AF389415

Energy Transfer in Alternative Vehicles

Energy Transfer in Alternative Vehicles

Editor

Wojciech Cieslik

MDPI • Basel • Beijing • Wuhan • Barcelona • Belgrade • Manchester • Tokyo • Cluj • Tianjin

Editor
Wojciech Cieslik
Division of Alternative Powertrains
Poznan University of Technology
Poznan
Poland

Editorial Office
MDPI
St. Alban-Anlage 66
4052 Basel, Switzerland

This is a reprint of articles from the Special Issue published online in the open access journal *Energies* (ISSN 1996-1073) (available at: www.mdpi.com/journal/energies/special_issues/Energy_Transfer_Alternative_Vehicles).

For citation purposes, cite each article independently as indicated on the article page online and as indicated below:

LastName, A.A.; LastName, B.B.; LastName, C.C. Article Title. *Journal Name* **Year**, *Volume Number*, Page Range.

ISBN 978-3-0365-7851-4 (Hbk)
ISBN 978-3-0365-7850-7 (PDF)

Cover image courtesy of Wojciech Cieslik

Contents

About the Editor

Wojciech Cieslik

DEng. Wojciech Cieslik was born in Poznan, Poland, in 1989. He received his engineer and master's degree in specialization of combustion engines at Poznan University of Technology. He defended his doctoral thesis on the conceptual combustion in the surrounding of non-combustible gases with awards from Polish Institute of Combustion and Polish Scientific Society of Combustion Engines.

Since 2015, he was employed as a research assistant in EU projects, and since 2018, he was employed as an assistant professor in the Institute of Combustion Engines and Powertrains. Currently, he is the author of more than 70 scientific articles achieving h-index = 8. His current research focuses on energy transfer in alternative vehicles including hybrids, electric and hydrogen vehicles. He is the associate supervisor for 3 PhD students. He is the supervisor of the PUT Powertrain scientific group, in which students learn about and create alternative propulsion analysis systems.

He is a member of the Polish Scientific Society of Combustion Engines and Polish Institute of Combustion. Work as an editor in journals: Combustion Engines and Marine Technology Bulletin (VETUS). From 2021, he is a Special Issue editor "Energy Transfer in Alternative Vehicles" and "Alternative Powertrains in Urban Mobility, Trends, Challenges and Opportunities in Energy Flow Analysis" in *Energies*.

He is currently working on the development of an alternative propulsion laboratory, and is making modifications to internal combustion engines for hydrogen-burning capabilities. His main interests also include water sports and building wooden boats.

Preface to "Energy Transfer in Alternative Vehicles"

This reprint is a collection of research that contribute to a better understanding of energy transfer related to alternative powertrains and their implementation in different types of vehicles. Hybridization of power units and alternative propulsion systems are the mainly developed technologies in the automotive field today. The modernization of conventional propulsion sources is justified by the increasingly stringent economic, ecological, and comfort requirements. The common link in any renewable energy source or propulsion system today is electric motors. Their correct use in propulsion offers the possibility of reducing or eliminating harmful emissions both in the form of toxic compounds and noise or unwanted vibrations. The research works were related to energy transfer in both hybrid and electric vehicles, and conversion of energy from renewable sources such as photovoltaic installations to power alternative vehicles. However, you will also find in this reprint papers that are related to modules of alternative propulsion such as drivetrain efficiency analysis, ways of energy accumulation in batteries, and influence of the charging method on the energy consumption of a vehicle during its life cycle. The search for solutions that reduce energy consumption and simultaneous methods of generating energy should go together for the global success of the energy transition. With the right transformation of electromobility, which is developing rapidly in all areas of transportation, starting with small personal vehicles and passenger cars through public transportation vehicles and ending with noticeable expansion in the area of urban transportation services will be possible to implement in near future. I encourage you to read all the articles dealing with the subject under discussion.

Wojciech Cieslik
Editor

Article

Research of Load Impact on Energy Consumption in an Electric Delivery Vehicle Based on Real Driving Conditions: Guidance for Electrification of Light-Duty Vehicle Fleet

Wojciech Cieslik [1],* and Weronika Antczak [2]

1   Department of Combustion Engines and Powertrains, Faculty of Civil and Transport Engineering, Poznan University of Technology, 60-965 Poznan, Poland
2   Faculty of Civil and Transport Engineering, Poznan University of Technology, 60-965 Poznan, Poland
*   Correspondence: wojciech.cieslik@put.poznan.pl

Abstract: Electromobility is developing rapidly in all areas of transportation, starting with small personal vehicles and passenger cars through public transportation vehicles and ending with noticeable expansion in the area of urban transportation services. So far, however, there is a lack of research determining how the effect of load weight defines the energy intensity of a vehicle under real conditions, especially in the areas of urban, suburban and highway driving. Therefore, this paper presents an analysis of a representative delivery vehicle and its energy consumption in two transportation scenarios where cargo weight is a variable. A survey was also conducted to determine the actual demand and requirements placed on the electric vehicle by transportation companies.

Keywords: electric light-duty vehicle; usability of an electric delivery vehicle; real energy consumption

Citation: Cieslik, W.; Antczak, W. Research of Load Impact on Energy Consumption in an Electric Delivery Vehicle Based on Real Driving Conditions: Guidance for Electrification of Light-Duty Vehicle Fleet. *Energies* **2023**, *16*, 775. https://doi.org/10.3390/en16020775

Academic Editor: Chunhua Liu

Received: 13 December 2022
Revised: 29 December 2022
Accepted: 4 January 2023
Published: 9 January 2023

1. Introduction

For a number of years now, road transport has gone through considerable alteration. Ongoing research on more efficient, yet environmentally friendly powertrains supports the EU's 2030 climate policy implementation [1]. However, most attention is paid to passenger car electrification, overlooking the need for heavy-duty (HDV) and light-duty (LDV) vehicle customization. In compliance with European Environment Agency insights, HDVs, i.e., buses, coaches, and trucks, account for approximately a quarter of the carbon dioxide (CO_2) emissions from road transport in the EU. Poland, as a holder of the largest truck fleet in EU, contributes to these emissions for the most part [2]. In order to launch a general road transport transformation, a framework for HDV and LDV electrification should be developed.

Profound discussion regarding electrification of heavy vehicles has considerable potential to direct the further development of electromobility in general. Research conducted in the past drew key conclusions that the crucial parameter influencing electric vehicles energy consumption is their mass [3–5]. In contrast to vehicles with conventional powertrains, where the engine power is the leading light in the fuel consumption rate, performance of vehicles driven by electric motors is not as reliant on the rated motor power. Such a feature, correlated with the high overall efficiency of electric powertrains, may shed light on the success of small EVs, notably evident in urban areas. Electro-micromobility is the central thread of pursuing change in the urban transport structure [6]. Cities worldwide have already experienced adjustments in the used and obtainable means of transport. This trend is intensely fostered through vehicle-sharing systems advancement and broadening the accessibility of electric means of microtransport, just as a couple of examples [7–9]. While micromobility is undoubtedly suitable for city residents' transportation, it will not meet the requirements of the transportation of mass goods. This need should not be omitted, as delivery vans are becoming a more and more common element of urban structures, due to the rapid rise in popularity of online shopping [10].

Scientists have already acknowledged this issue, beginning to conduct examinations verifying the actual suitability of electric vans for delivery and transportation companies. Most of these studies are focused on analyzing a variety of delivery scenarios by means of algorithms and thus proposing the best charging stations' arrangement to overcome battery capacity limitations [11–16]. The conclusions and proposed adjustments are fundamental for reasonable planning of road infrastructure modification; however, they mostly do not consider real driving conditions, including traffic, variable loads, and driving style.

One of the possibilities to conduct research that takes account of the wide range of potential variables is to carry out road tests performed in compliance with the RDE (real driving emissions) procedure. It allows the counting of different road infrastructures, traffic, road slopes, and driving behaviors, simultaneously assuring compliance with European Union regulations test procedures [17]. Thanks to the prescriptive test method, results obtained for the different vehicle and powertrain models can be compared and submitted for analysis. While the RDE tests are focused on reflecting the vehicle's impact on the environment, research that draws on their procedures but omits exhaust emission analysis is referred to as testing RDC (real driving conditions) [18,19]. Such examinations have been performed in the past; however, they regarded combustion engine, hybrid, and electric passenger cars for the most part [20–27]. Thereby, the electrification potential of HDVs and LDVs remains an insufficiently researched subject.

According to the data presented in Table 1, the number of both passenger EVs and large EVs registered in Poland has doubled year on year since 2019. While such an increasing rate is highly desirable, it is clearly seen that large EVs still play a minor role in the global service sector. To reach the goal of climate neutrality, immediate changes in this area are needed.

Table 1. Number of registered electric vehicles in Poland in recent years (based on [28]).

Type of Vehicle	2019	2020	2021	November 2022
Passenger EV	5091	10,041	18,795	28,386
Passenger PHEV	3546	8834	19,206	28,540
Large EV	**519** [1] 224 [2]	**839** [1] 430 [2]	**1657** [1] 651 [2]	**2638** [1] 790 [2]
Small EV [3]	6450	9308	11,091	16,541
FCEV	1	0	79	124

[1] **Delivery vans and trucks**; [2] buses; [3] electric bikes, scooters.

Light-duty vehicles, being so far challenging for electrification, remain a common element of the urban landscape and thus have an input into cities' noise and air pollution. Moreover, the demand for the road freight transport has been continuously growing. The necessity to undertake further actions aiming to decarbonize the transport sector has encouraged authors to pursue research on the actual usage potential of electric delivery vans. With the aim of conducting tests possibly akin to real-life driving conditions, a survey has been created. The questionnaire focuses on gathering data that portray the expectations imposed on the delivery vehicles, as well as actions that could possibly spur users to turn to electromobility.

This article consists of six main sections. The first one describes the general research problem and presents the current state of the Polish fleet of various EVs. The second chapter is covers the survey that was disseminated among transport companies and other relevant businesses and data gathered on their demand for electromobility. The third part describes the research objective: details regarding vehicles chosen for tests and the software utilized for data gathering. In the fourth section, a description of the routes driven during examinations and weather conditions on the measurement days can be found. The fifth chapter is a comprehensive presentation of results and data analysis, while the sixth one contains a conclusion and guidelines for light-duty fleet electrification.

The information gathered throughout the series of driving tests combined with the interviewees' answers may serve as a comprehensive guide for electrification of the light-duty vehicle fleet.

2. Survey Assessment: Guidelines and Demand of Transport Companies for Electromobility

2.1. Questionnaire Design

For familiarization with the actual working conditions and the scale of electric delivery vehicle usage, a questionnaire was created. The target group consisted of individuals performing professional duties with the aid of delivery vehicles or representing relevant companies, i.e., delivery and transportation companies, as well as self-employment. The survey investigated interviewees' perspectives on the usefulness of electric delivery vehicles at their current technological level. Moreover, interviewees' perceptions on the ongoing projects fostering electromobility and expectations associated with them were assessed.

2.2. Data Collection

The questionnaire was designed with the aid of the Google Forms platform and disseminated online. It was distributed predominantly on media platforms among professional groups and through mailing lists to selected businesses. Video footage promoting the study and encouraging receivers to take part in the survey had been released on YouTube [29] and further advertised on social networks. Additionally, business cards containing references to the questionnaire were produced and spread around university and cooperating car dealers. The survey was available in Polish and disseminated among relevant companies and individuals across Poland. The data were collected from mid-July to mid-August 2022.

2.3. Data Analysis

A total of 51 responses were gathered. Initially, interviewees were asked about the type of propulsion with which the delivery vehicle is equipped. Obtained data show the predominant usage of CI engines in large cars, having been declared by 83% of the respondents. Delivery cars driven by electric motors were used by only 3% of the respondents, pointing to their still-modest use (Figure 1).

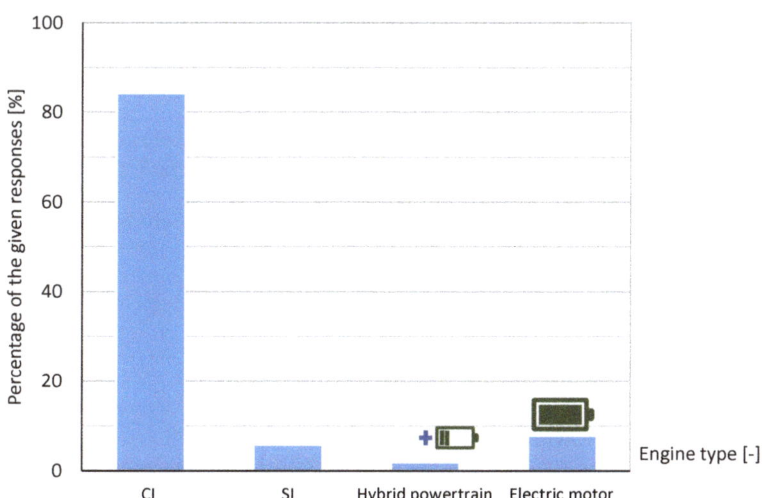

Figure 1. Current status of the surveyed fleets in terms of their propulsion system (based on the survey).

The number of collected survey responses represented virtually 1000 delivery vans. Both small fleets, beginning from two delivery vans, as well as large fleets comprising more than 200 vehicles, are included herein (Figure 2).

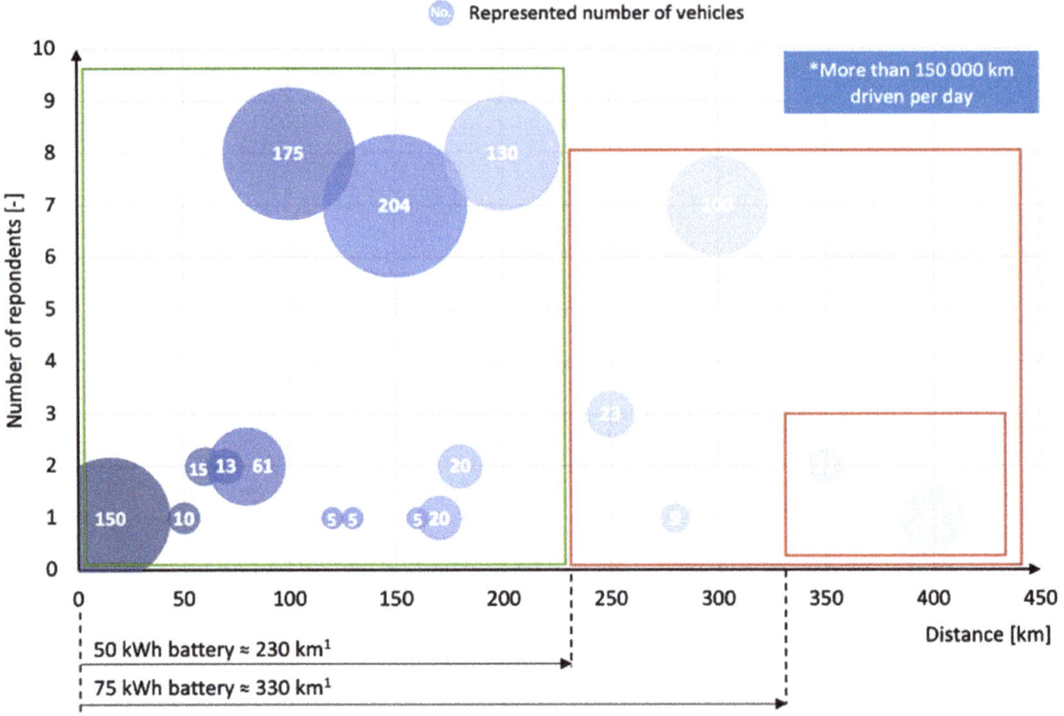

Figure 2. Number of vehicles managed by each fleet and their average daily distance, indicating the potential of different versions of a battery capacity to meet vehicle range requirements as well as the range of electrification potential in particular companies (based on the survey). [1] Electric range ensured by the manufacturer for Toyota Proace Electric with 16″ steel wheels. It is highlighted that these figures may not reflect real driving conditions (RDCs). Electric range depends on accessories package, driving style, conditions, speed, load, etc. [30].

As shown in Figure 2. daily distance covered by the delivery vehicle declared by almost a third of respondents exceeds the range declared by the manufacturer for the Toyota Proace Electric with a 50 kWh battery. Greater battery capacity naturally enhances the range, but still does not meet all potential users' needs. It should be highlighted that the theoretical range ensured by the producer is given for the unloaded car with the basic accessories package. Thus, it may be assumed that the predominant part of the electric delivery vans executing delivery or transportation services in real driving conditions, that is, with additional load and greater daily distance covered, will demand recharging during the workday, i.e., while loading or unloading. However, this solution requires infrastructure customization.

Of the examined group, 77% chose cargo vans. That statistic alone points out the reasonableness of the car model choice for the examination. Frequently chosen by users' car bodies included Luton and city vans as well, both being represented by 22% of the answers (Figure 3). This confirms the appropriateness of the choice of research object.

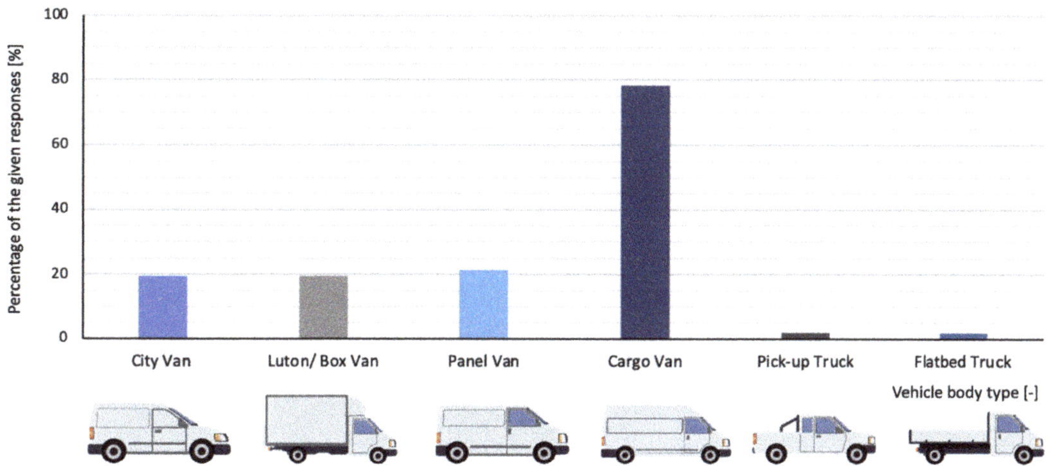

Figure 3. Type of light-duty vehicle used in the company (based on own survey).

Commercial vehicles are characterized by a wide variation of construction depending on their intended use, with many models of currently available vehicles available in a variety of bodies. Both passenger and cargo versions are observed, with open or closed cargo area. Selected cars available on the Polish market have been compared with each other by the basic parameters declared by manufacturers. In this way, a summary was created indicating selected electric vehicles and their range according to the WLTP test, depending on battery capacity (Figure 4).

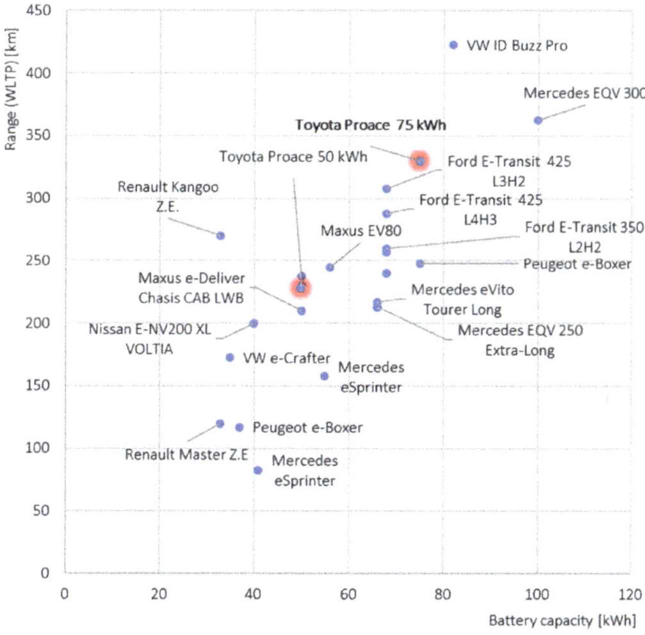

Figure 4. List of battery capacities of an electric LDV available on the Polish market, combined with their catalogue range based on the WLTP test [31] (marked with a red, color two versions of the test vehicle varying in battery capacity).

The examined group was roughly equally divided in terms of car loading at the beginning of a workday. Merely 4% of respondents declared to take less than 100 kg of goods (Figure 5). This information should be considered notably, as load is one of the factors affecting an electric car range.

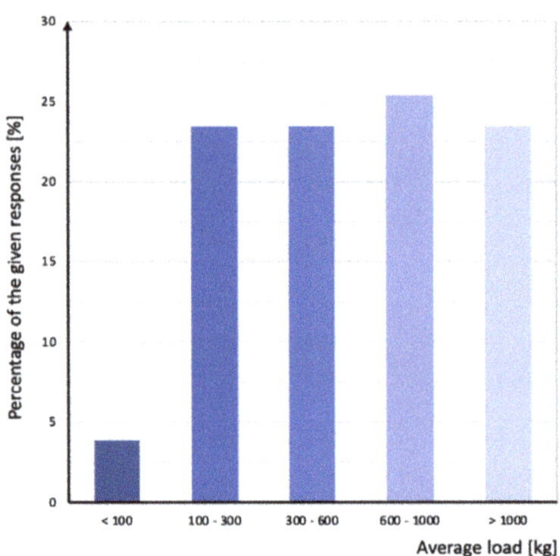

Figure 5. The average load at the beginning of a workday (based on own survey).

Interviewees were requested to indicate which of the ongoing projects fostering electromobility could encourage them to modify the fleet of vehicles into electric-powered models (Figure 6). Government subsidy for purchasing electric cars proved to be the most efficient way to raise the interest in electromobility. Moreover, free charging stations, as well as allowance for bus lanes usage, turned out to play an important role for potential electric delivery vehicle users.

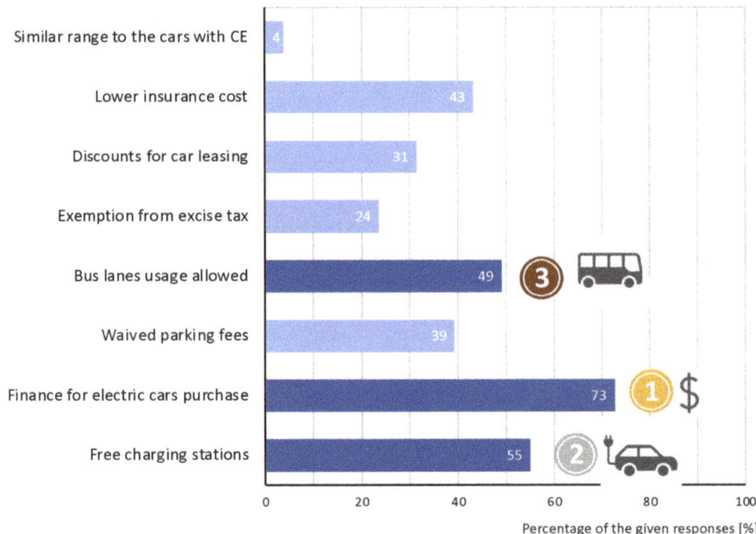

Figure 6. Factors deciding willingness to modify a fleet vehicle to electric-powered models (based on own survey).

2.4. Conclusions

Building on the survey answers, the following conclusion can be drawn:

- Light-duty vehicles are currently predominantly equipped with CI Engines. The usage of electric motor is still modest—declared by only 3% of the respondents.
- Almost a third of the respondents declared the average daily distance covered being greater than 230 km, which is the range declared by the manufacturer for the Toyota Proace Electric with a 50 kWh battery.
- The cargo van is the most frequently chosen light-duty vehicle body type.
- Government subsidies for purchasing electric cars, free charging stations and allowance for bus lane usage proved to be the most efficient ways to raise interest in electromobility.

3. Research Object

The electric vehicle under test was a cargo van-type body structure. Its GVW is 3055 kg, and in the provided version of the equipment for the road tests conducted, the weight of the vehicle was 2115 kg (the weight limits related to the tested vehicle are shown in Figure 7; these values are presented on the basis of the registration certificate, which is an approval document showing the parameters of a specific model, and on the basis of the manufacturer's data). Based on these values, the vehicle's loading ranges were determined, defining the carrying capacity.

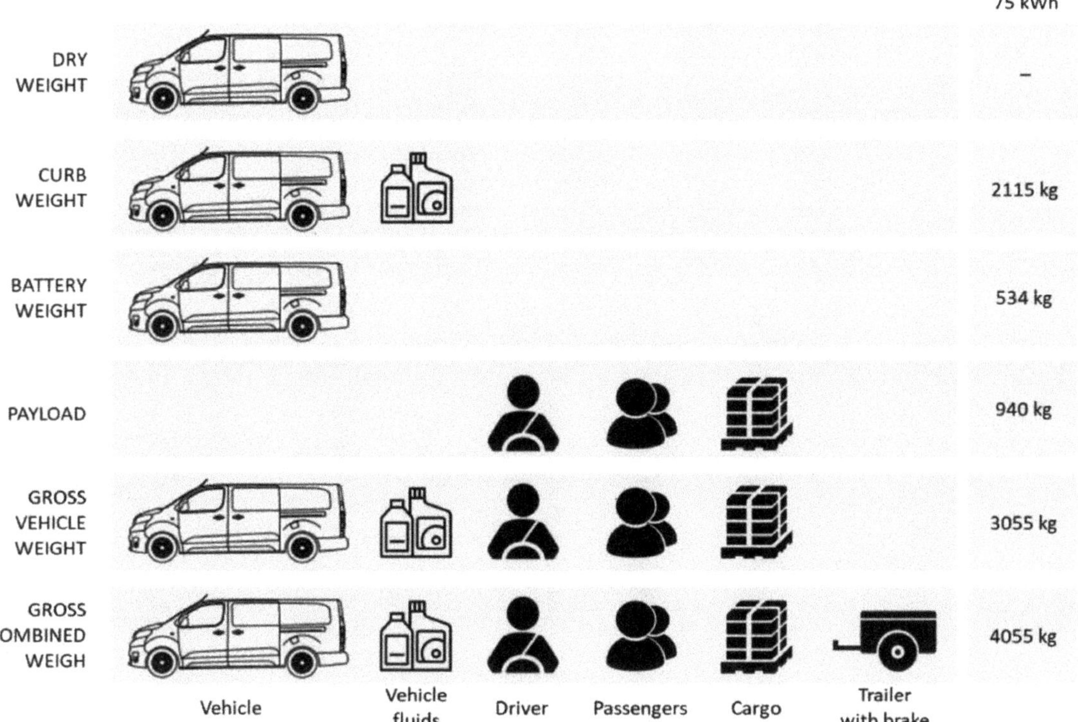

Figure 7. Mass limits of the tested vehicle (values read from the registration certificate of used in research vehicle) based on [32].

During the tests, the vehicle was equipped with a diagnostic system consisting of a diagnostic computer and a GPS signal recorder (Figure 8). The vehicle was equipped with a 75 kWh battery (optionally, the vehicle can also be equipped with a smaller 50 kWh battery capacity). Despite the available space, larger battery capacities are not available, which is determined by the maximum allowable weight of the vehicle. Data were collected with the use of an OBD diagnostic system and GPS module. The OBD system allowed the gathering of parameters related to the powertrain or the high-voltage battery performance. The frequency of data collection equaled 2 Hz.

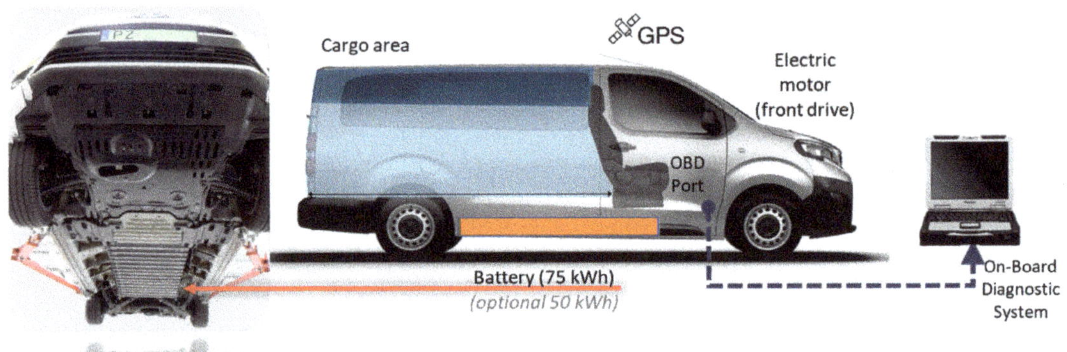

Figure 8. Schematic of the measurement system including a view of the location of the traction battery [32].

In its current configuration, the vehicle can be loaded with a weight of 940 kg. This weight represents both the weight of the load space goods and the weight of the driver and passengers. Therefore, when taking into account the maximum possible loading weight, it is necessary to take into account the weight of the vehicle's users as well (in the tested version, the homologation specifies three people in the passenger compartment). The research reported in the current work concerns the analysis of the maximum loading weight, that is, the weight of the driver (90 kg) and a cargo weight of 850 kg (Figure 9).

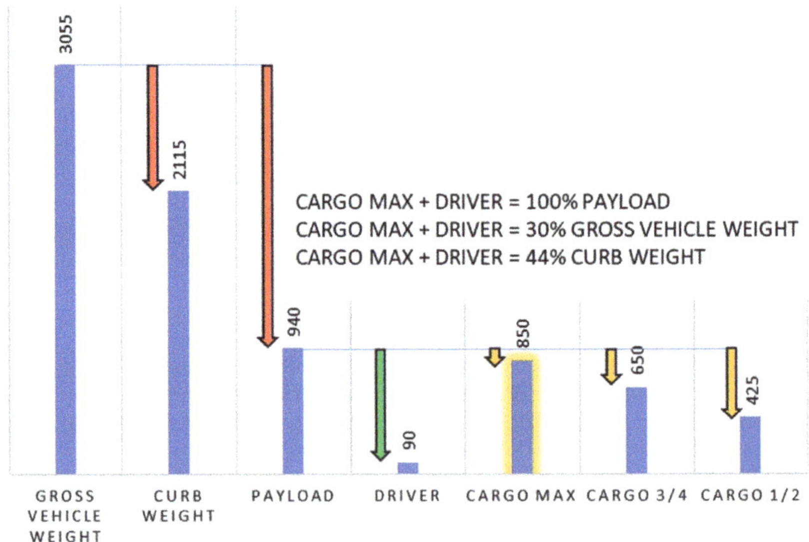

Figure 9. Weights taken into account during road surveys conducted (based on the data from the vehicle registration certificate—Figure 7).

The realized research aims to assess the actual power consumption of an electric delivery car both in real traffic conditions and real driving conditions that include the influence of a load on the powertrain performance. An electric vehicle (EV), contrary to a hybrid (HEV) or a fuel cell (FCEV) vehicle, is characterized by considerably fewer powertrain operating modes. Considering the drive phase only, two modes can be differentiated: drive mode, during which the high voltage battery is discharged and deceleration mode, when the battery is recharged (Figure 10). The high-voltage battery may by charged with the aid of an external power source; however in this research, the charging process analysis is regarded as not a critical element.

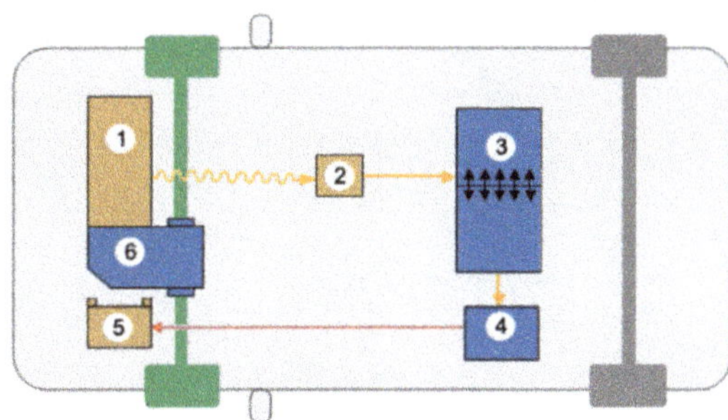

Figure 10. Drive train operating modes while driving (①—electric drive motor, ②—inverter, ③—traction battery, ④—on-board charger/DC-DC voltage transformer, ⑤—ancillary battery, ⑥—reduction gear) [32].

4. Measurement Route and Conditions

Research was conducted in compliance with RDC (real driving conditions) test requirements, which are shown in Table 2 in detail, and effective traffic regulations.

Table 2. Real driving conditions shorter test requirements [33].

Selected RDE/RDC Test Requirements	Urban	Rural	Motorway
Cycle repetition (+/− 10%) [%]	29 < ratio ≤ 34	33	←
Speed [km/h]	< 60	$60 \leq V \leq 90$	V > 90
Max. speed [km/h](+/− 15 km/h for less than 3% of driving time)	-	-	145
Average speed (stops included) [km/h]	$15 \leq V \leq 30$	-	-
Minimum travelled distance [km]	16	←	←
Altitude difference (beginning/end) [m]	100	←	←
Maximum slope [m/100 km]	1200 m/100 km	←	←

The marked route, depicted in Figure 11, met the RDC test requirements imposed by the European Union regulations. Thereby it consisted of the urban, rural and motorway sections, closely selected in accordance with the requirements imposed for the particular route sectors (Figure 12).

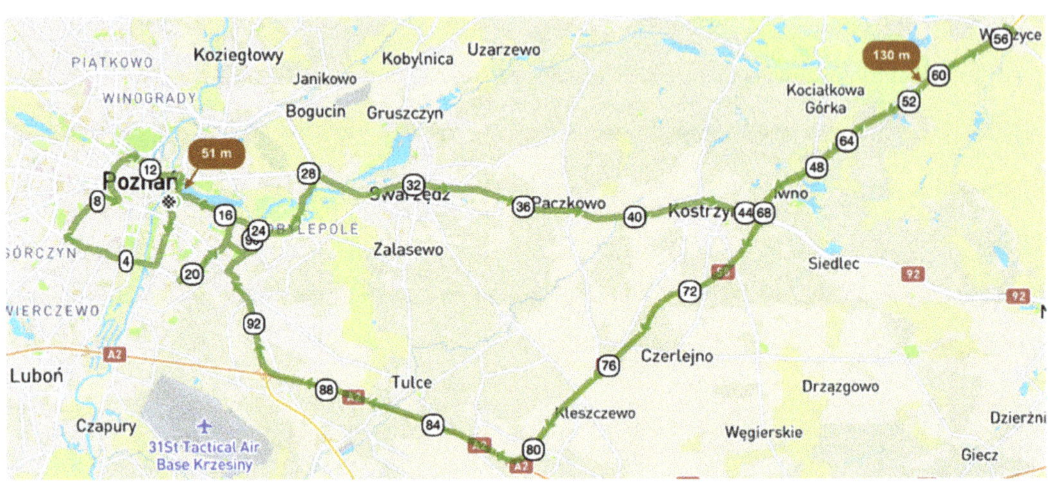

Figure 11. Route driven during the examinations.

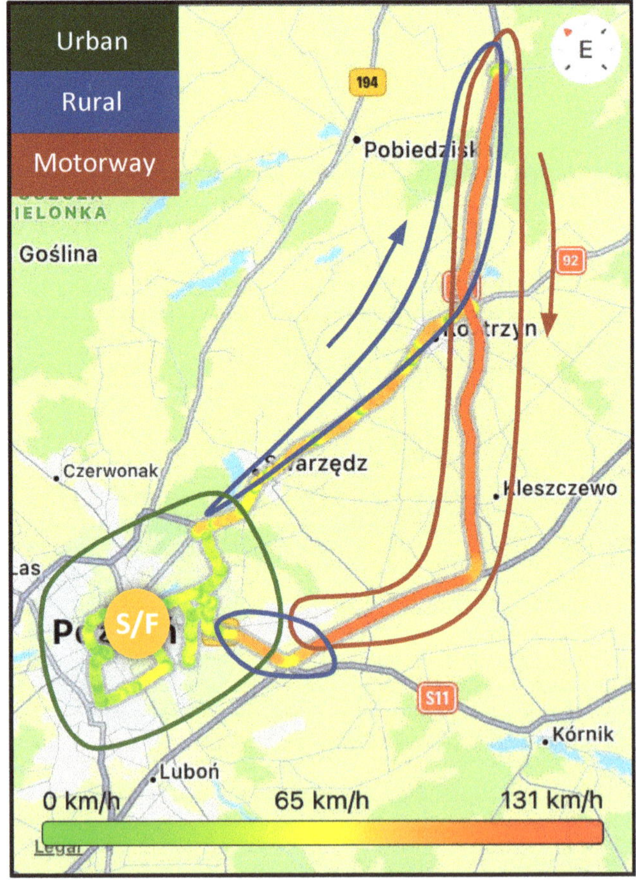

Figure 12. Route divided into particular sections (S\F—Start/Finish).

Tests were performed with the aid of an electric delivery van in Poznan (Poland) and its vicinities. The car was driven by only one driver throughout the tests, thus eliminating the influence of the driving manner on the gathered data. The length of the route averaged 100 km. The highest elevation, amounting to 131 m, was reached on the highway, while the lowest point, equal to 51 m, was encountered in the urban area (Figure 13). Thereby, the general elevation difference totaled 80 m.

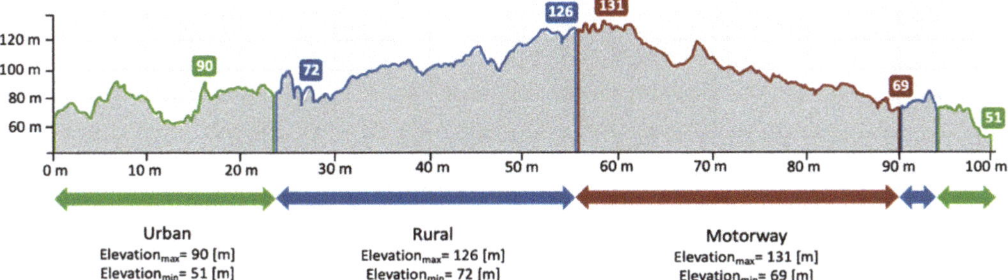

Figure 13. The elevation pattern throughout the route.

Each drive included stopovers determined by the infrastructure of the particular route's sectors. Naturally, drives along the motorway inheld no stops. However, due to the traffic lights and intersections encountered in the urban and rural sectors, in each drive several dozen stops were registered. As an example, during one of the tests 40 stops were enforced in the urban area (Figure 14) and 6 ones in the rural route, giving eventually the total of 46 stopovers along the route.

Figure 14. Stops enforced by the infrastructure in the urban section along the marked route.

Drives were carried out on the working days at the hours of moderate traffic. They were realized in July 2022. Although the measurements were conducted at an interval of more than two weeks, during the period of varying temperatures, the temperature circled around 25 °C when the tests were conducted (Figure 15). During the recordings, the settings of the comfort systems including air conditioning were set at the same level.

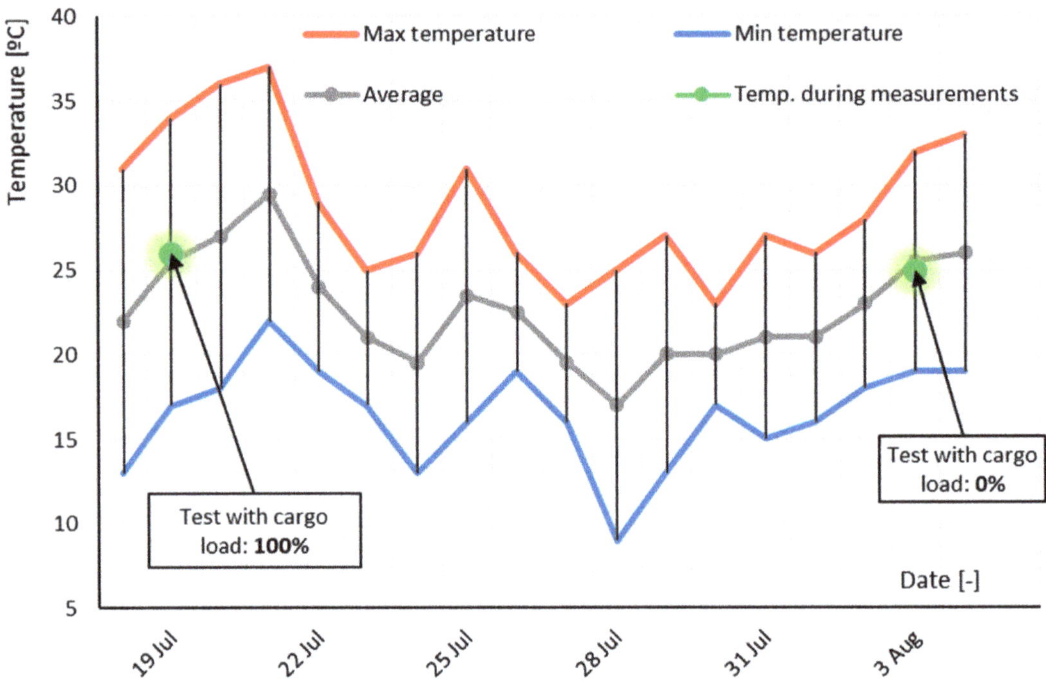

Figure 15. Ranges on measurement days [based on [34] and authors' own measurements].

5. Research Results

The presented test results are representative of the two examined cases that differ in the cargo load set in the vehicle's rear compartment. The fundamental research question of this research paper was to define the load impact on the energy consumption in the real driving conditions. For this reason, two extreme cases have been considered. The first one assumed 100% of the maximum load (the addition of the driver weight and the cargo weight equal to 940 kg). As a matter of the second case the cargo area was emptied, thus the only loading constituted the driver's weight. For both cases, full battery charge at the beginning of the drive was assured.

Both rides met the requirements of the test under real traffic conditions (guidelines shown in Table 2). The basic parameters for the proportion of the road in the urban, rural and motorway route are shown in Figure 16. Some differences can be seen in the two runs, consisting especially in the varying values and characteristics of maintaining a constant speed in freeway driving, but this did not adversely affect the fulfillment of the test requirements. Varying driving conditions, traffic volumes are taken into account in the test procedure allowing the two measurements to be compared.

The general number of stops differs between the drives as well as between the particular sections of the route. Such state is a direct result of a road infrastructure, that is, the number of junctions, and naturally of the current traffic intensity. The number of stops during the drives equals respectively: for the test with a 100% of a maximum cargo load—38 stops, and for the test with a 0% of a maximum cargo load—40 stops.

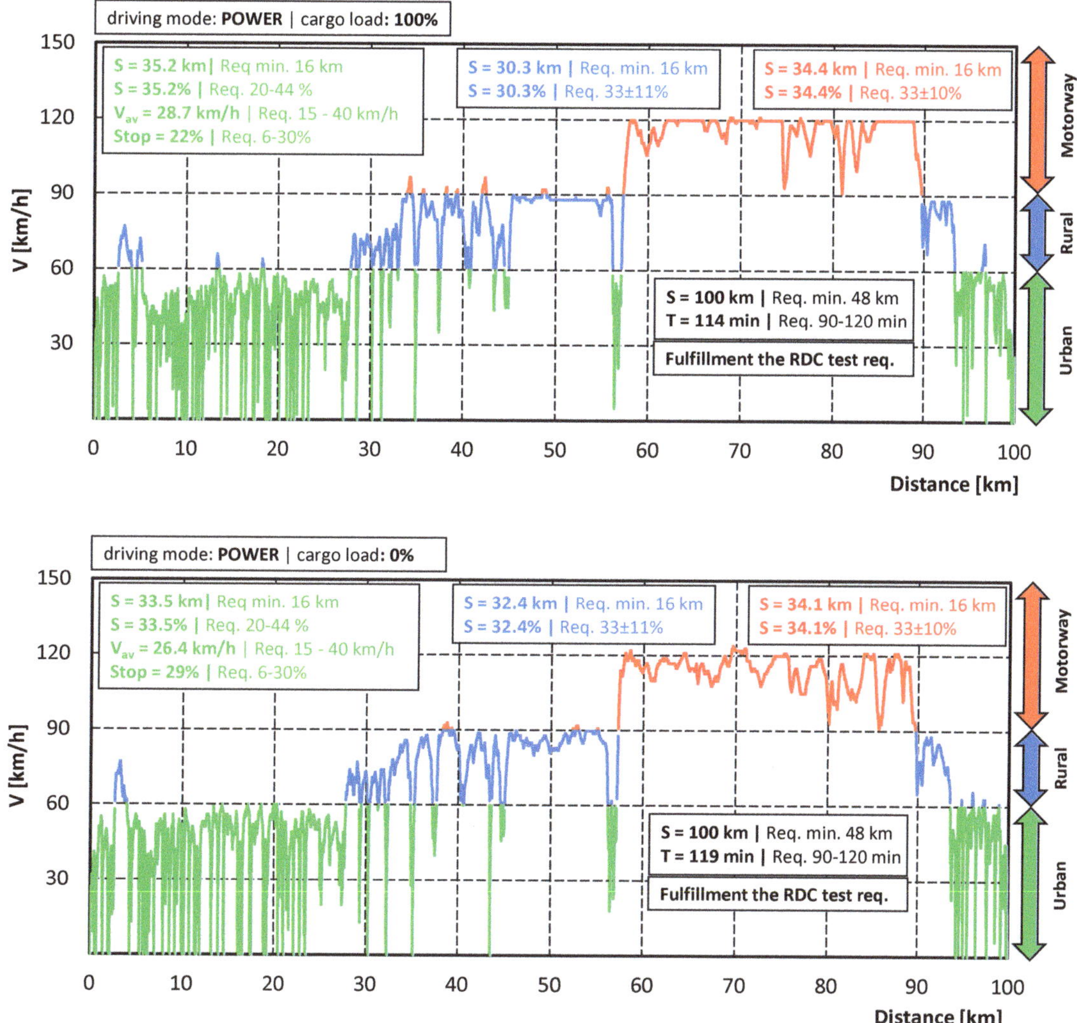

Figure 16. The course of RDC test in different driving mode with defining basic parameters for meeting test requirements.

By analyzing parameters essential to determine the energy flow in a vehicle, the authors compiled tests parameterization in terms of a particular drive phase share (acceleration -a_+, constant speed—a_0, deceleration—a_-) and a stopover share. For both drives these values, depicted in Figure 17, are similar and thereby allow us to make a reliable assessment regarding the actual load influence on the powertrain's energy consumption.

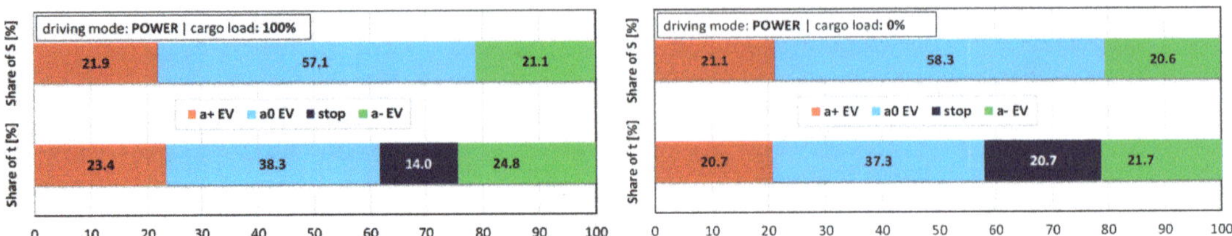

Figure 17. Comparison of the phase motion share during the RDC test regarding the variable cargo load.

An electric vehicle is characterized by two states of drivetrain operation: the drivetrain consumes or generates energy from/to a high voltage battery. With respect to the route parameters in the real driving condition test, therefore, the time and distance intervals in which the vehicle consumes or generates energy were determined—presented in Figure 18 (periods in which the vehicle does not consume energy from the battery during standstill are not recorded—in most cases, at standstill, the battery is also discharged for vehicle comfort purposes—air conditioning).

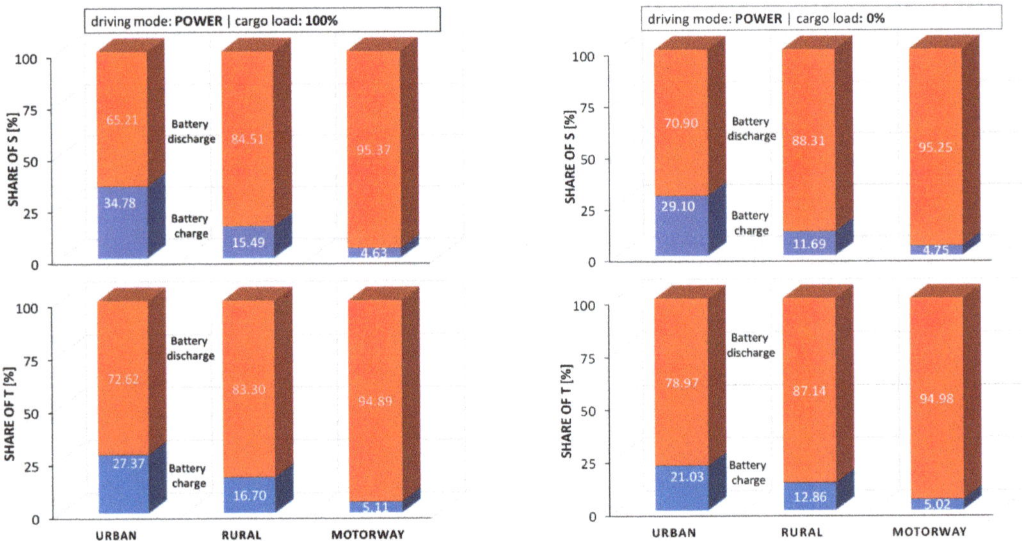

Figure 18. Parameterization of time- and route-dependent energy consumption and recovery for individual phases of the RDC test.

For the both measurement runs, the highest energy recovery values are determined for urban driving conditions, but it should be pointed out that the procedure counts energy recovery/consumption with respect to vehicle speed, so any braking from highway or suburban route speeds automatically enters into the sum of energy recovery for suburban and urban routes, respectively. Nevertheless, it should be noted that in each case the loaded vehicle recorded higher shares of both time and distance of energy recovery, which is confirmed by the energy flow results shown in Figure 19. Greater time or distance of energy recovery resulting also in higher values of recovered energy to the battery did not result in lower energy consumption, as higher vehicle load results in higher energy consumption in the acceleration phase. The higher the speed of the vehicle, the smaller this difference is between runs. In city driving conditions, energy consumption is almost 20% higher for a loaded vehicle than for an empty one. When driving on the highway, this difference decreases to about 8%. However, taking into account the higher energy recovery for a vehicle loaded with a mass of cargo, these differences decrease in the total energy flow.

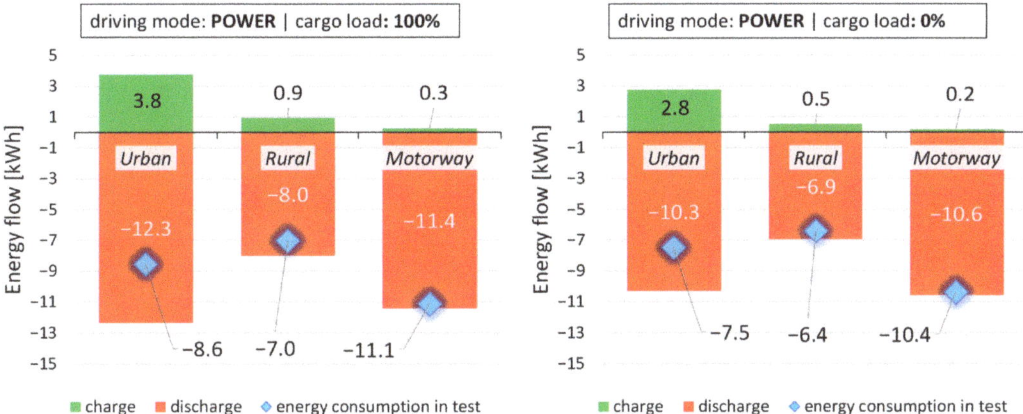

Figure 19. Energy consumption balance in terms of different cargo load.

A loaded vehicle is characterized not only by the total amount of energy consumed at a higher level than a vehicle moving unloaded. Also, the individual operating points (energy intensity of the drivetrain) reach higher values, which is noted in area 1 in Figure 20. It can also be seen that the temporary energy recovered to the battery during braking is higher (area 2 in Figure 20) compared to the unloaded vehicle. Despite the unloaded vehicle reaching higher speeds on the highway, the energy flow is at a lower level (area 3 in Figure 20).

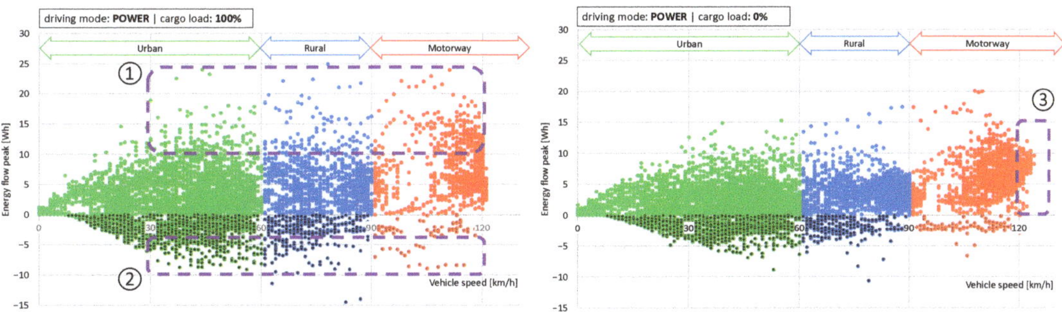

Figure 20. Energy flow areas segmented by speed ranges for different vehicle payload levels.

Based on the above short-term energy consumption maps, the maximum ongoing energy consumption was determined, which indirectly determines the use of the propulsion system. Due to the limitations of the OBD system's measurement monitor, the drive engine's operating parameters were not determined in the current work (this will be done in future work). Instead, the maximum values presented indicate that the energy consumption from the battery in practically every speed range is higher for a loaded vehicle (Figure 21).

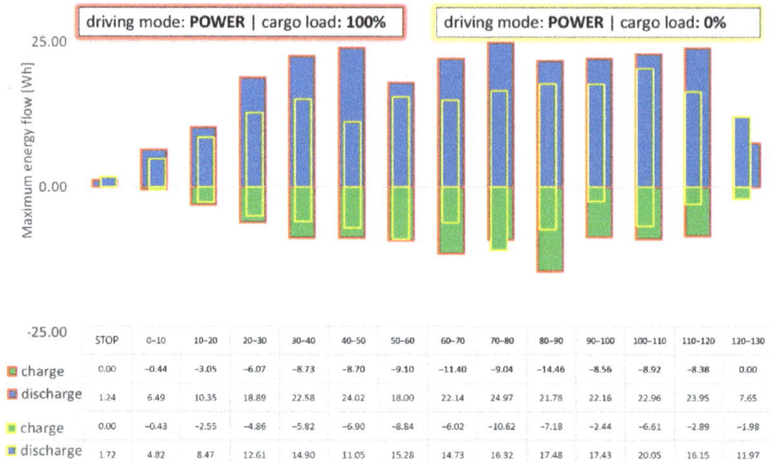

	STOP	0–10	10–20	20–30	30–40	40–50	50–60	60–70	70–80	80–90	90–100	100–110	110–120	120–130
charge	0.00	−0.44	−3.05	−6.07	−8.73	−8.70	−9.10	−11.40	−9.04	−14.46	−8.56	−8.92	−8.38	0.00
discharge	1.24	6.49	10.35	18.89	22.58	24.02	18.00	22.14	24.97	21.78	22.16	22.96	23.95	7.65
charge	0.00	−0.43	−2.55	−4.86	−5.82	−6.90	−8.84	−6.02	−10.62	−7.18	−2.44	−6.61	−2.89	−1.98
discharge	1.72	4.82	8.47	12.61	14.90	11.05	15.28	14.73	16.32	17.48	17.43	20.05	16.15	11.97

Figure 21. Maximum values of energy flow in speed intervals steps of 10 km/h.

The greatest energy recovery is realized in urban speed intervals, but this is due to the frequency of braking in urban driving conditions. Instantaneous maximum energy recovery values are highest in the 60–90 km/h speed range. Energy recovery for speeds below 10 km/h is marginal, so below this speed it is necessary to use the vehicle's conventional braking system.

6. Summary

The presented research made it possible to determine the usability of an electric delivery vehicle in real traffic conditions with extreme load options. The conclusions of the presented work are presented below:

- The maximum range of the LDV in actual traffic conditions differs from the declared by the manufacturer (330 km WLTP), both in unloaded and fully loaded trips (by 15% and 22%, respectively).
- The vehicle's weight increased by 850 kg of loading affects the range reduction as summarized in Figure 22. The largest decrease (by almost 14%) in range was recorded for the urban route, due to increased energy consumption during acceleration.
- The impact of route type (average speed and proportion of recuperation) significantly affects the vehicle range—differences between highway and urban routes reach 25–30%.

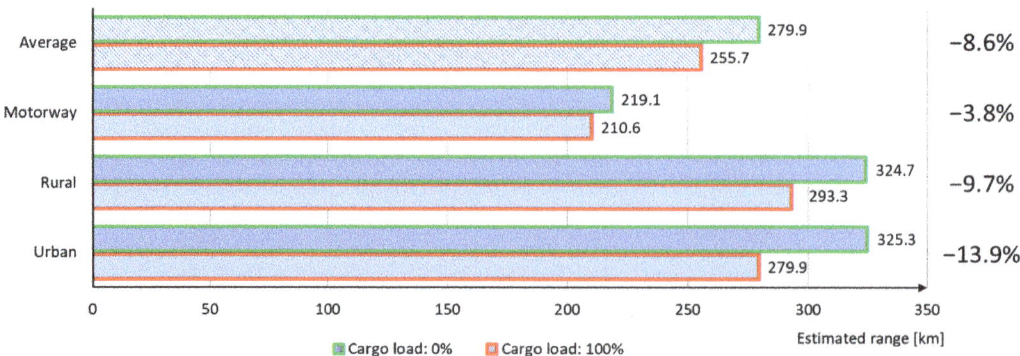

Figure 22. Effect of cargo weight on estimated range in different phases of the RDC test.

Based on the research presented, guidelines and conclusions were determined for transportation companies interested in modernizing their vehicle fleets:

- In an urban application, an electric delivery vehicle will meet most of the transportation requirements among surveyed entrepreneurs.
- The delivery vehicle should be adapted to the daily operation of the company. This may allow to reduce the battery capacity (reduce the purchase price) or increase the battery charging intervals. This applies to companies that declare a daily distance in the range of up to 100 km.
- Variable loading has an impact on the maximum range of the vehicle during the day, and proper planning of unloading from the heaviest goods can greatly increase the range of the vehicle.
- The delivery electric vehicle should be used especially in urban transportation, as the energy recuperation significantly reduces the energy consumption of the vehicle.

Author Contributions: Conceptualization, W.C. and W.A.; methodology, W.C.; formal analysis, W.C. and W.A.; investigation, W.C. and W.A.; writing—original draft preparation, W.C. and W.A.; visualization, W.C. and W.A.; writing—review and editing, W.C. and W.A.; supervision, W.C. All authors have read and agreed to the published version of the manuscript.

Funding: This research was funded by Poznan University of Technology, grant 0415/SBAD/0337.

Data Availability Statement: The data presented in this study are available on request from the corresponding author.

Acknowledgments: The authors of this article would like to thank Toyota Professional Bońkowscy for the provision of a vehicle for research.

Conflicts of Interest: The authors declare no conflict of interest.

Abbreviations

BEV	battery electric vehicle
CAN	controller area network
EV	electric vehicle
FCEV	fuel cell electric vehicle
HEV	hybrid electric vehicles
ICE	internal combustion engine
LCA	life-cycle assessment
NEDC	New European Driving Cycle
PHEV	plug-in hybrid electric vehicles
RDC	real driving conditions
SOC	state of charge
TTW	tank to wheel
WLTP	Worldwide Harmonised Light Vehicles Test Procedure

References

1. Clean and Sustainable Mobility. Available online: https://www.consilium.europa.eu/en/policies (accessed on 27 November 2022).
2. Carbon Dioxide Emissions from Europe's Heavy-Duty Vehicles. Available online: https://www.eea.europa.eu/publications (accessed on 27 November 2022).
3. Weiss, M.; Cloos, K.C.; Helmers, E. Energy efficiency trade-offs in small to large electric vehicles. *Environ. Sci. Eur.* **2020**, *32*, 46. [CrossRef]
4. Zhao, C.; Gong, G.; Yu, C.; Liu, Y. Research on Key Factors for Range and Energy Consumption of Electric Vehicles. SAE Technical Paper 2019-01-0723. 2019. [CrossRef]

5. Bull, M. Mass Reduction Performance of PEV and PHEV Vehicles. 22nd Enhanced Safety of Vehicle Conference, Washington, 2011. Available online: https://www-esv.nhtsa.dot.gov/Proceedings/22/ (accessed on 10 November 2022).

6. Ziemba, P.; Gago, I. Compromise Multi-Criteria Selection of E-Scooters for the Vehicle Sharing System in Poland. *Energies* **2022**, *15*, 5048. [CrossRef]

7. Turoń, K.; Kubik, A.; Chen, F. What Car for Car-Sharing? Conventional, Electric, Hybrid or Hydrogen Fleet? Analysis of the Vehicle Selection Criteria for Car-Sharing Systems. *Energies* **2022**, *15*, 4344. [CrossRef]

8. Turoń, K.; Kubik, A.; Ševčovič, M.; Tóth, J.; Lakatos, A. Visual Communication in Shared Mobility Systems as an Opportunity for Recognition and Competitiveness in Smart Cities. *Smart Cities* **2022**, *5*, 802–818. [CrossRef]

9. Aiello, G.; Quaranta, S.; Certa, A.; Inguanta, R. Optimization of Urban Delivery Systems Based on Electric Assisted Cargo Bikes with Modular Battery Size, Taking into Account the Service Requirements and the Specific Operational Context. *Energies* **2021**, *14*, 4672. [CrossRef]

10. More Customers Are Shopping Online Now Than at Height Of Pandemic, Fueling Need For Digital Transformation. Available online: https://www.forbes.com/sites/blakemorgan/2020/07/27/more-customers-are-shopping-online-now-than-at-height-of-pandemic-fueling-need-for-digital-transformation/?sh=768cdf076bb9 (accessed on 24 November 2022).

11. Sánchez, D.G.; Tabares, A.; Faria, L.T.; Rivera, J.C.; Franco, J.F. A Clustering Approach for the Optimal Siting of Recharging Stations in the Electric Vehicle Routing Problem with Time Windows. *Energies* **2022**, *15*, 2372. [CrossRef]

12. Erdelić, T.; Carić, T. Goods Delivery with Electric Vehicles: Electric Vehicle Routing Optimization with Time Windows and Partial or Full Recharge. *Energies* **2022**, *15*, 285. [CrossRef]

13. Iwan, S.; Nürnberg, M.; Bejger, A.; Kijewska, K.; Małecki, K. Unloading Bays as Charging Stations for EFV-Based Urban Freight Delivery System—Example of Szczecin. *Energies* **2021**, *14*, 5677. [CrossRef]

14. Martins, L.d.C.; Tordecilla, R.D.; Castaneda, J.; Juan, A.A.; Faulin, J. Electric Vehicle Routing, Arc Routing, and Team Orienteering Problems in Sustainable Transportation. *Energies* **2021**, *14*, 5131. [CrossRef]

15. Wang, L.; Wu, Z.; Cao, C. Integrated Optimization of Routing and Energy Management for Electric Vehicles in Delivery Scheduling. *Energies* **2021**, *14*, 1762. [CrossRef]

16. Baek, D.; Chen, Y.; Chang, N.; Macii, E.; Poncino, M. Battery-Aware Electric Truck Delivery Route Exploration. *Energies* **2020**, *13*, 2096. [CrossRef]

17. Zardini, A.; Bonnel, P. *Real Driving Emissions Regulation: European Methodology to Fine Tune the EU Real Driving Emissions Data Evaluation Method, EUR 30123 EN*; Publications Office of the European Union: Luxembourg, 2020; ISBN 978-92-76-17157-7. [CrossRef]

18. Szałek, A.; Pielecha, I.; Cieslik, W. Fuel Cell Electric Vehicle (FCEV) Energy Flow Analysis in Real Driving Conditions (RDC). *Energies* **2021**, *14*, 5018. [CrossRef]

19. Al-Wreikat, Y.; Sodre, J. Evaluating the Energy Consumption of an Electric Vehicle Under Real-World Driving Conditions. SAE Technical Paper 2022-01-1127. 2022. [CrossRef]

20. Cieślik, W.; Szwajca, F.; Golimowski, J. The possibility of energy consumption reduction using the ECO driving mode based on the RDC test. *Combust. Engines* **2020**, *182*, 59–69. [CrossRef]

21. Pielecha, I.; Cieslik, W.; Szalek, A. Impact of Combustion Engine Operating Conditions on Energy Flow in Hybrid Drives in RDC Tests. SAE Technical Paper 2020-01-2251. 2020. [CrossRef]

22. Andrych-Zalewska, M.; Chlopek, Z.; Merkisz, J.; Pielecha, J. Comparison of Gasoline Engine Exhaust Emissions of a Passenger Car through the WLTC and RDE Type Approval Tests. *Energies* **2022**, *15*, 8157. [CrossRef]

23. Pielecha, J.; Skobiej, K.; Gis, M.; Gis, W. Particle Number Emission from Vehicles of Various Drives in the RDE Tests. *Energies* **2022**, *15*, 6471. [CrossRef]

24. Pielecha, I.; Szałek, A.; Tchorek, G. Two Generations of Hydrogen Powertrain—An Analysis of the Operational Indicators in Real Driving Conditions (RDC). *Energies* **2022**, *15*, 4734. [CrossRef]

25. Selleri, T.; Melas, A.D.; Franzetti, J.; Ferrarese, C.; Giechaskiel, B.; Suarez-Bertoa, R. On-Road and Laboratory Emissions from Three Gasoline Plug-In Hybrid Vehicles—Part 1: Regulated and Unregulated Gaseous Pollutants and Greenhouse Gases. *Energies* **2022**, *15*, 2401. [CrossRef]

26. Melas, A.; Selleri, T.; Franzetti, J.; Ferrarese, C.; Suarez-Bertoa, R.; Giechaskiel, B. On-Road and Laboratory Emissions from Three Gasoline Plug-In Hybrid Vehicles—Part 2: Solid Particle Number Emissions. *Energies* **2022**, *15*, 5266. [CrossRef]

27. Pielecha, J.; Skobiej, K.; Kubiak, P.; Wozniak, M.; Siczek, K. Exhaust Emissions from Plug-in and HEV Vehicles in Type-Approval Tests and Real Driving Cycles. *Energies* **2022**, *15*, 2423. [CrossRef]

28. Electromobility Meter: On Polish Roads in 2021. Available online: https://www.pzpm.org.pl/pl/Rynek-motoryzacyjny/Licznik-elektromobilnosci/Rok-2021/Grudzien-2021 (accessed on 24 November 2022).

29. Electric Vans—Will They Manage to Meet the Demands of the Modern World—Toyota Proace EV. Available online: https://youtu.be/Y6kQzUP2CBw (accessed on 10 November 2022).

30. Toyota Proace Electric. Available online: https://www.toyota.co.uk/new-cars/proace-ev (accessed on 24 November 2022).

31. Electric Vehicle Database. Available online: https://ev-database.org/ (accessed on 10 November 2022).
32. Proace EV Service Manuals. Available online: https://www.techdoc-toyota.com (accessed on 10 November 2022).
33. Gis, W.; Pielecha, J.; Merkisz, J.; Kruczyński, S.; Gis, M. Determining the route for the purpose light vehicles testing in Real Driving Emissions (RDE) test. *Combust. Engines* **2019**, *178*, 61–66. [CrossRef]
34. Weather Archive. Available online: https://www.weatheronline.pl/weather/maps/city (accessed on 10 November 2022).

Article

Fuel Cell Electric Vehicle (FCEV) Energy Flow Analysis in Real Driving Conditions (RDC)

Andrzej Szałek [1], Ireneusz Pielecha [2,*] and Wojciech Cieslik [2]

1 Toyota Motor Poland, ul. Konstruktorska 5, 02-673 Warszawa, Poland; andrzej.szalek@toyota.pl
2 Faculty of Civil and Transport Engineering, Poznan University of Technology, ul. Piotrowo 3, 60-965 Poznan, Poland; wojciech.cieslik@put.poznan.pl
* Correspondence: ireneusz.pielecha@put.poznan.pl; Tel.: +48-61-224-4502

Abstract: The search for fossil fuels substitutes forces the use of new propulsion technologies applied to means of transportation. Already widespread, hybrid vehicles are beginning to share the market with hydrogen-powered propulsion systems. These systems are fuel cells or internal combustion engines powered by hydrogen fuel. In this context, road tests of a hydrogen fuel cell drive were conducted under typical traffic conditions according to the requirements of the RDE test. As a result of the carried-out work, energy flow conditions were presented for three driving phases (urban, rural and motorway). The different contributions to the vehicle propulsion of the hydrogen system and the electric system in each phase of the driving route are indicated. The characteristic interaction of power train components during varying driving conditions was presented. A wide variation in the contribution of the fuel cell and the battery to the vehicle's propulsion was identified. In urban conditions, the share of the fuel cell in the vehicle's propulsion is more than three times that contributed by the battery, suburban—7 times, highway—28 times. In the entire test, the ratio of FC/BATT use was more than seven, while the energy consumption was more than 22 kWh/100 km. The amounts of battery energy used and recovered were found to be very close to each other under RDE test conditions.

Keywords: hydrogen vehicle; energy flow; hybrid powertrain; real driving conditions

Citation: Szałek, A.; Pielecha, I.; Cieslik, W. Fuel Cell Electric Vehicle (FCEV) Energy Flow Analysis in Real Driving Conditions (RDC). *Energies* **2021**, *14*, 5018. https://doi.org/10.3390/en14165018

Academic Editor: Woojin Choi

Received: 19 July 2021
Accepted: 11 August 2021
Published: 16 August 2021

Publisher's Note: MDPI stays neutral with regard to jurisdictional claims in published maps and institutional affiliations.

1. Introduction

The use of hydrogen for energy production can be particularly important for industries that are difficult to convert to electric power. This is especially relevant for transportation and industrial production. Currently, most hydrogen is produced from natural gas without CO_2 capture during production (CCS—Carbon Capture and Storage). Beyond 2030, hydrogen production from this source with CO_2 capture is not expected to increase significantly, as this process will only become cost-competitive when CO_2 emission fees are around USD 90 per ton. In contrast, hydrogen from renewable electricity is and will only be cost-effective if low-cost excess electricity is used. Furthermore, it is assumed that in major hydrogen-consuming regions, hydrogen production from biomass will only play a minor role [1].

Depending on the raw materials used in hydrogen production and the amount of CO_2 emissions accompanying this process, the produced hydrogen is labeled by colors. Gray hydrogen is produced from fossil fuels, and the associated CO_2 is released into the atmosphere. When a process is used to capture CO_2 that is infused, for example into a mine shaft, the hydrogen is referred to as blue. If renewable energy and a CO_2-free process are used to produce hydrogen, the resulting hydrogen will be referred to as green hydrogen.

For producing hydrogen from fossil fuels, steam reforming and gasification processes are used. The efficiency of these processes, their mass scale of production and the inexpensive price of raw materials result in a low price of hydrogen. On the other hand, however, they require additional hydrogen purification processes. For hydrogen produced from fossil fuels to have a purity above 99%, it must be purified in an enrichment step. This technology is currently used on an industrial scale primarily as a pressure swing adsorption (PSA). The contamination of hydrogen with hydrogen sulfide has a huge impact on the durability of fuel cells, while the carbon monoxide content affects the voltage generated by the cell.

Current hydrogen production is for chemical applications where it is fully consumed. The demand generated by transport using hydrogen will be covered by the newly established production facilities. The processes used there qualify the product as green hydrogen, giving the desired effect of zero-emission transport at the same time.

A dominant process in producing green hydrogen is electrolysis. As a result of water electrolysis, hydrogen and oxygen are extracted with very high purity, above 99.999%. This fact is used in direct hydrogen production on the spot of the hydrogen fueling station for fuel cell vehicles, without the need for further purification. However, to obtain such pure hydrogen, the water must be preconditioned. The current energy efficiency of hydrogen production by electrolysis is about 75%.

According to the International Energy Agency [1], the structure of world hydrogen production consists of about 48% hydrogen produced from natural gas, 30% from oil and 18% from coal. The remaining 4% is produced by the electrolysis of water.

Hydrogen has a significantly higher energy density value than batteries (in terms of mass and volume), which benefits the vehicle storage capacity and affects the driving range of the vehicle. Taking these advantages into account, hydrogen fuel cells or internal combustion engines powered by hydrogen can be used in passenger cars, vans, trucks buses and other means of transport (Figure 1).

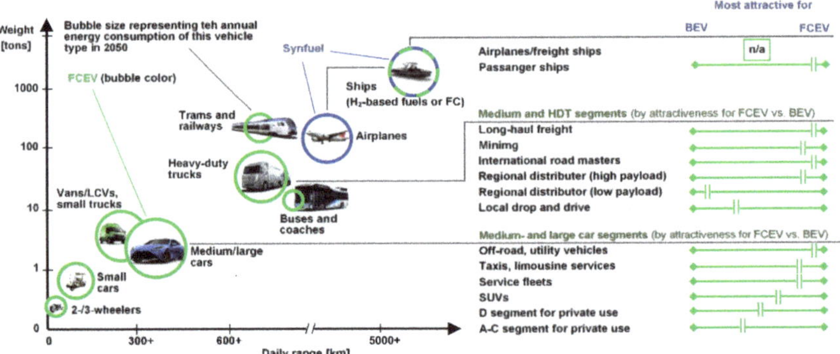

Figure 1. Comparison of range and payload for hydrogen and battery technology in means of transportation [2].

The BEV systems can be used in small passenger vehicles where the daily mileage limit is quite low. With respect to trucks, the use of FCEVs starts to be very beneficial. Fuel cells require far fewer raw materials in the production stage than electrochemical batteries. An additional advantage is the lack of the use of cobalt and the limited use of platinum (compared to internal combustion engine vehicles).

The current price of hydrogen for end-user transport in Europe ranges from EUR 5 to EUR 9.5 per kg, depending on the region. The lowest price is due to the fact that it is produced as a waste product in industrial chemical processes, while the highest price is a contractual price intended to equate the cost of operating a fuel cell car with a spark ignition (SI) engine.

According to the Hydrogen Council [3], the price of hydrogen for fuel cells will decrease by about 60% for the end user over the next decade. This will occur in regions with access to cheap natural gas and the ability to store captured CO_2. In addition, with an increased demand for hydrogen, the cost of hydrogen supply over the coming decade could decrease by as much as 70%. As a result, the cost of distributed hydrogen in 2030 could be in the range of USD 4.5–6 per kg. A comparative analysis of refueling vehicles with conventional fuels, hydrogen and recharging batteries at the filling station shows (Figure 2) that this time is significantly shorter for conventional fuels and hydrogen than electric vehicles, while hydrogen provides a much greater driving range than in the case of electric vehicles. Additional factors favoring hydrogen as a fuel are the investment costs associated with the construction and size of refueling stations.

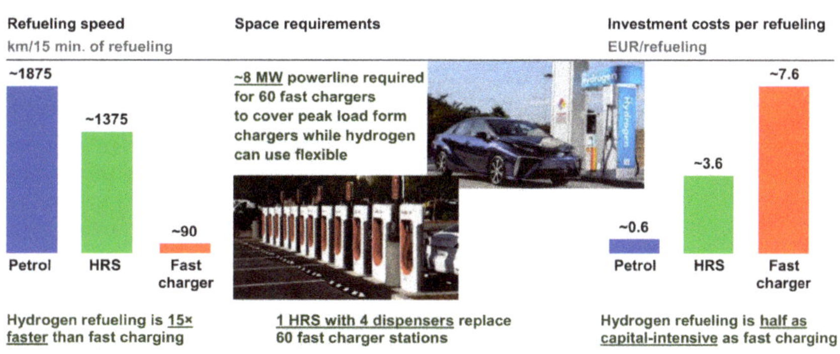

Assumptions: average mileage of passenger car = 24,000 km; number of PCs in EU in 2050: ~180 million; ICE: range = 750 km/ref., refueling time = 3 min.; FCEV: range: 600 km/ref., refueling time = 5 min., fast charger = 1080 km^2; BEV: range = 470 km/ref., refueling time = 75 min., gas station = 1080 m^2; WACC 8%; fast charger: hardware = USD 100,000, grid connection = USD 50,000, installation costs = USD 50,000, lifetime = 10 years; HRS: capex (1,000 kg daily) = EUR 2,590,000, lifetime = 20 years, refueling demand/car = 5 kg; gas: capex = EUR 225,750, lifetime = 30 years, 1 pole per station.

Figure 2. Comparison of refueling performance and investment rates for traditional and near-future fuels [2].

According to the Hydrogen Council, the CO_2 emission of hydrogen pathways (the well-to-tank stage) from natural gas via SMR (steam methane reforming) was ~75 g/km, accounting for ~60% of the total CO_2 emissions of a FCEV from lifecycle perspective [4]. Most hydrogen today is produced from fossil fuels and emits carbon (grey hydrogen). For producing low-carbon hydrogen from natural gas with CCS, the following two technology options exist: steam methane reforming (SMR) and autothermal reforming (ATR) [3]. SMR combines natural gas and pressurized steam to produce syngas, which is a blend of carbon monoxide and hydrogen. ATR combines oxygen and natural gas to produce syngas. This process can easily capture up to 95% of CO_2 emissions. ATR technology is typically used for larger plants compared with SMR technology. Based on the data presented in [5], CO_2 emission during hydrogen production is (kg CO_{2-e}/kg H_2): coal gasification (no CCS)—12.7–16.8; coal gasification and CCS (best case)—0.71; SMR (no CCS)—8.5; SMR and CCS (best case)—0.76.

The data compiled by the International Council on Clean Transportation Europe [6] show that the average real-world fuel (kg) and electricity consumption (kWh/100 km) values for lower medium and SUV segment cars registered in the European Union are, respectively, BEV: 20.6 and 21.9 kWh and FCEV: 1.0 and 1.2 kg.

In a study, the authors of [6] stated that the life cycle for GHG emissions of average gasoline- and diesel-powered ICEVs (internal combustion engine vehicle) are very similar and range from 226–227 g CO_2 eq./km for small, 245–246 g CO_2 eq./km for lower medium and 266–288 g CO_2 eq./km for SUV segment cars. The emissions from FCEV vehicles looks similar to the following: of medium segment, 202 g CO_2 eq./km (from natural gas) and 55–60 g CO_2 eq./km (from renewable hydrogen). However, the data presented in [7] show that in some European countries, the amounts of the average carbon footprint over a lifetime (segment D) are significantly different: Germany—426 g CO_2 eq./km and France—112 g CO_2 eq./km. This is mainly due to the way in which hydrogen is produced.

When the electrolysis is powered by 100% renewable energy, the gain in emissions from hydrogen production makes it possible to reach the BEV level (80 g CO_2 eq./km).

As hydrogen in gaseous form has a very low density (0.089 kg/m^3) and is significantly lighter than air, it is usually stored compressed. In vehicle propulsion applications, to increase the energy density, two standards are usually used for hydrogen storage pressure, that is 35 MPa, which corresponds to a density of 23 kg/m^3 and 70 MPa, which corresponds to a density of 38 kg/m^3. For a pressure of 35 MPa, the volumetric energy density of hydrogen is 2.8 MJ/dm^3, while for a pressure of 70 MPa, the energy density is 4.7 MJ/dm^3.

Hydrogen fuel cell power cars started to reach the US consumer market already in 2014. Official U.S. sales of Hyundai's cars began in June 2014, Toyota's in October 2015 and Honda's in December 2016 (primarily in the state of California).

"Global Market for Passenger Hydrogen Fuel Cell Vehicles" conducted a study at the beginning of the HFCV sale, which projected that by the end of 2020, global sales would amount to more than 27,500 passenger cars powered by hydrogen fuel cells. In 2020 alone, 8500 units were sold [8]. A barrier to the development of this drive has been identified as the lack of available hydrogen refueling infrastructure. The report also indicated that sales of hydrogen-powered cars and SUVs will increase in the coming years. Last year, more than 8500 of such vehicles were sold, which was the highest annual sales rate ever recorded. It should be noted that such high sales in 2020 were achieved despite the huge economic slowdown experienced by the automotive industry during the SARS-CoV-2 pandemic. Consequently, sales of passenger cars and SUVs, light commercial vehicles and full-size trucks and buses are expected to grow very rapidly in the coming years [8].

In 2020, the global number of hydrogen refueling stations was 553; it is planned that in 2021 this number will increase by another 221 stations. In Europe, there are 200 stations (including 100 in Germany), in Asia—275 and in North America—75. The most dynamic development of this technology is observed in Germany, China, Korea and Japan [9].

2. Analysis of Hydrogen Usage in Internal Combustion Engines and Fuel Cells

The automotive deployment of hydrogen is currently in two application areas (a) as a fuel in internal combustion engines and (b) as a fuel in fuel cells.

Research on hydrogen-powered internal combustion engines began in the 1930s [10]. A broad spectrum of categorization of a hydrogen internal combustion engine (HICE) based on typical injection and ignition strategies was presented by Yip et al. [11]. Typical solutions involve the indirect injection of hydrogen into the intake tract. The second technical solution is the direct injection of hydrogen into the cylinder [12] or the use of both variants [13]. It is realized in spark-ignition [14] and compression-ignition engines as well.

Smirnov and Nikitin [15] conducted studies of hydrogen ignitability in closed chambers. Models of hydrogen combustion were proposed and verified with reference to pre-mixed and non-premixed combustion and detonation models.

One form of using hydrogen in an internal combustion engine is its co-combustion with the following other fuels (dual–fuel systems): gasoline [16], diesel [17], natural gas [18,19], methanol [20,21] or as an additive to other fuels (butanol [22], natural gas [18] or fuel mixtures [23]).

Simulation studies of co-combustion of hydrogen with diesel fuel by Babayev et al. [17] indicate that (a) compressed ignition hydrogen reacting jets are fundamentally different from diesel jets, (b) both the free-jet and the global mixing modes govern the compressed ignition hydrogen combustion cycles and (c) jet-mixing combustion is more effective and should be maximized in compressed ignition H_2 engines.

A common solution is to use fuel cells together with internal combustion engines in Range Extender systems. Such a solution presented by Chubbock and Clague [24] involves a package of two fuel cells with a total power of 7.8 kW (the FC power mass index is 0.2 kW/kg) and a tank for storing 1.5 kg of hydrogen. The system uses a three-cylinder internal combustion engine with a displacement of 660 cm^3 and a power output of 30 kW (operating as a power generator system).

The fuel cells used in the first prototype vehicles (in 2002) achieved a volumetric power factor of 1.0 kW/dm^3 with a mass power factor of 0.75 kW/kg [25]. In the FCHV model (in 2008), these ratios were 1.45 kW/dm^3 and 0.9 kW/kg, respectively. The first-generation Toyota Mirai had values of 3.1 kW/dm^3 and 2 kW/kg, while the new generation of the Mirai vehicle achieves 5.4 kW/dm^3 (4.4 kW/kg excluding end plates) and 5.4 kW/kg, respectively [26].

Honda used 130 kW fuel cells in the Clarity model, for which the volumetric and mass power factors were 3.1 kW/dm^3 and 2.0 kW/kg [27,28].

Although the parameters and metrics of fuel cells are known, there is a lack of analysis on energy consumption in typical road tests by fuel cell powered automotive vehicles. There are few publications on different vehicles [29] or other research tests [30]. Therefore, the aim of this paper is to fully analyze the energy flow in the Real Driving Condition test (based on the Real Driving Emissions test) and, additionally, to analyze the use of fuel cells, a high-voltage battery and an electric motor.

3. Materials and Methods

3.1. Research Objects

The research was conducted using a Toyota Mirai first-generation vehicle (Table 1). The vehicle uses components from Toyota's hybrid vehicle models mass-produced since 1997. These components consisted of the vehicle's power management unit, known as the power controller and the voltage converter, both used from a third-generation Prius model; the traction electric motor was taken from a Lexus RX 450h hybrid model; and the high-voltage battery was taken from a Toyota Camry model.

Table 1. Toyota Mirai powertrain system [31,32].

Component	Parameter	Value
Vehicle	weight	1850 kg
	top speed	179 km/h
	acceleration 0 to 60 mph	9.6 s
	range (homologation cycle)	approx. 483 km
Fuel cell	type	PEM (polymer electrolyte)
	power	114 kW
	power density	2.0 kW/kg; 3.1 kW/dm^3
	number of cells	370
	humidificiation method	internal circulation system
Electric motor	type	permanent magnet synchronous
	peak power	123 kW at 4500 rpm
	maximum torque	335 N·m
	total speed reduction ratio	3.542
Battery	type	Nickel Metal Hydride (NiMH)
	capacity	6.5 Ah
	nominal voltage	244.8 V (7.2 × 34)
hydrogen storage	internal volume	122.4 dm^3
	nominal/filling pressure	70 MPa/87.5 MPa
	mass	approx. 5.0 kg
	refueling time	3 min.

The vehicle was equipped with a stack of 370 fuel cells creating a 114-kW power output. Two hydrogen tanks with a pressure of 70 MPa were used in the vehicle (Figure 3). This produced the highest unit mass power density of compressed hydrogen. The voltage from the fuel cell stack is converted to 650 V and powers an AC electric motor.

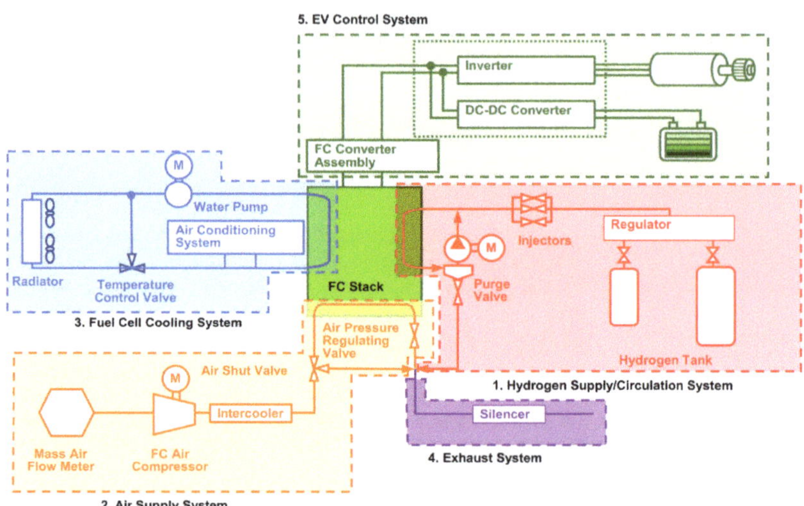

Figure 3. Toyota Mirai hydrogen system component layout (based on [33]).

The generation of the fuel cell used in this model has a volumetric energy density 28 times higher than that of the first generation used by Toyota. The first generation had a volumetric energy density of only $0.11\ \mathrm{kW/dm^3}$, while the one used in the tested Mirai model is $3.1\ \mathrm{kW/dm^3}$. This was achieved, partly due to the design of the cell stack with a configuration that allows internal self-humidification, using the circulation of water produced in each cell. This feature eliminated the need for a humidifier, significantly reducing the volume of the entire cell stack.

The vehicle has two tanks of 120 and 122 $\mathrm{dm^3}$, holding a total of 5 kg of hydrogen (Table 2). The calculated ratio of the mass of hydrogen, at maximum hydrogen filling, to the empty mass of the tanks is 5.7%.

Table 2. Parameters of hydrogen tanks used in Toyota Mirai [34].

Component	Parameter	Value
Compressed hydrogen tanks	Number of tanks	2
	storage density	5.7 wt%
Front tank	Capacity/weight	$60\ \mathrm{dm^3}/42.8\ \mathrm{kg}$
Rear tank	Capacity/weight	$62.4\ \mathrm{dm^3}/44.7\ \mathrm{kg}$

The use of the Fuel Cell Boost Converter (FDC) from the hybrid model in the Mirai enabled the use of an inverter and an electric traction motor, already used in series-produced hybrid vehicles. In addition to these components, an Intelligent Power Module (IPM) was also used [33].

The high-voltage battery used in this model has the same function as in the hybrid-drive models. Its main role is to accumulate the energy regenerated during braking [35]. The energy stored in the battery is used to power the powertrain during vehicle startup and during acceleration [36]. This keeps the instantaneous hydrogen consumption at a very low, or zero, level compared to if the energy was generated by the fuel cell alone. Since the mass of the vehicle is 1850 kg, the designers used a 244.8 V traction battery to ensure adequate performance. Since Toyota uses nickel-metal hydride batteries for a majority of its hybrid models, the same battery was used for the model Mirai. The battery structure contains 34 modules with 6 cells of 1.2 V each.

3.2. Research Equipment and Methodology of Determination of the Energy Flow

The measurements were performed using a specialized, dedicated diagnostic tester utilizing the OBD (on-board diagnostics) connector. The research used data provided by one of the vehicle systems—the hybrid control. This system operates using selected vehicle data, fuel cell stack, parameters of the electric motor and the parameters of the high-voltage battery. The vehicle driving conditions were determined based on the measurements of the vehicle speed and the data sampling time. The resolution was 1 Hz.

The assessment of the energy flow was carried out based on the measurements of the engine speed and load, the speed and torque of the electric motors/generators, the battery voltage and the current (including the boost voltage).

Using the above measurement data, the following quantities were determined:

- energy flow (urban, rural, motorway):

$$\Delta E_i = \int_{t=0}^{t=t_{max}} U_{BAT} \cdot I_{BAT} dt \tag{1}$$

the instantaneous energy flow values ΔE_i were divided in accordance with the following criteria:

- discharging (urban, rural, motorway):

$$\Delta E_{dis} = \int_{t=0}^{t=t_{max}} U_{BAT} \cdot I_{BAT} dt \; (\text{if } \Delta E_i < 0), \tag{2}$$

- charging (urban, rural, motorway):

$$\Delta E_{ch} = \int_{t=0}^{t=t_{max}} U_{BAT} \cdot I_{BAT} dt \; (\text{if } \Delta E_i > 0 \text{ and } T_{reg} \geq 0), \tag{3}$$

- regenerative braking (urban, rural, motorway):

$$\Delta E_{reg} = \int_{t=0}^{t=t_{max}} U_{BAT} \cdot I_{BAT} dt \; (\text{if } \Delta E_i > 0 \text{ and } T_{reg} < 0), \tag{4}$$

where U_{BAT}—voltage (V), I_{BAT}—current (A), dt—time interval (h), T_{reg}—braking torque (Nm);
- boost value (urban, rural, motorway):

$$boost = \frac{U_{HV}}{U_{LV}} \tag{5}$$

where U_{LV}—low voltage side (V) and U_{HV}—high voltage side (V).

4. Results

4.1. Driving Test Evaluation

The main problem of a constantly developing industry is its negative impact on the environment. Transportation is one of the most rapidly changing industries, and it significantly affects the concentration of hazardous substances in the air. To reduce the impact of vehicles on the environment, increasingly stringent standards for exhaust emissions are being introduced and solutions are being developed to minimize vehicle emissions. Exhaust emission standards are set to control the pollutants emitted from automotive vehicles around the world. Exhaust emission values are measured under conditions in an established type of the approval test. This part of the vehicle certification process is responsible for the environmental performance of the vehicle and is the same for all passenger cars. The course of the test corresponds to the most likely road conditions, and the tests performed, which are the same for all vehicles, authorize the comparison of emission results between them. Nowadays, the focus is more and more on road testing, i.e., testing under real driving conditions. These tests have now been integrated into European Union regulations under the name RDE (Real Driving Emissions) [37,38]. These are made to best reflect the actual operating conditions of the vehicle in terms of environmental aspects. The research presented in this paper omits the analysis of exhaust emissions (which, in a fuel cell vehicle, is zero), focusing on the analysis of the energy consumption of a modern propulsion system based on RDE test standards. With this in mind, the authors refer to this test by the acronym RDC (Real Driving Conditions) [39,40]. Due to increasing electrification of vehicles, comparative work on the energy consumption of propulsion is extremely important for the development of the transportation field. The RDC (RDE) test procedure is universal within the European Union and can be carried out on selected sections of a road that meet the basic requirements. The route is divided into three sections corresponding to the speed of urban, rural and motorway driving conditions. The test was performed with the FC vehicle in Warsaw (Poland) and met all the requirements, as shown in Figure 4.

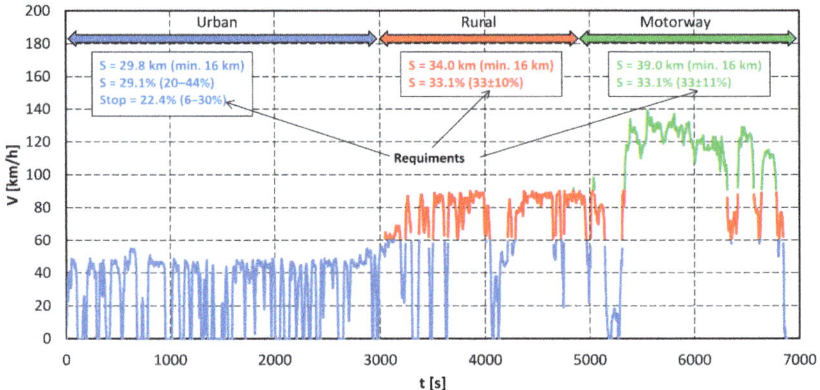

Figure 4. Course of the RDC test with characteristics phases (S = 102.8 km, t = 114.5 min).

4.2. State of Charge (SOC) Change Analysis

The drive system of the Toyota Mirai is equipped with a high-voltage nickel-metal hydride battery. During driving, the battery is charged and discharged due to the characteristic parameters of the route, such as the amount of acceleration and braking in a given section of the route. The energy recovered during braking can be reused to power the vehicle, as is the case in hybrid and electric vehicles. The study identified the areas with the highest and lowest average battery charge levels (Figure 5a).

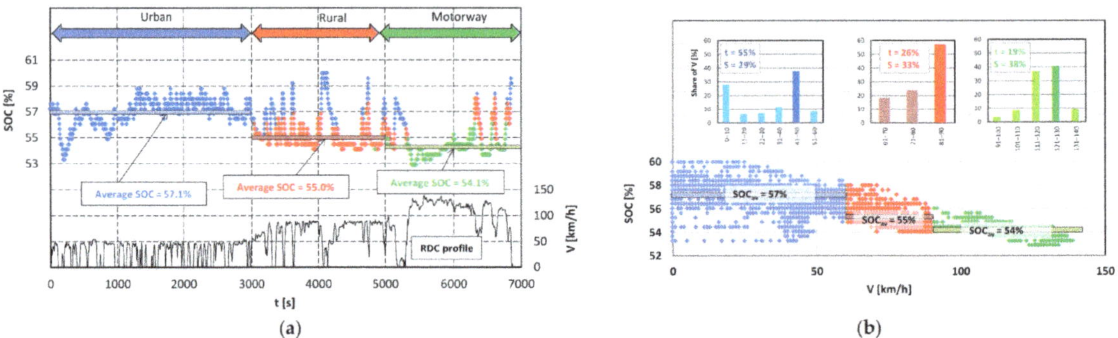

Figure 5. Changes in battery SOC: (**a**) with averaged values for travel phases; (**b**) in relation to travel speed.

As the driving speed increases, resulting in less braking zone, the high-voltage battery reaches lower average charge levels. Of course, this value is dependent on the characteristics of the route, since a single slowing of the vehicle results in a significant increase in recovered energy. Despite the direct response of the level of charge to vehicle driving parameters, the total SOC fluctuation area is in the range of 53–60%, which is a small range of the full battery capacity of 1.6 kWh. Slight variations of the SOC are characteristic of hybrid powertrains, where the engine (in the case of the vehicle under study, a fuel cell) is the main source of propulsion. The battery is supposed to mainly store energy from recovery modes or excess energy production by the powertrain.

The largest changes in battery charge are recorded in the range of driving speeds up to 60 km/h (which corresponds to urban driving speeds), where the highest average charge was recorded at the same time (Figure 5b). The change in battery charge level oscillates in the range of ΔSOC = 7.06% over the entire test interval.

4.3. Powertrain Performance Evaluation

In a hybrid vehicle, the energy for propulsion comes from two sources. The range of the propulsion power source in the intervals of each stage of the RDC test route was determined in the tested vehicle. The battery and fuel cell power consumption conditions shown in Figure 6a,b indicate areas of battery-only operation and areas of dual power source cooperation in the vehicle drive. At low vehicle speeds in the 0–10 km/h range, the propulsion energy comes from the battery. Higher vehicle speeds result in the fuel cell starting to work in the power generation process. An increase in vehicle speed increases both the instantaneous maximum values of the powertrain energy demand and the average values in each speed window with an interval of every 10 km/h (Figure 7).

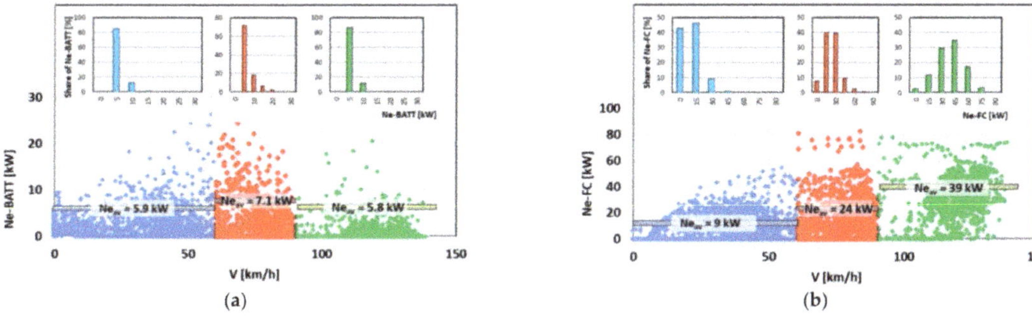

Figure 6. Powertrain usage conditions: (**a**) battery, (**b**) fuel cell; with a specific subdivision of the test phases.

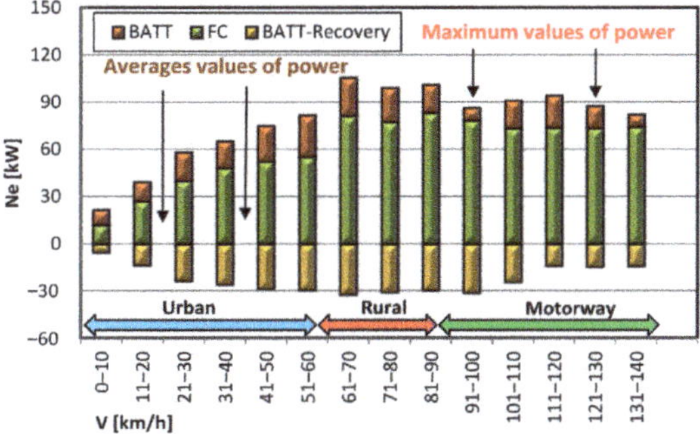

Figure 7. Average and maximum changes in battery (discharge and regeneration) and fuel cell contributions during each phase of the RDC test.

Due to the individual driving parameters, some speed ranges recorded lower energy consumptions, however, the justification for these results should be found in the temporary driving conditions, for example, the lowest average energy values were obtained in the speed range 41–50 km/h, this speed usually occurs only in transition states between acceleration and deceleration to the maximum speed of the urban speed range. Over a wide range of speed ranges, similar energy recovery values were recorded for both the urban and rural sections, indicating the versatility of energy recovery at varying speeds.

4.4. Evaluation of Electric Drive Operating Conditions and Energy Consumption

The energy to operate the vehicle comes mainly from the fuel cell, transient situations caused by acceleration additionally consume energy from the high-voltage battery, but in total, due to energy recovery to the battery during braking, the energy flow from this source becomes neutral. The small changes between EC_ALL and EC_FC are due to the inclusion of changes in battery discharge (EC_BATT) and battery recharge due to recovery regenerative braking (EC_REC), according to the following equation:

$$EC\_ALL = EC\_FC - EC\_BATT + EC\_REC \tag{6}$$

In this way, the difference between EC_ALL and EC_FC is not significant because the battery usage and recharge is close to zero. These conditions are illustrated in Figure 8.

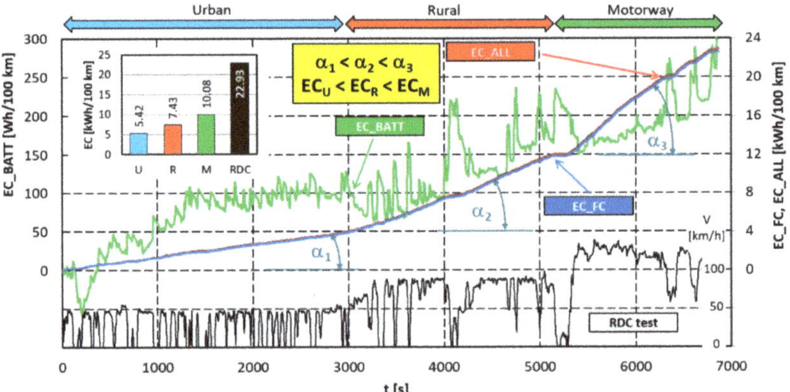

Figure 8. Energy consumption conditions in the RDC test phases (EC_U—energy consumption in urban phase, EC_R—energy consumption in rural phase, EC_M—energy consumption in motorway phase).

Confirmation of the above statement regarding the main use of the fuel cell for vehicle propulsion is provided by an analysis of the energy flow shares (discharge and charge) per vehicle speed for the battery (Figure 9). It can be concluded that in selected speed ranges, the regenerated energy is equal to the energy consumed during driving or acceleration. However, it should be remembered that this graph only shows the energy flow from the battery and the fuel cell is also used for propulsion, which generates much more energy to drive the vehicle. The drive characteristics also indicate that braking energy recuperation only occurs until 8 km/h, below which the vehicle decelerates using the conventional braking system. The highest energy consumption from the battery was recorded for accelerating the vehicle from a standstill when the fuel cell is not yet generating the required propulsion power. In the presented single RDC test, the total energy recovered is 322 Wh more than the energy consumed; therefore, the energy recovery system is highly efficient (the battery does not require an external source of charge to obtain the energy needed to drive the vehicle—part of the energy also comes from the operation of the fuel cell, what is indicated later in this article).

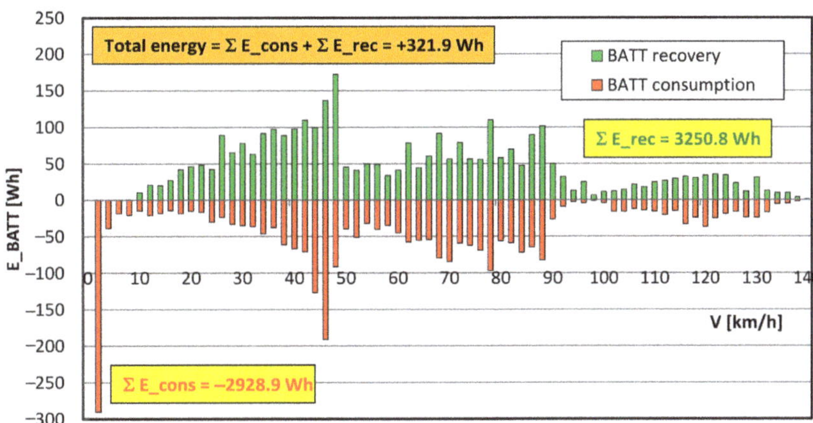

Figure 9. Shares of battery energy use during discharge and charge in relation to driving speed.

Complementing the presented summary evaluation of the powertrain during the RDC test is a discussion of energy flow on selected instantaneous single vehicle acceleration and deceleration states. Based on Figure 10, it is possible to describe the following characteristic operating points of the drive system:

1. during vehicle acceleration, the initial energy input from the battery is visible until the fuel cell starts producing energy (for further analysis see Figure 11);
2. during braking, the fuel cell operation is shut down and the energy is recovered to the high-voltage battery;
3. depending on the vehicle's acceleration rate, larger amounts of energy are consumed from both the battery and the fuel cell sources;
4. when the vehicle is standing still, the energy requirements of the on-board systems (comfort, entertainment) are fulfilled by the high-voltage battery.

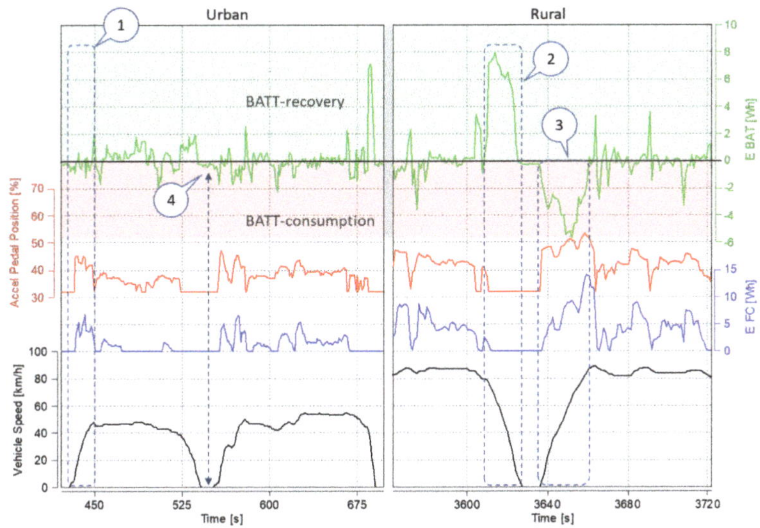

Figure 10. Interaction of powertrain components during driving of Toyota Mirai (selected events on road).

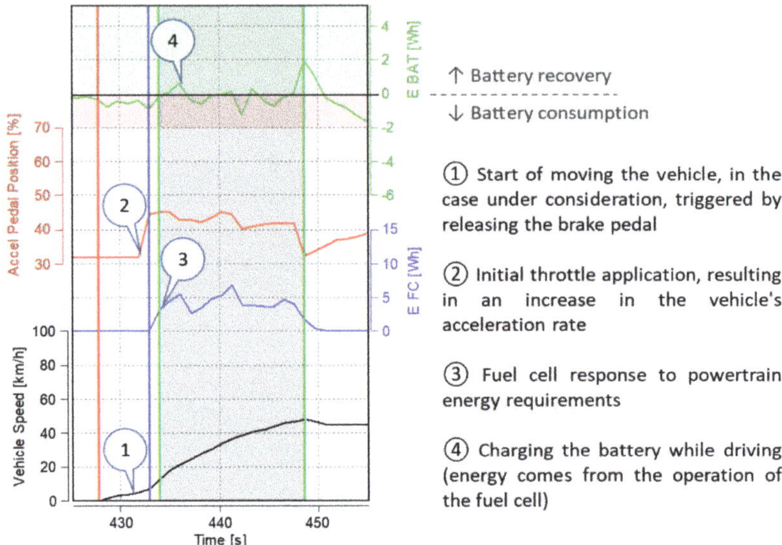

Figure 11. Analysis of a selected single acceleration process of a FC vehicle.

Regardless of the energy source, the wheel propulsion is provided entirely by the electric motor. It is, therefore, important to recognize the operating conditions of the drivetrain (Figure 12). During the entire RDC test, the powertrain generates 83.7% of the vehicle's propulsion and 16.3% of the energy is recovered during braking. During both propulsion and braking of the vehicle, the powertrain operates within the voltage range of mainly 300–350 V—achieving more than 60% of the total time share. Individual operating points generate higher voltages in the range of 350–652 V; however, the total share of these values is much lower (these are noted at intervals of higher vehicle speeds or higher powertrain loads). The maximum torque achieved during the test was 216 Nm, which is 65% of the max torque claimed by the manufacturer; therefore, the drivetrain in the RDC test does not require maximum torque to complete the run. The regenerative braking characteristics indicate a constant braking torque in the electric motor speed range 1500–9000 rpm at the level of 15 Nm. Regenerative braking is only possible down to a speed of about 7 km/h.

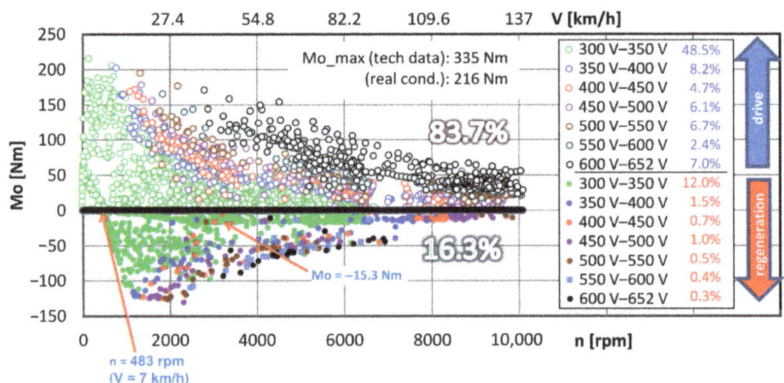

Figure 12. Characteristics of an electric motor in relation to its operating voltage.

The charging and discharging conditions of the high-voltage battery are shown in Figure 13. Although the nominal voltage is reported as 244.8 V, its operating conditions indicate much larger positive fluctuations around this value. The minimum value fluctuates around 240 V, but the maximum value far exceeds 300 V. The operating conditions of the battery indicate that it is possible to receive about 26 kW of power when discharged. During its charging, much higher power values were obtained (more than 32 kW), indicating slightly different charging and discharging conditions (with similar current values); the voltage variations are around 20 V.

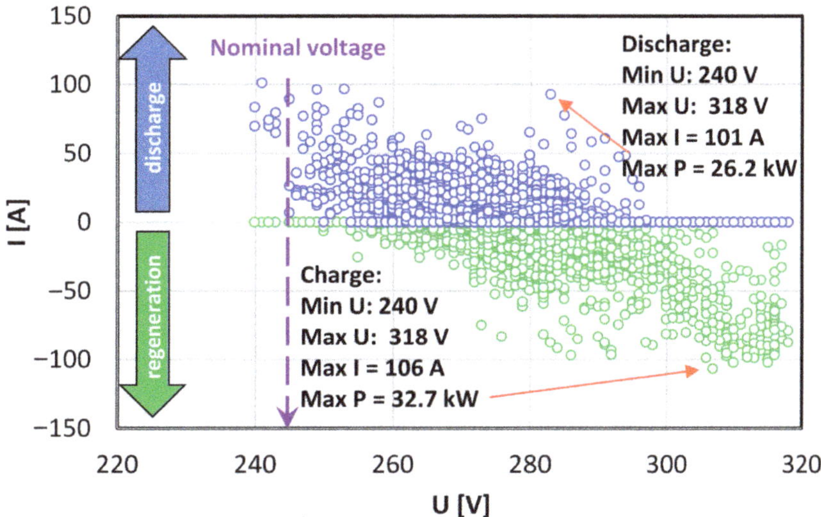

Figure 13. Characteristics of a high voltage battery.

The version of the first-generation Toyota Mirai powertrain presented in this paper includes a Fuel Cell Boost Converter. This is a significant change from the powertrain presented in 2008 (designated as Toyota FCHV adv). This means that the fuel cell using boost can largely self-power the vehicle's electric motor. The operating conditions in Figure 14 indicate that at a fuel cell current value of about 100 A, the converter maintains a voltage of 650 V. For current values from about 200 A, the converter converts voltage from the cell to the maximum level of 650 V. The voltage–current characteristics of the fuel cell indicate voltage values of 200–300 V at no load to about 200 V at maximum current values of over 350 A. The power characteristics of the fuel cell do not have a typical maximum; therefore, increasing the current increases its power. Thus, applying voltage amplification at a specific current value, in this case at about 100 A, effectively increases the power directed to the electric motor.

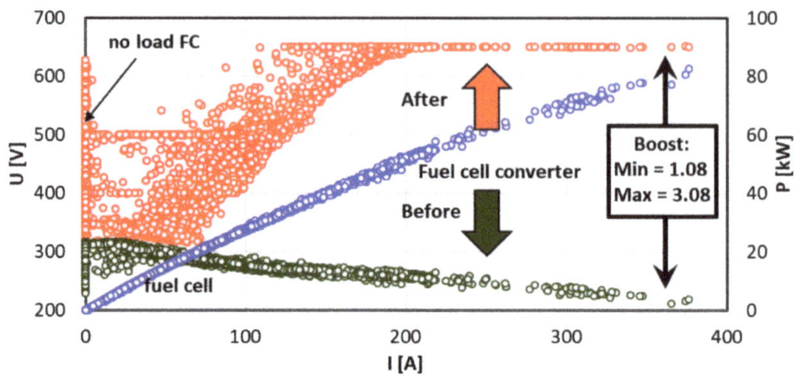

Figure 14. Current–voltage characteristics of a fuel cell and its voltage converter.

An evaluation of the total energy flow in the Toyota Mirai drive system is shown in Figure 15. The contribution of the HV battery in the individual phases of the RDC test is not large—as the driving speed increases, the contribution decreases. For the entire test, the share of battery utilization is about 13%. The share of energy recovery to the battery (regenerative braking and fuel cell charging) is slightly higher. The larger values of energy recovered to the battery apply to each phase of the test—approximately 0.6 percentage points on average. This shows that in the entire driving test, slightly more energy was supplied to the battery than was used. Energy flow analysis shows the vehicle consumed 22.285 kWh in the test. However, the fuel cell "produced" 22.60 kWh, which is indicated in Figure 15—as 101.5% of the total energy consumed. This difference is due to the energy recovery to the battery. Summarizing the energy flow in the RDC test, it should be stated that the contribution of the fuel cell to the vehicle's propulsion is more than 70%. The rest is half, indicating the battery's contribution and energy recuperation.

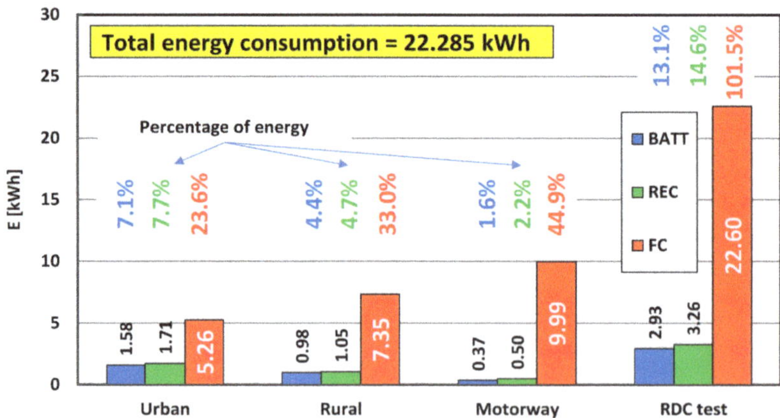

Figure 15. Energy consumption conditions for the RDC test phases and the entire RDC test.

5. Conclusions

Modern hybrid drives that use a fuel cell instead of an internal combustion engine as the main source of propulsion are now a trend in the development of future zero-emission automotive. The advantage of fueling the vehicle quickly is a key advantage over electric vehicles, which take much more time to charge depending on the charging type. Fuel cells are a multipurpose source of electricity that can be converted to electric drive in basically any type of propulsion system; therefore, the research presented above is important because of the recognition of the energy balance of such vehicles. The widespread use of fuel cells in automobile, truck and even maritime transport brings significant benefits. However, the global use of this type of propulsion depends on the development of a hydrogen re-fueling infrastructure. The study determined at what ranges the drive is realized with a fuel cell, and in what ranges a high-voltage battery is engaged. The operating conditions of these systems have been specified.

The above analysis of the operating conditions of the hydrogen vehicle propulsion system under real traffic conditions (according to the RDE test procedure) indicates the following:

- in most cases, the high voltage battery is charged only from energy regeneration during braking; however, there are also situations where the battery is charged from the energy generated by the fuel cell (in the test, the battery charge from the fuel cell reached 0.315 kWh—Figure 15);
- as the vehicle speed increases (in other words as the RDC test interval changes), the battery energy consumption decreases and the fuel cell energy consumption increases (Figure 15);
- the vehicle is initially started from standstill by using a high-voltage battery, only after a certain time, depending on the load, is the fuel cell activated (Figure 11).

Author Contributions: Conceptualization, I.P. and A.S.; methodology, A.S., I.P. and W.C.; validation, I.P., W.C. and A.S.; formal analysis, I.P., W.C. and A.S.; investigation, I.P.; resources, A.S.; data curation, I.P. and A.S.; writing—original draft preparation, I.P., W.C. and A.S.; visualization, I.P., W.C. and A.S.; supervision, I.P. and A.S.; project administration, I.P. and A.S. All authors have read and agreed to the published version of the manuscript.

Funding: This research received no external funding.

Institutional Review Board Statement: Not applicable.

Informed Consent Statement: Not applicable.

Data Availability Statement: The data presented in this study are available on request from the corresponding author.

Acknowledgments: The authors wish to thank Toyota Motor Poland for providing the vehicles for the road tests.

Conflicts of Interest: The authors declare no conflict of interest.

References

1. Technology Roadmap. Hydrogen Fuel Cell. International Energy Agency. Paris, 2015. Available online: www.iea.org (accessed on 21 June 2021).
2. *Hydrogen Roadmap Europe: A Sustainable Pathway for the European Energy Transition, Fuel Cells and Hydrogen 2 Joint Undertaking*; Publications Office of the European Union: Luxembourg, 2019. [CrossRef]
3. Path to Hydrogen Competitiveness. A Cost Perspective. Hydrogen Council, 2020. Available online: www.hydrogencouncil.com (accessed on 21 June 2021).
4. Hydrogen Scaling Up. A Sustainable Pathway for the Global Energy Transition. Hydrogen Council, November 2017. Available online: https://hydrogencouncil.com/wp-content/uploads/2017/11/Hydrogen-scaling-up-Hydrogen-Council.pdf (accessed on 4 August 2021).
5. Australia's National Hydrogen Strategy. COAG Energy Council Hydrogen Working Group. Available online: https://www.industry.gov.au/sites/default/files/2019-11/australias-national-hydrogen-strategy.pdf (accessed on 4 August 2021).
6. Bieker, G. A Global Comparison of the Life-Cycle Greenhouse Gas Emissions of Combustion Engine and Electric Passenger Cars. International Council on Clean Transportation Europe. July 2021. Available online: https://theicct.org/sites/default/files/publications/Global-LCA-passenger-cars-jul2021_0.pdf (accessed on 1 August 2021).
7. Amant, S.; Meunier, N.; de Cosse Brissac, C. Road Transportation: What Alternative Motorisations are Suitable for the Climate? A Comparison of the Life Cycle Emissions, in France and Europe. Available online: https://www.carbone4.com/wp-content/uploads/2021/02/Road-transportation-what-alternative-motorisations-are-suitable-for-the-climate-Carbone-4.pdf (accessed on 4 August 2021).
8. Hydrogen Fuel News. Hydrogen News Daily–Alternative Energy Reports. Available online: https://www.hydrogenfuelnews.com/fuel-cell-passenger-vehicles/8546147 (accessed on 20 June 2021).

9. Ludwig-Bölkow-Systemtechnik GmbH. Available online: www.lbst.de (accessed on 21 June 2021).
10. Das, L.M. Hydrogen engines: A view of the past and a look into the future. *Int. J. Hydrogen Energy* **1990**, *15*, 425–443. [CrossRef]
11. Yip, H.L.; Srna, A.; Yuen, A.C.Y.; Kook, S.; Taylor, R.A.; Yeoh, G.H.; Medwell, P.R.; Chan, Q.N. A review of hydrogen direct injection for internal combustion engines: Towards carbon-free combustion. *Appl. Sci.* **2019**, *9*, 4842. [CrossRef]
12. Yu, X.; Li, D.; Yang, S.; Sun, P.; Guo, Z.; Yang, H.; Li, Y.; Wang, T. Effects of hydrogen direct injection on combustion and emission characteristics of a hydrogen/Acetone-Butanol-Ethanol dual-fuel spark ignition engine under lean-burn conditions. *Int. J. Hydrogen Energy* **2020**, *45*, 34193–34203. [CrossRef]
13. Zareei, J.; Rohani, A.; Mazari, F.; Mikkhailova, M.V. Numerical investigation of the effect of two-step injection (direct and port injection) of hydrogen blending and natural gas on engine performance and exhaust gas emissions. *Energy* **2021**, *231*, 120957. [CrossRef]
14. Younkins, M.; Boyer, B.; Wooldridge, M. Hydrogen DI dual zone combustion system. *SAE Int. J. Engines* **2013**, *6*, 45–53. [CrossRef]
15. Smirnov, N.N.; Nikitin, V.F. Modeling and simulation of hydrogen combustion in engines. *Int. J. Hydrogen Energy* **2014**, *39*, 1122–1136. [CrossRef]
16. Karthikeyan, S.; Periyasamy, M. Impact on the power and performance of an internal combustion engine using hydrogen. *Mater. Today Proc.* **2021**, in press. [CrossRef]
17. Babayev, R.; Andersson, A.; Dalmau, A.S.; Im, H.G.; Johansson, B. Computational characterization of hydrogen direct injection and nonpremixed combustion in a compression-ignition engine. *Int. J. Hydrogen Energy* **2021**, *46*, 18678–18696. [CrossRef]
18. Zareei, J.; Haseeb, M.; Ghadamkheir, K.; Farkhondeh, S.A.; Yazdani, A.; Ershov, K. The effect of hydrogen addition to compressed natural gas on performance and emissions of a DI diesel engine by a numerical study. *Int. J. Hydrogen Energy* **2020**, *45*, 34241–34253. [CrossRef]
19. Oni, B.A.; Sanni, S.E.; Ibegbu, A.J.; Aduojo, A.A. Experimental optimization of engine performance of a dual-fuel compression-ignition engine operating on hydrogen-compressed natural gas and Moringa biodiesel. *Energy Rep.* **2021**, *7*, 607–619. [CrossRef]
20. Gong, C.; Li, Z.; Sun, J.; Liu, F. Evaluation on combustion and lean-burn limitof a medium compression ratio hydrogen/methanol dual-injection spark-ignition engine under methanol late-injection. *Appl. Energy* **2020**, *277*, 115622. [CrossRef]
21. Wang, H.; Ji, C.; Shi, C.; Ge, Y.; Wang, S.; Yang, J. Development of cyclic variation prediction model of the gasoline and n-butanol rotary engines with hydrogen enrichment. *Fuel* **2021**, *299*, 120891. [CrossRef]
22. Meng, H.; Ji, C.; Wang, S.; Wang, D.; Yang, J. Optimizing the idle performance of an n-butanol fueled Wankel rotary engine by hydrogen addition. *Fuel* **2021**, *288*, 119614. [CrossRef]
23. Zhen, X.; Li, X.; Wang, Y.; Liu, D.; Tian, Z. Comparative study on combustion and emission characteristics of methanol/hydrogen, ethanol/hydrogen and methane/hydrogen blends in high compression ratio SI engine. *Fuel* **2020**, *267*, 117193. [CrossRef]
24. Chubbock, S.; Clague, R. Comparative analysis of internal combustion engine and fuel cell range extender. *SAE Int. J. Alt. Power* **2016**, *5*, 175–182. [CrossRef]
25. Konno, N.; Mizuno, S.; Nakaji, H.; Ishikawa, Y. Development of compact and high-performance fuel cell stack. *SAE Int. J. Alt. Power* **2015**, *4*, 123–129. [CrossRef]
26. Toyota Motor Europe. Toyota Service and Repair Information. Available online: www.toyota-tech.eu (accessed on 24 June 2021).
27. Yoshizumi, T.; Kubo, H.; Okumura, M. Development of high-performance FC stack for the new MIRAI. In *SAE Technical*; SAE International: Warrendale, PA, USA, 2021. [CrossRef]
28. Kimura, K.; Kawasaki, T.; Ohmura, T.; Atsumi, Y.; Shimizu, K. Development of new fuel cell vehicle Clarity Fuel Cell. *Honda RD Tech. Rev.* **2016**, *28*, 1–7.
29. Matsunaga, M.; Fukushima, T.; Ojima, K. Powertrain system of Honda FCX Clarity fuel cell vehicle. *World Electr. Veh. J.* **2009**, *3*, 820–829. [CrossRef]
30. Tazelaar, E.; Bruinsma, J.; Veenhuizen, B.; Van den Bosch, P. Driving cycle characterization and generation, for design and control of fuel cell buses. *World Electr. Veh. J.* **2009**, *3*, 812–819. [CrossRef]
31. Lohse-Busch, H.; Stutenberg, K.; Duoba, M.; Iliev, S. Technology Assessment of a Fuel Cell Vehicle: 2017 Toyota Mirai, ANL/ESD-18/12, Argonne National Laboratory. 2018. Available online: https://www.anl.gov/argonne-scientific-publications/pub/144774 (accessed on 20 June 2021).
32. Outline of the Mirai. Key Specifications, Mirai–Toyota Europe. Available online: www.toyota-europe.com (accessed on 21 June 2021).
33. Yumiya, H.; Kizaki, M.; Asai, H. Toyota Fuel Cell System (TFCS). *World Electr. Veh. J.* **2015**, *7*, 85–92. [CrossRef]
34. Hasegawa, T.; Imanishi, H.; Nada, M.; Ikogi, Y. Development of the fuel cell system in the Mirai FCV. In *SAE Technical*; SAE International: Warrendale, PA, USA, 2016. [CrossRef]
35. Pielecha, I.; Cieślik, W.; Szałek, A. Energy recovery potential through regenerative braking for a hybrid electric vehicle in a urban conditions. *IOP Conf. Ser. Earth Env.* **2019**, *214*, 012013. [CrossRef]
36. Pielecha, I.; Cieślik, W.; Szałek, A. The use of electric drive in urban driving conditions using a hydrogen powered vehicle–Toyota Mirai. *Comb. Eng.* **2018**, *172*, 51–58. [CrossRef]
37. The European Parliament and the Council of the European Union. Regulation (EU) 2019/631 of the European Parliament and of the Council of 17 April 2019 setting CO_2 emission performance standards for new passenger vehicles and for new light commercial vehicles, and repealing Regulations (EC) No 443/2009 and (EU) No 510. *Off. J. Eur. Union* **2019**, *111*, 13–534.

38. Gis, W.; Gis, M.; Pielecha, J.; Skobiej, K. Alternative exhaust emission factors from vehicles in on-road driving tests. *Energies* **2021**, *14*, 3487. [CrossRef]
39. Szałek, A.; Pielecha, I. The Influence of engine downsizing in hybrid powertrains on the energy flow indicators under actual traffic conditions. *Energies* **2021**, *14*, 2872. [CrossRef]
40. Cieslik, W.; Szwajca, F.; Golimowski, W.; Berger, A. Experimental analysis of residential photovoltaic (PV) and electric vehicle (EV) systems in terms of annual energy utilization. *Energies* **2021**, *14*, 1085. [CrossRef]

Article

Experimental Analysis of Residential Photovoltaic (PV) and Electric Vehicle (EV) Systems in Terms of Annual Energy Utilization

Wojciech Cieslik [1], Filip Szwajca [1], Wojciech Golimowski [2,*] and Andrew Berger [3]

[1] Department of Combustion Engines and Powertrains, Faculty of Civil and Transport Engineering, Poznan University of Technology, sq. M. Sklodowskiej-Curie 5, 60–965 Poznan, Poland; wojciech.cieslik@put.poznan.pl (W.C.); filip.szwajca@put.poznan.pl (F.S.)

[2] Department of Agroengineering and Quality Analysis, Faculty of Engineering and Economics, Wroclaw University of Economics and Business, Komandorska 180/120 Street, 53–345 Wroclaw, Poland

[3] Department of Physics and Engineering, University of Scranton, Scranton, PA 18510, USA; berger@scranton.edu

* Correspondence: wojciech.golimowski@ue.wroc.pl; Tel.: +48-71-368–0288

Abstract: Electrification of powertrain systems offers numerous advantages in the global trend in vehicular applications. A wide range of energy sources and zero-emission propulsion in the tank to wheel significantly add to electric vehicles' (EV) attractiveness. This paper presents analyses of the energy balance between micro-photovoltaic (PV) installation and small electric vehicle in real conditions. It is based on monitoring PV panel's energy production and car electricity consumption. The methodology included energy data from real household PV installation (the most common renewable energy source in Poland), electric vehicle energy consumption during real driving conditions, and drivetrain operating parameters, all collected over a period of one year by indirect measuring. A correlation between energy produced by the micro-PV installation and small electric car energy consumption was described. In the Winter, small electric car energy consumption amounted to 14.9 kWh per 100 km and was 14% greater than summer, based on test requirements of real driving conditions. The 4.48 kW PV installation located in Poznań produced 4101 kWh energy in 258 days. The calculation indicated 1406 kWh energy was available for EV charging after household electricity consumption subtraction. The zero-emission daily distance analysis was done by the simplified method.

Keywords: energy consumption; real driving conditions; electric vehicle; solar panels; energy flow; renewable energy

Citation: Cieslik, W.; Szwajca, F.; Golimowski, W.; Berger, A. Experimental Analysis of Residential Photovoltaic (PV) and Electric Vehicle (EV) Systems in Terms of Annual Energy Utilization. *Energies* **2021**, *14*, 1085. https://doi.org/10.3390/en14041085

Academic Editor: Anastasios Dounis

Received: 5 January 2021
Accepted: 15 February 2021
Published: 19 February 2021

Publisher's Note: MDPI stays neutral with regard to jurisdictional claims in published maps and institutional affiliations.

1. Introduction

The introduction of environmentally friendly and highly efficient energy conversion systems represents one of the biggest challenges in the development of vehicle drivetrain systems [1]. Considering global energy consumption, 44% of global transport energy is consumed by light-duty vehicles, the next 26% by heavy-duty vehicles [2], and 30% by others [3]. Over 99.8% of transport means are still powered by combustion engines, impeding fast transport decarbonization [3,4].

The mechanical energy generated by petroleum product's combustion processes produces problematic carbon dioxide and other toxic exhaust compounds [5,6]. The EU legislation pays special attention to CO_2 emission and, starting in 2021, has limited it to 95 g/km [7,8]. The CO_2 emission is tightly bound with fuel consumption and related directly to engine efficiency in tank to wheel (TTW) calculations. A wide variety of powertrains systems, including internal combustion engines (ICE) and electric vehicles (EV), dedicated to vehicles demand not only tank to wheel analysis but also life cycle assessment (LCA). It provides detailed information about the environment's energetic impacts [9].

LCA analysis carried out by Message et al. [10] shows the most significant climate change, expressed in gram CO_2/km, for conventional vehicles using fossil fuels, particularly Petrol Euro 4, and the lowest for EVs, highlighting the significance of energy source. The elementary difference in LCA between conventional and electric powertrain relies on Well-to-Tank (WTT) and TTW share. EV characterizes the majority of the WTT energy conversion share compared to conventional ICE, where TTW is dominant [10]. Respectively, other studies indicate EV lifetime relevance [11]. The attractiveness of EVs in terms of global warming potential increases with its lifetime.

An electric vehicle's advantage is that it consumes energy from various energy sources, such as fossil fuels, renewable energy (RES), bioenergy, or nuclear [12]. The trends in EV technology field development provide a necessity for energy market analysis. The study presented by Xu et al., 2020 [13] discusses four different charging strategies, and how they will influence greenhouse gas (GHG) emissions in 2050, i.e., when high decarbonization levels and large RES shares are expected. The analysis included electricity mix type and reduced energy consumption from gas-fired power stations, and increased energy share from RES by controlling the charging process. Other predictions, up to 2040, indicate the benefits of implementing the ClimPol scenario. It significantly increases energy sharing from RES, reducing kilogram carbon dioxide equivalence per kWh to just above 0.2, from the global case of slightly more than 0.8 [14]. The trend is primarily influenced by an increased share of wind, solar biomass, and nuclear power in energy generation and EV battery charging, with results depending on the EU country [15]. Overall, the lowest charging effect has been achieved in France in 2015. Forecasting shows a 19% increasing RES share in 28 EU countries, along with a 17% decrease of share of electricity from solid fuels between 2020–2050.

Rising demand for RES energy led to fast photovoltaic infrastructure development, which became competitive in terms of low cost and high efficiency. The flexibility of the design of PV systems allows energy production in a wide range of voltage, from systems with power above 100 MWp to household applications, most often below 15 kWp [16,17]. Integration of household PV systems and electric vehicle use are a promising solution for global GHG reduction and locally lower fuel costs [17,18]. Energy analysis from Kyoto, with limited areas intended for RES, indicated 74% CO_2 emission and 37% cost reduction from the power and transport sectors by applying photovoltaic rooftop systems and electric vehicles [19]. Results from 12 stands at an EV charging station equipped with photovoltaic panels (48 kW) and Li-ion 100 kW battery energy storage show the possibility of achieving 100% renewable electricity using appropriate control modes [20]. Modeling [21] of 400 combinations shows an attractive solution: cooperation of stationary battery (EV) with household PV infrastructure. The electric vehicle, being mobile energy storage, can effectively replace traditional battery storage. Using the battery in an EV as energy storage in such vehicle to grid (V2G) combinations, the self-sufficiency of solar self-consumption of household residentials increases [22]. In relation to V2G Technology, Wu Y. et al. proposes a real-time energy management system (EMS), that allows for a 29–55% reduction of the total cost of a Photovoltaic assisted charging station [23].

Energy stored in lithium-ion batteries has many advantages [24]. However, significant limitations are the dependence of the distance range on charging infrastructures, battery capacity, and drive quality [25,26]. Hence, it is important to monitor the high voltage battery state of charge to better understand energy flow phenomena and distance range prediction [27]. The range–distance prediction can be estimated based on various data collection methods. Zhang J. et al. obtained data from fifty EV taxis driving in Beijing [28]. In another study, testing was carried out by 32 electric busses traveling four routes under different working conditions [29]. Many studies about energy consumption prediction used data from experimental tests to validate new approaches [30–32]. In order to determine energy consumption by an electric vehicle, some researchers [33,34] used the real driving emission (RDE) test procedure as a suitable method to compare results with conventional ICE vehicles.

In recent years, Poland has noticed a rapid growth in interest in photovoltaic energy development [35]. The installed generated power increased by approximately 30% between the end of 2019 and May 2020 and achieved more than 1950 MW. The high growth rate places Poland in the top five EU countries in terms of new power. Most solar energy is produced by PV micro installations, representing more than 70% of all Polish PV markets in 2019. New regulations and supporting programs cause changes in the Polish renewable energy market [36]. The number of new registered battery electric vehicles has grown parallel to the number of household PV installations. Within a year, Poland's quantity of EVs increased by 80% to almost 7300 vehicles [37]. In effect, electric energy consumption will constantly be rising [38].

In the analysis of electric vehicle research, some major fields of experimental study can be quoted:

- Energy consumption by electric vehicle [28,39],
- Energy production by household photovoltaic installations [40,41],
- EV charging process analysis and optimization with reference to PV source of electricity [42–44].

Regarding the high rate of changes in the energy market and non-ICE vehicle development, the authors decided to investigate the energy balance between energy production from the micro photovoltaic system and the energy consumption by small passenger cars equipped with a battery electric powertrain system. The approach of combining real objects (vehicle, residential building) in the energy balance consideration, as proven in the literature, is a commonly analyzed issue. In terms of the studied field, the novelty is the connection between the following approaches: annual balance, changing of ambient conditions, and road approved electric vehicle energy consumption evaluation based on a real driving conditions (RDC) test.

This study's aim is to assess household micro-photovoltaic systems' self-sufficiency in connection with battery electric vehicle use. The research goals are:

- How much energy does a city electric vehicle consume during its intended operation?
- Is the 4.48 kW photovoltaic installation capable of satisfying the energy demand in the assumed scenarios of driving an electric vehicle?

The extent of this study included the energy flow analysis from a PV system, then charging, followed by RDC testing to discover the amount of energy consumed by an electric vehicle. It covered a one-year period with energy generation analysis from photovoltaic panels being carried out during the same periods as the vehicle tests, with average energy consumption during driving estimated for both winter and summer measurements.

2. Methods of Analysis

The aim of this study was to evaluate the energy flow generated by the solar panels, followed by analysis of the energy flow during vehicle's charging and the energy consumption in the RDC test. The analysis of energy generation from photovoltaic panels was carried out during the same periods as the vehicle tests, i.e., in winter and summer periods in 2020 on the territory of Poznan city in Poland. In the same periods, the average driving energy consumption was also estimated. The following questions were posed: How much energy does a city electric vehicle consume during its intended operation periods? Is the applied 4.48 kW photovoltaic installation able to guarantee an electric vehicle's energy demand in the assumed driving scenarios? The scope of the research included two test runs compliant with the RDC test procedure in urban, rural, and motorway cycles in winter and summer conditions of 2020. The measurements were made in the ECO driving mode, which in earlier studies [34] showed a beneficial reduction in energy consumption by the vehicle. The measurements made in the same driving mode allowed the estimation of the impact of weather conditions on the overall energy consumption of the vehicle on selected road sections. The data from the test run were recorded in real-time based on the information pulled from the vehicle controller area network (CAN) by a dedicated on-board diagnostics system (OBD) scan tool. The main parameters that were recorded during the measurements in the actual traffic conditions of the vehicle were those describing the operation of the electric motor (rotational speed, value of the current and voltage, and torque and vehicle speed) and the parameters concerning the accumulation of energy in the battery (state of charge (SOC), and power).

The characteristics of the electric vehicle used in road tests are shown in Table 1. The vehicle used in the tests—ŠKODA CITIGO^e iV—is supplied with an electric drive, allowing different driving modes and variable intensity of regenerative braking. The 61 kW ŠKODA CITIGO^e iV powertrain used a Li-Ion battery of 36.8 kWh full capacity and 32.3 kWh useable capacity. The location of the batteries have also been shown in Table 1.

Table 1. Technical data of the analyzed powertrain fitted in ŠKODA CITIGO^e iV [34].

Electric Motor/Car		Battery	
Parameter	Value	Parameter	Value
Max. voltage	360 V	Type	Li-ion
Max. power output	61 kW	Capacity total	36.8 kWh
Max. torque	212 Nm	Capacity usable	32.3 kWh
Maximum speed	130 km/h	Charge port	AC-Type 2 DC-CCS2
Operating weight	1235 kg	Charge power	AC-7.2 kW DC-40 kW

This article presents an analysis of three stages of energy conversion with emphasis on the vehicle's energy consumption under real traffic conditions (Figure 1). The comparison of the stages of energy conversion from the PV energy generation to vehicle charging, allowed us to develop a compilation of the possibility of driving the vehicle using energy from renewable sources only.

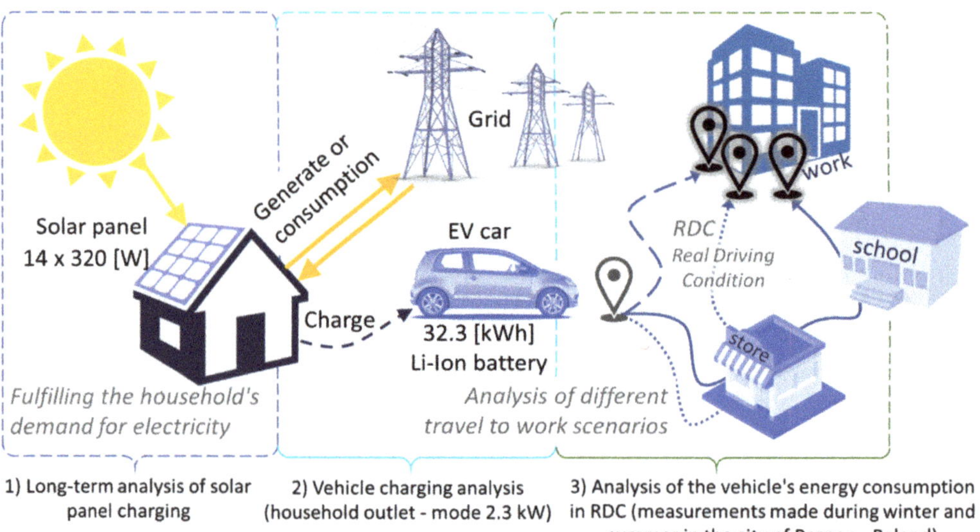

Figure 1. Division of analysis carried out in the discussed studies.

2.1. Long-term Analysis of Solar Panel Household Power Generation

Long term analysis of producing electric energy has been based on monitoring working the parameters of household micro-PV installation located in the northern part of Poznan and in service since February 2020. The system has been working in on-grid mode with solar electric power used by household and the surplus sold to the grid. The PV installation contains fourteen monocrystalline rooftop PV panels, DC/AC inverter, fuses, and assembly parts. The 320 W panels' total power output is 4.48 kW. The installation also includes three-phase 5.2 kW inverter SMA Sunny Tripower 5.0 and a two-way energy meter. The real-time view of the energy data (± 1 Wh) is possible through dedicated web and mobile apps.

The daily average energy consumption has been calculated using Equation (1) below.

$$E_{houshold_{usage}}avg = \frac{\left(E_{from\ the\ grid} + E_{direct\ use\ from\ solar\ panels}\right)}{\sum analyzed\ days} \tag{1}$$

$$E_{houshold_{usage}}avg = \frac{2293 + 402}{258} = 10.4 \left[\frac{kWh}{day}\right]$$

The calculation was based on 258 days with 2293 kWh of energy taken from the grid and 402 kWh of energy supplied by PV installation.

Available energy dedicated to an electric vehicle charging has been estimated using Equation (2).

$$E_{available\ for\ EV\ vehicle\ charging} = E_{from\ solar\ panels} - \left(E_{from\ the\ grid} + E_{direct\ use\ from\ solar\ panels}\right) \tag{2}$$

$$E_{available\ for\ EV\ vehicle\ charging} = 4101 - (2293 + 402) = 1406\ [kWh]$$

This additional energy of 1406 kWh is available for EV charging (taking into account the billing period of 258 days). Due to the analyses carried out in different periods of the year, it is possible to change the flow of energy from the source in the form of photovoltaic panels and the power plant. The trending differences are shown in Figure 2.

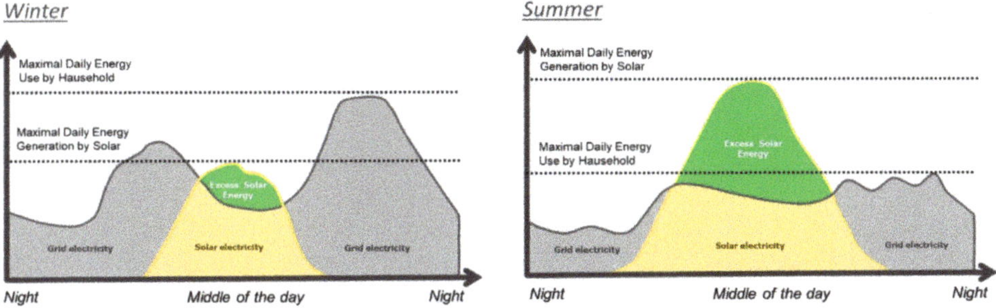

Figure 2. Characteristic correlations in electricity production from photovoltaic panels in the summer and winter months (based on [45]).

2.2. Electrical Vehicle Charging Modes

Small passenger EV used in this investigation could be charged following the modes below:

- 2.3 kW (AC) Type 2 from the household grid with dedicated converter supplied by the manufacturer,
- 7.2 kW (AC) Type 2 from the wall outlet or a public charging station,

- 40.0 kW (AC/DC) combined charging system (CCS) from the rapid charging station.

The combined charging system (CSS) can, within one hour, charge up to 80% of the car battery capacity.

The full charge time increases with decreased charging power. Charging profile of the first mode—2.3 kW (AC)—was observed over one full cycle. The charging level ($\pm 0.1\%$), voltage (± 1 V), and current (± 0.01 A) were sampled with 1.3 Hz and were registered from vehicle's CAN in real time. The observation setup is shown below in Figure 3.

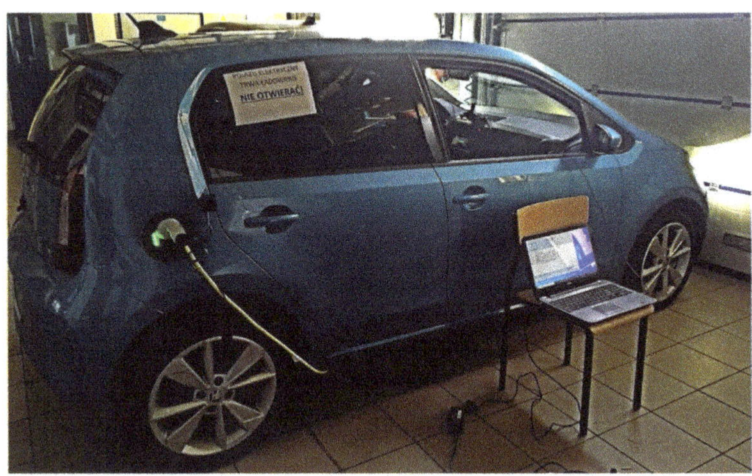

Figure 3. Vehicle charging setup.

2.3. Vehicle Energy Consumption in RDC Test

The test route was proposed in [34,46] and determined to lead through the city of Poznań and its surrounding areas. It covered urban, rural, and motorway conditions. The maximum motorway legal speed is 140 km/h. Selected test requirements related to the course of the test run have been presented in Table 2. The duration of all the test runs exceeded 90 min, and the total length of the track did not change.

Table 2. Real driving conditions test requirements with map of the route traveled during the measurements [34].

Route Pattern Followed in Research	Selected RDE/RDC Test Requirements	Urban	Rural	Motorway
	Cycle repetition [%] ($\pm$ 10%)	29 < ratio $\leq$ 34	33	$\leftarrow$
	Speed [km/h]	< 60	60 $\leq$ V $\leq$ 90	V > 90
	Max. speed [km/h] ($\pm$ 15 km/h for less than 3% of driving time)	-	-	145
	Average speed (stops included) [km/h]	15 $\leq$ V $\leq$ 30	-	-
	Minimum travelled distance [km]	16	$\leftarrow$	$\leftarrow$
	Altitude difference (beginning/end) [m]	100	$\leftarrow$	$\leftarrow$
	Maximum slope [m/100km]	1200 m/100 km	$\leftarrow$	$\leftarrow$

The main problem of constantly developing industry is its negative impact on the environment. One of the most dynamically changing sectors of industry is transport, which significantly affects the concentration of hazardous substances in the air. In order to reduce the impact of vehicles on the environment, increasingly restrictive emission standards are being introduced, and solutions are being sought to minimize the emission of exhaust fumes from vehicles. Exhaust emission standards are set to control the pollutants emitted from automotive vehicles around the world. Exhaust emission values are measured under conditions in an established type approval test. This part of the vehicle certification process is responsible for the environmental performance of the vehicle and is the same for all passenger cars. The course of the test corresponds to the most likely road conditions, and the tests performed, which are the same for all vehicles, authorize the comparison of emission results between them. However, currently, more and more attention is paid to road tests, i.e., tests performed in real driving conditions. At present, these tests have been included in the European Union regulations under the name RDE (real driving emissions) [47,48]. They are performed in order to best reflect the actual vehicle operation conditions in terms of ecological aspects. Such tests must be performed with specific requirements, the main assumptions of which are presented in Table 2. The winter and summer runs performed in this research met the requirements specified in the RDE test directive (Table 3). However, due to the lack of exhaust emission measurement, these tests are named RDC.

Table 3. Meeting real driving emissions (RDE) test requirements for summer and winter performed measurements [47,48].

Test Specification	Result Winter Conditions	Result Summer Conditions	Requirement	
Urban component [km]	22.3	23.4	>16	
Rural component [km]	27.2	23.6	>16	
Highway component [km]	28.3	30.0	>16	
Total route length [km]	77.8	77.0	>48	
Urban component [%]	29.7	30.3	29−44	
Rural component [%]	34.9	30.7	33 ± 10	
Highway component [%]	35.4	39.0	33 ± 10	
Average speed on urban route [km/h]	20.9	24.0	15–40	
Duration of stops on urban route [%]	28.9	28.24	6–30	
Trip duration at more than 100 km/h [min]	14.2	15.2	>5	
Maximum speed [km/h]	117.0	116	<160	
Trip duration at more than 145 km/h during the highway component [%]	0.0	0.0	<3	
Maximum single stop duration [s]	103.0	99.9	<180	
Trip duration [min]	99.5	92.8	90–120	
Urban: data set no. $a_i > 0.1$ m/s^2	1146	1024	>150	
Rural: data set no. $a_i > 0.1$ m/s^2	263	269	>150	
Highway: data set no. $a_i > 0.1$ m/s^3	158	188	>150	
Urban: 95. percentile $V{\cdot}a_{pos}$ [m^2 /s^3]	17.0	15.0	<17.2 [1]	<17.7 [2]
Rural: 95. percentile $V{\cdot}a_{pos}$ [m^2 /s^3]	24.7	14.8	<24.9 [1]	<24.6 [2]
Highway: 95. percentile $V{\cdot}a_{pos}$ [m^2 /s^3]	25.9	24.7	<27.3 [1]	<27.3 [2]
Urban: RPA [m/s^2]	0.323	0.268	>0.142 [1]	>0.137 [2]
Rural: RPA [m/s^2]	0.099	0.085	>0.047 [1]	>0.053 [2]
Highway: RPA [m/s^2]	0.065	0.078	>0.025 [1]	>0.025 [2]

[1] and [2]—Specific value determined for each trip, taking into account vehicle speed parameters based on European Union regulations. [1] Data calculated for winter real driving conditions (RDC) test. [2] Data calculated for summer RDC test.

2.4. Energy Supply and Demand for Selected Driving Scenarios

The choice of a means of transport for many users is motivated by economic factors (total cost of vehicle use). Literature sources of daily commutes provided average distances in the EU. Commutes within cities ranged between 4–25 km. Commutes from suburban areas can be significantly longer (Figure 4). In the next part of this section, we analyzed and presented three scenarios of daily commute.

Based on described trends [50] of dependency of distance and travel frequency and information about average distances covered in Poland [49], three different distances have been selected to analyze (Figure 5). Scenario 1 assumes a 15 km distance per day focusing on an urban area. Consequently, scenarios 2 and 3 concern the suburban areas where the residence is a farther distance from the workplace.

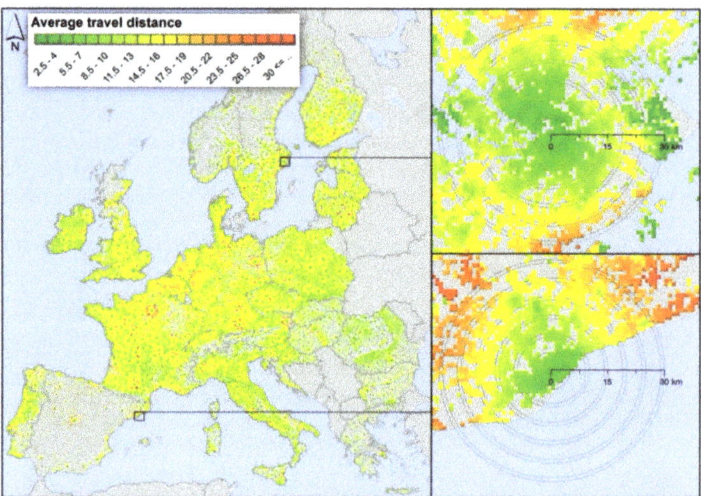

Figure 4. Average travel distances (km) in Europe, with focus on Stockholm and Barcelona, showing the dependence of the average distance on the distance from the metropolitan area [49].

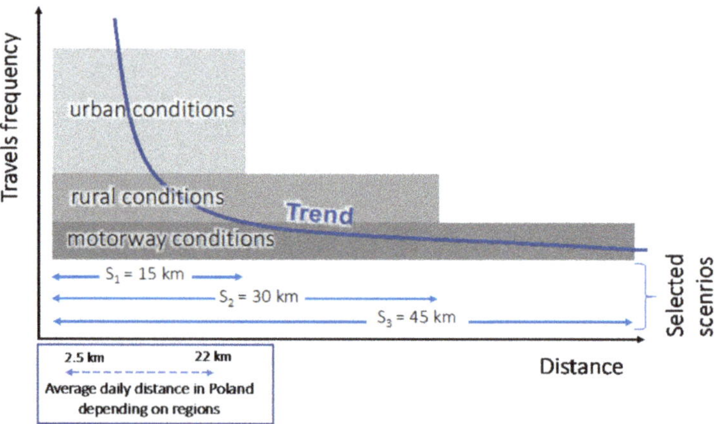

Figure 5. Typical distribution of frequency of trips vs. distance covered in Italy [50], supplemented by proposed scenarios and statistical data on average distances covered in Poland [49].

3. Results

3.1. Long Term Analysis of PV Energy Production

In Poland, the solar radiation falls consistently within 1050–1160 kWh/m^2/year, with highest values observed in the central part (Poznan or Warsaw) and in the south of the country (Krakow), as shown in Figure 6. The photovoltaic systems produce similar amounts of energy throughout the country. The Institute of Renewable Energy report shows that, in Poland, about 70% of currently installed photovoltaic sources are micro installations with an increasing trend. In 2019, there were 640 MW of new power installed, three times more than in 2018. Such a rapidly growing branch of power industry presents real possibilities for power self-sufficient households with excess energy to be dedicated to zero-emission transport.

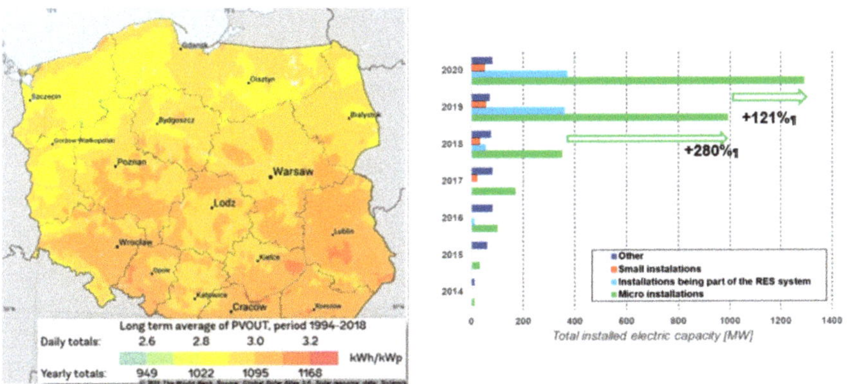

Figure 6. Average annual sum of photovoltaic (PV) power potential [51] and the development of micro installations in Poland [35].

Measurements of an existing single-family building with a 4.48 kWp photovoltaic installation were made. The Sunny Tripower 5.0 model STP5.0–3AV–40 424 inverter enabled real-time measurement of the generated power by the photovoltaic system, with monitoring performed by a dedicated application (SMA Smart Connected) and data archiving by the Sunny Portal service. The inverter's parameters are shown in Table 4 and the values of the energy obtained are presented in Figure 7.

Table 4. Characteristics of the inverter Sunny Tripower 5.0 used in the tested home installation [52].

Input (DC)		Output (AC)		Efficiency	
Parameter	**Value**	**Parameter**	**Value**	**Parameter**	**Value**
Max. PV array power	9000 Wp	Rated power (at 230 V, 50 Hz)	5000 W	Max. efficiency	98.2%
Max. input voltage	850 V	Max. output current	3×7.7 A	European efficiency	97.4%

The differences between energy generation in the summer and winter months were significant. Shortening the time of solar radiation of the panels by 30% reduced the maximum power generated by 19% (Figure 7), resulting in a total reduction in the share of accumulated energy by about 40% (energy gain in June was 664.5 kWh while in March only 406.7 kWh). Such large differences in the total values of energy produced raises doubts about the ability to meet the energy demand for both the power supply to the building and the electric vehicle.

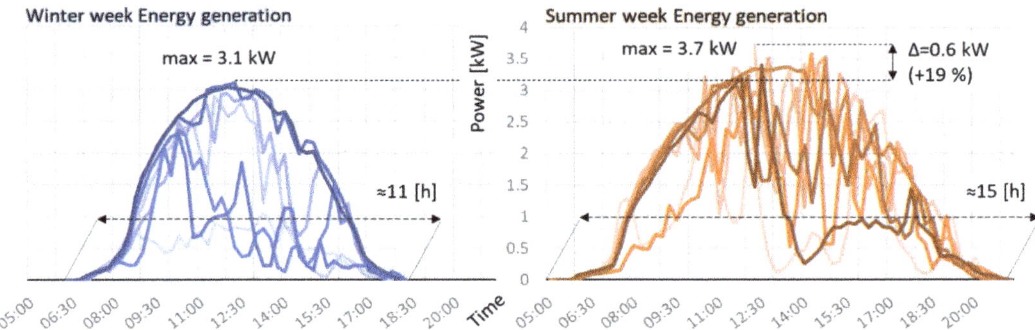

Figure 7. The data from 4.48 kWp solar energy installation located in Poznan city in Poland—energy generated during the winter and summer weeks, collected by the Sunny Tripower 5.0 data acquisition system.

The compiled daily electricity production characteristics (Figure 7) were averaged and presented in relation to the daily usage pattern of the electric vehicle (Figure 8). Part of the energy produced during the day (around the afternoon hours) was transferred to the power plant, because at this point, the energy was not used for the household's needs. The graph shows the maximum vehicle charging time with a discharged battery. Charging time, in this case, is long also because of the choice of basic charging technology. In the absence of energy production by photovoltaic panels, the energy needed to charge the vehicle (as well as other home usage) was drawn back from the power plant.

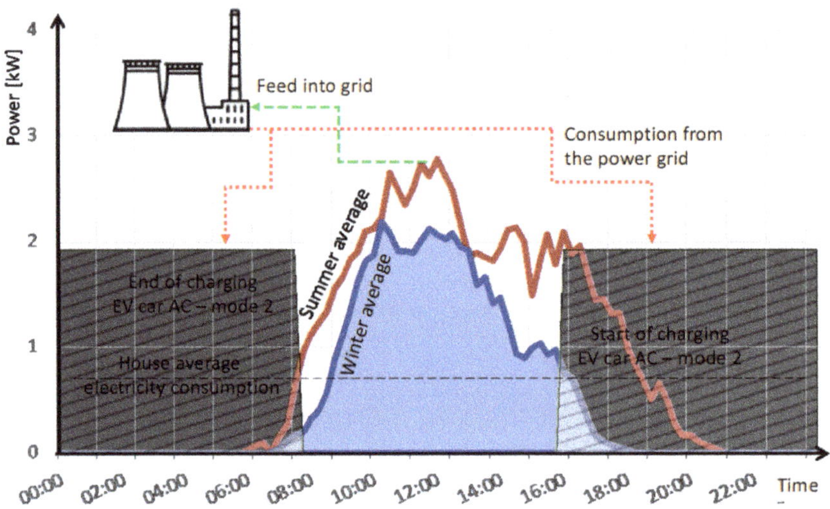

Figure 8. Electric vehicle charging scenario (using charging mode 2) in relation to the average generation of electricity from the photovoltaic installation (red—summer, blue—winter).

3.2. Vehicle Charging Analysis

The charging profiles, represented by SOC, power, and voltage, are shown in Figure 9. The authors did not optimize the charging process. Algorithms implemented by the vehicle manufacturer controlled the charging. Battery charging from 15.2% to 95.6% lasted almost sixteen hours. The charging process was carried out by constant current (CC) mode, i.e., with approximate constant current and variable voltage using the original manufacturer household charger. The voltage rose from 296.25 V to 351.75 V at the end of charging.

At the same time, the current value was changing significantly. At 96% of SOC, current reduced to slightly above 0, and with it, large voltage variations were observed. When the car was not being used and had a full battery, the charger ensured stand by energy consumption. The battery delivered 30.16 kWh energy after one charging cycle.

3.3. The Impact of Atmospheric Conditions on the Energy Consumption of an Electric Vehicle

The driving cycles, realized in compliance with the RDC procedures, were started at 100% battery SOC level (software readout). The test runs were performed by a single driver to assure consistency in the driving style. Compared to the previous analysis results of the driving energy consumption in the RDC test [34], this article focuses on the determination of the total energy consumption by route sections: urban, rural, and motorway under different atmospheric conditions. The vehicle velocity and relative SOC profile during the RDC tests are shown in Figure 10. Presented curves marked in blue and red colors represent different ambient conditions. Vehicle speed and state of charge are represented by solid and dotted lines, respectively. In the tests with similar conditions achieved during the measurement journeys, the indications of the vehicle speed in relation to the distance travelled showed a high similarity to the journey, both in the urban part and in the sections with increased speeds. Some differences in the speed on the suburban and highway routes were dictated by road conditions, and there was no possibility to repeat them.

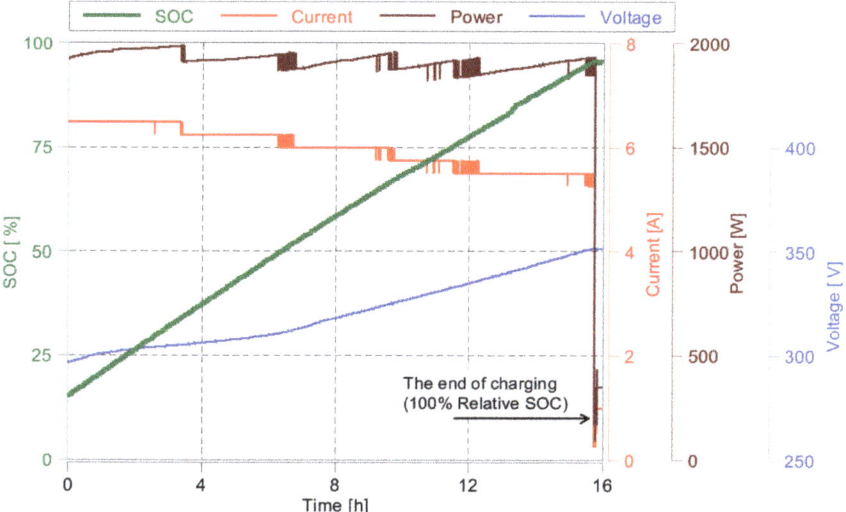

Figure 9. Single cycle charging profile carried out by household dedicated converter in constant current (CC) mode.

The driving cycles, realized in compliance with the RDC procedures, were started at battery level SOC 100% (software readout). The test runs were performed by a single driver to avoid inconsistency in the driving style. During the test run, the vehicle speed and battery level were recorded. Due to running the vehicle in ECO mode, the maximum speed was limited by the drivetrain controller; both runs were comparable speeds in the given test intervals, and nevertheless, the energy consumption was about 11% more in the winter period. For this reason, further work identifies the intervals of route split and vehicle speed affecting the increased energy consumption.

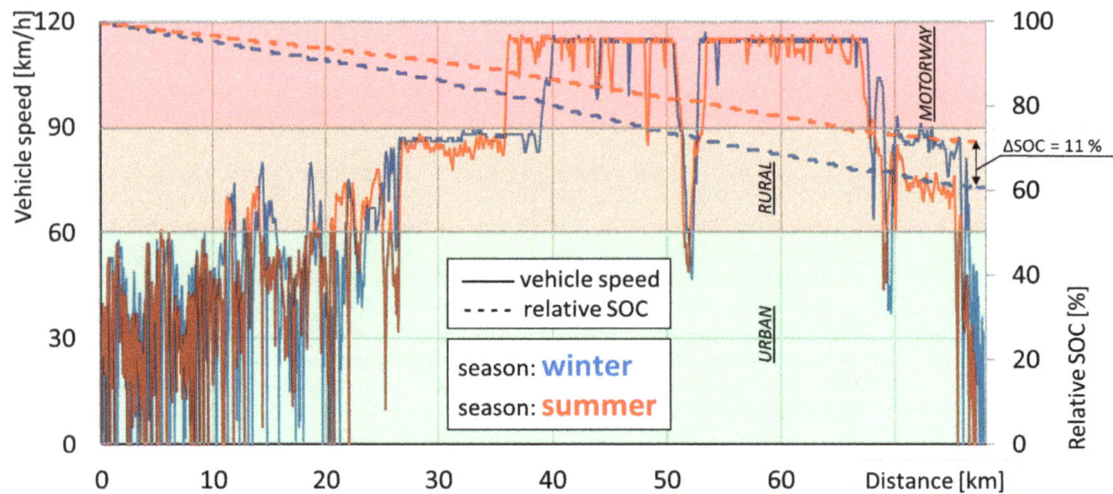

Figure 10. Comparison of the real driving conditions (RDC) test for the performed test runs in winter and summer conditions.

The flow of the energy ΔE was determined based on the flow of current (I_{BAT}) and voltage (U_{BAT}) of the battery as a result of its discharge and regenerative braking charge during driving of the car:

$$\Delta E_i = \sum_{t=0}^{t=t_{max}} U_{BAT} \times I_{BAT}\, dt \tag{3}$$

- discharging:

$$\Delta E_{dis} = \sum_{t=0}^{t=t_{max}} U_{BAT} \times I_{BAT} dt \; (when\; \Delta E_i < 0) \tag{4}$$

- energy recovery (regenerative braking):

$$\Delta E_{reg} = \sum_{t=0}^{t=t_{max}} U_{BAT} \times I_{BAT} dt \\ (when\; \Delta E_i > 0\; and\; M_{reg} < 0) \tag{5}$$

In order to determine the individual electric powertrain operating conditions, road portions were specified where the system operated in these individual conditions. On this basis, the operating modes were divided into individual phases: driving, acceleration, standstill, and braking, during operation of the electric drive. The adopted criteria have been shown in Table 5 and Figure 11.

Table 5. Vehicle motion phase criteria.

Parameter	Key Assumptions		
Drive	$a = 0$	$v > 0$	
Acceleration	$a > 0$	$v > 0$	
Standstill	-	$v = 0$	
Regenerative braking	$a < 0$	$v > 0$	$I_{BAT} > 0$

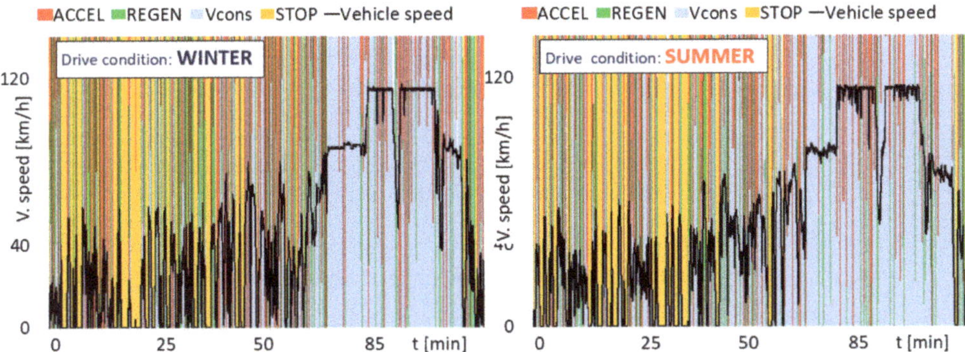

Figure 11. Distribution of motion phases in winter and summer conditions.

The energy balance as a function of the type of the road is shown below (Figure 12). Regardless of the weather conditions, similar energy recovery values were recorded. However, due to more vehicle stops during winter measurements, higher energy recovery was achieved in both the urban and suburban parts during winter measurements. The greatest amount of energy was recovered in the urban cycle due to the high number of brakings, compared to the smooth traffic road portions.

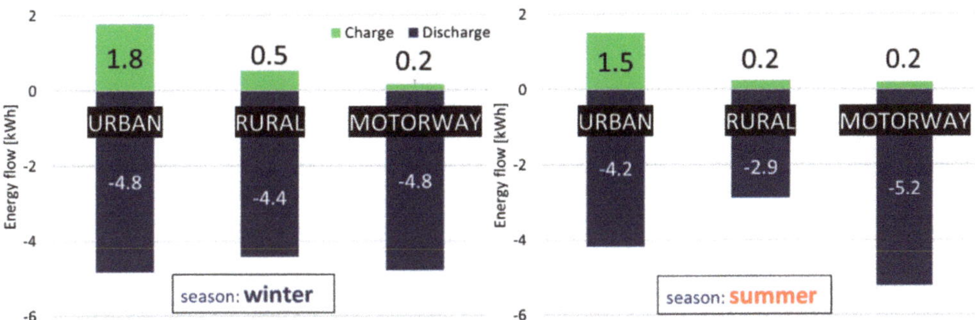

Figure 12. Energy consumption balance in terms of road and weather conditions.

The specification of the assumptions used in the energy flow summation (Figure 12) are specified in Table 6. The assumptions for the urban, rural, and motorway segmentation of the route are consistent with the RDC test assumptions presented earlier (Tables 2 and 3). The value of power delivered or generated from/to the battery was recorded during the measurements, with the following assumptions, the summed energy flows presented earlier were determined. These assumptions also apply to the energy consumption totals in Figure 14.

Table 6. Vehicle motion phase criteria [34,35].

General Form of the Equation	Energy Flow	Speed Requirements		
		Urban	Rural	Motorway
$\Delta E_{char/dischar} = \sum\limits_{t=0}^{t=t_{max}} P_{BAT} \times dt$	Charge $P_{BAT} > 0$ Discharge $P_{BAT} < 0$	$v \leq 60$	$60 > v \geq 90$	$v \geq 90$

During the RDC test, the vehicle's energy consumption was dependent on the road conditions. Increased energy consumption of the vehicle was noticeable in the higher speed ranges. The bar charts of Figure 13 show the energy flow and the share of energy consumption for different speed ranges. The marked points represent the total energy flow of both recovered (green) and consumed (red) energy in a specific speed range. The largest amount of energy was recovered in the 20–50 km/h range. The amount of energy recovered in the urban speed range allowed us to increase the vehicle range. However, the energy consumption in each speed range was higher than the recovered energy. The highest energy consumption was recorded in the intervals of increased vehicle speed, both during summer and winter driving conditions; in the speed range 110–120 km/h, the vehicle consumes more than 4 kWh (the distance covered at this speed is almost 25% of the entire RDC test). The shares of each speed interval in the test indicate nearly identical driving conditions in the urban route speed range (0–60 km/h) and in the motorway route range (v > 90 km/h).

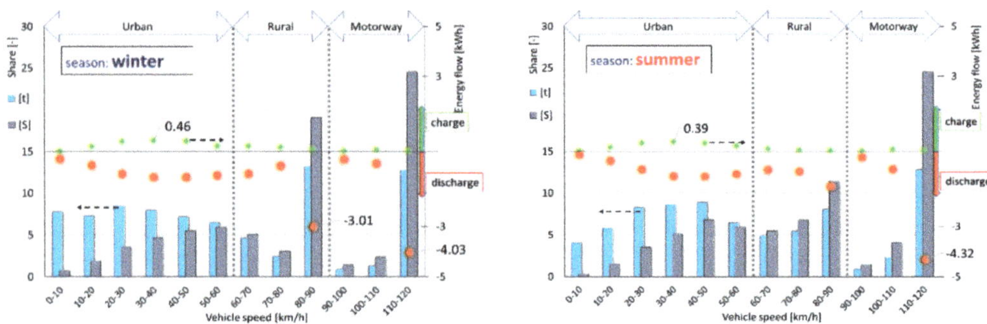

Figure 13. Energy flow characteristic and the share of energy consumption for different speed ranges in the RDC tests.

A summary of the vehicle's energy consumption during the RDC test for sections of the route (urban, rural, and motorway) is shown below in Figure 14 and Table 7 (1, 2 and 3). It compares both the energy consumption without recovery and the reduction of energy consumption after taking into account the recovery of energy from braking, which is shown in Table 7 (1', 2' and 3'). The graph shows the energy flow characteristics for both winter and summer conditions. The energy consumption was then calculated for 100 km of the sections under consideration (urban, rural, and motorway), thus obtaining the total energy consumption of the vehicle in winter conditions (14.9 kWh/100 km) and in summer conditions (13.1 kWh/100 km). Averaging the total energy consumption of a vehicle, without division into sections of the route and weather conditions, gives 14 kWh/100 km. The assumptions presented below are affected by some simplifications, but the paper is intended to undertake a preliminary analysis of the possibility of supplying an electric vehicle from a renewable source. Due to the variety of drivers and routes taken, these calculations are not applicable to every type of vehicle or every road with their own unique characteristics; nevertheless, the authors estimated the average energy consumption for a small urban vehicle, which was then compiled together with the assumed travel scenarios of the electric vehicle user.

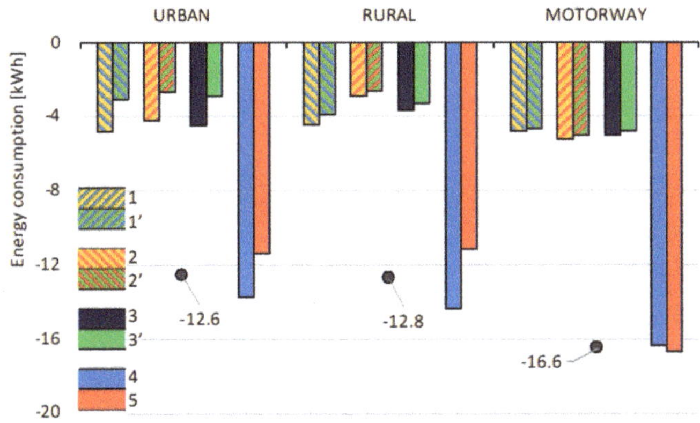

Figure 14. Energy flow characteristics and the share of energy consumption for different speed intervals in the RDC tests.

Table 7. Energy consumption of an electric vehicle in the RDC test of Figure 13.

[-]	Period	Explanation	Urban [kWh]	Rural [kWh]	Motorway [kWh]
1	Winter condition	Energy consumption of the vehicle in the selected road section **without considering energy recovery**	−4.83	−4.433	0.79–4.79
1'		Energy consumption of the vehicle in the selected road section **with considering energy recovery (energy recovery during vehicle braking)**	−3.06	−3.91	−4.63
2	Summer condition	Energy consumption of the vehicle in the selected road section **without considering energy recovery**	−4.16	−2.88	−5.23
2'		Energy consumption of the vehicle in the selected road section **with considering energy recovery (energy recovery during vehicle braking)**	−2.67	−2.64	−5.03
3	Average from research periods	Energy consumption of the vehicle in the selected road section **without considering energy recovery**	−4.51	−3.65	−5.01
3'		Energy consumption of the vehicle in the selected road section **with considering energy recovery (energy recovery during vehicle braking)**	−2.87	−3.27	−4.83
4	Winter condition	Estimated energy consumption for 100 km in a selected section of the tested route (including energy recovery from braking characteristic of the route)	−13.73	−14.36	−16.39
5	Summer condition	Estimated energy consumption for 100 km in a selected section of the tested route (including energy recovery from braking characteristic of the route)	−11.43	−11.15	−16.72

The above average values, without division into atmospheric conditions, were used for the analysis of the electric vehicle driving scenarios and are presented below.

3.4. Energy Supply and Demand for Selected Scenarios of Driving a Vehicle

The analysis of the average distance travelled by passenger vehicles in 2.4 has been used to develop a theoretical list of three scenarios (Table 8) in which the distance, together with the share of individual route sections, is a variable. Scenario (S1)—the minimum analyzed commuting distance is 15 km and covers urban driving conditions only. Scenario (S2) assumes a one-way distance of 30 km with 15 km in urban and 15 km in suburban driving conditions. Scenario (S3) assumes the participation of all three sections of the route and is 45 km total in one direction. Increasing the distance makes it necessary to charge the vehicle more often. This frequency was estimated based on calculations of the distance of a particular route compared with the energy consumption of the vehicle presented in the previous sections. According to the investigation's assumptions about charging, the vehicle must ensure an adequate charge to cover the entire route planned during the day. An overview of the charging frequency is presented in Figure 15.

Table 8. Analyzed scenarios of distances covered by an electric vehicle.

Scenarios Under Investigation	Daily Distance Work	Travel Conditions			Total Distance in Week	Energy Usage in the Scenario	Frequency of Vehicle Charging	Average Energy Demand Per Month
		Urban	Rural	Motorway				
[-]	[km]	[km]	[km]	[km]	[km]	[kWh]	[-]	[kWh]
S1	15	15	-	-	210	3.78	8.54→8	113.4
S2	30	15	15	-	420	7.62	4.24→4	228.6
S3	45	15	15	15	630	12.6	2.56→2	378

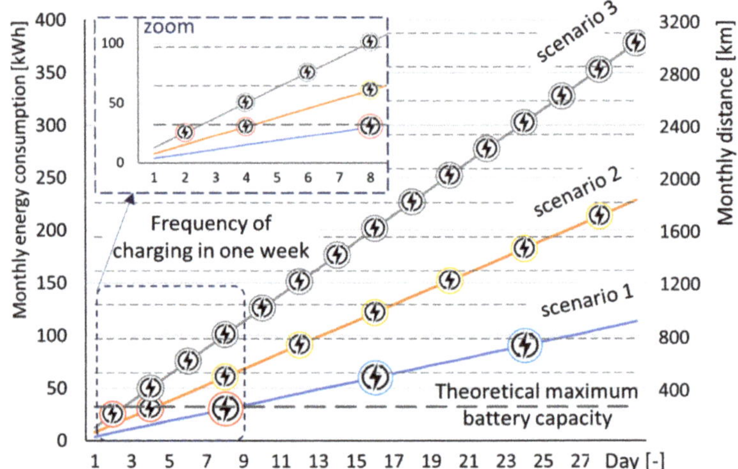

Figure 15. Frequency of charging an electric vehicle in the presented scenarios depending on the distance covered by a user.

The summary of energy production to power an electric vehicle, Figure 16, shows the energy produced by the photovoltaic installations in the months from March to October, marked as green. The calculated energy consumption of the vehicle in all scenarios shows it can be met by the production of electricity from solar panels in all months except October. However, if we also consider the current average electricity consumption of household appliances, Scenario 3 is not possible in any month. In such cases, the energy to power the vehicle in scenario 3 would come directly from the power plant. The solution to reduce electricity costs in such cases should be increasing the number of photovoltaic panels.

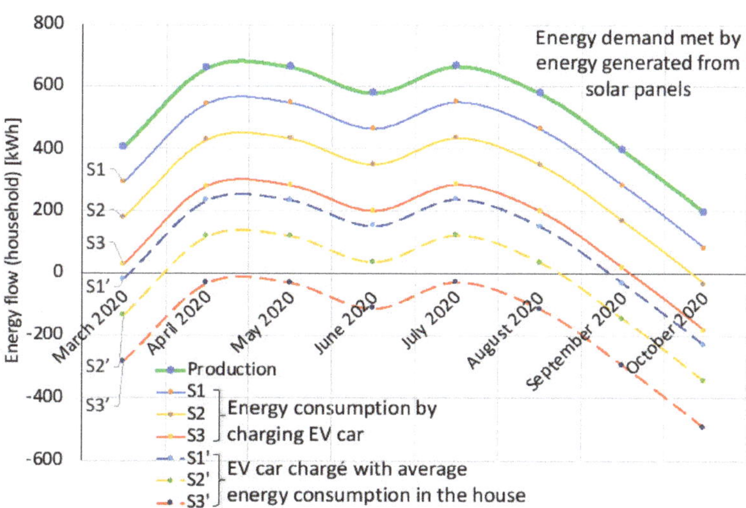

Figure 16. Energy balance of the presented scenarios of driving an electric vehicle in relation to the energy produced and consumed by a household with a 4.48 kW photovoltaic installation.

4. Discussion

This paper focused on an experimental assessment of the energy flow between household PV installation and a small size battery electric vehicle. Karkosiński et al., 2018 [53] conducted a similar approach without RDC test requirements and household energy consumption. The results indicated a significant impact of the sun's shining state during the day for energy generation, obtaining $100\ \mathrm{W/m^2}$ sun irradiance on a cloudy day and almost $250\ \mathrm{W/m^2}$ on a clear day in January. In effect, 11.4 kWh energy was obtained on an average sunny day in January. The study of [54], with polycrystalline and monocrystalline PV modules, showed an 80% difference between the amount of energy produced in summer and winter over a three year period. The results of this study confirmed the above trends through detailed analyses of PV energy production during the week of ride test cycles. In addition, a 19% increase in peak power has been observed in summer with the effective energy production time during the day raised by approximately 4 h.

EV energy consumption was examined with RDC tests. The average energy consumption in considered cases was 13.1 kWh/100 km in summer and 14.9 kWh/100 km in winter. The elevated energy consumption was likely caused by interior comfort system energy demand. For reference, in [33] where energy consumption was determined also with the RDC test, the average energy consumption was 19.6 kWh/100 km, calculated from two repeated rides with the difference between the rides of 1.3 kWh.

Comparison of three types of powertrain: gasoline, hybrid, and electric, in terms of an RDC test, was performed by Pielecha J. et al., 2020 [55]. During the RDE test, the lowest accumulated energy demand was achieved for the electric vehicle, approximately 30% lower than the combustion engine and 10% lower than a plug-in hybrid powertrain. In the research presented, like in [55], energy consumption was analyzed separately for each road type and in terms of ambient climate conditions, not powertrain type. In [36], the influence of drive mode and braking strategy on energy flow in small-sized EV was investigated within the RDC test requirements.

During the summer, smaller energy consumption was observed, 17% for urban conditions and 22% for rural residents, including braking energy recovery. The result of another investigation [56] shows decrease of possible driving range from 150 km at 20 °C to 85 km at 0 °C. Doyle A. et al., 2019 [57] indicated that interior thermal comfort systems consume an average of 14% of the total trip's energy by the cooling system and 18% by a heating operation.

The last part of the investigation included analyses of energy balance between household PV installation and EV. To assess sufficiency of PV installations, three scenarios of distance covered by EV were considered. In effect, it was possible to use an EV car charged by surplus energy from PV installation. The mentioned aspect of cooperation between electric vehicles and renewable energy sources is key to effective electric powertrain future developments.

5. Conclusions

The article presents an analysis of energy flow from the stage of production of electricity from a renewable source in the form of solar energy (PV panels), through the charging of the electrical vehicle, and the subsequent consumption of this energy while driving. Charging of electric vehicles, especially in areas with limited access to charging points, can be difficult. Therefore, the estimates of both the energy consumption of the vehicle and the necessary frequency of charging the vehicle are shown here. The energy consumption of the vehicle has been recorded for driving conditions during both winter and summer periods. The influence of the type of route (urban, rural, or motorway) and distance covered have significant impact on the vehicle's energy consumption. The presented scenarios are a stage of preliminary elaboration by the authors of the mechanism of simulating the energy consumption of electric vehicles considering various road conditions. In addition, oversupply of energy produced by the residential PV system used in this study indicates the possibility of reliance of charging of the EV from that source only.

The study specific conclusions:

- The electric vehicle's (urban type) consumed energy during the RDC test:
    - In winter conditions: 11.39 kWh/RDC test (estimated at 100 km = 14.9 kWh)
    - In summer conditions: 10.35 kWh/RDC test (estimated at 100 km = 13.1 kWh)
- The 4.48 kW PV installation can guarantee sufficient EV energy demand:
    - For all three scenarios in March–September period without energy demand by household appliances
    - For Scenario 1 and 2 in April–August period with household appliances
- For Scenario 3, the PV installation cannot guarantee the total energy demand while also powering household appliances. This case will be the subject of further research by the authors.

Author Contributions: Conceptualization, W.C., F.S., W.G., and A.B.; methodology, W.C., F.S., W.G., and A.B.; formal analysis, W.C., F.S., W.G., and A.B.; investigation, W.C. and F.S.; writing—original draft preparation, W.C., F.S., W.G., and A.B.; writing—review and editing, W.C., F.S., and W.G.; visualization, W.C. and F.S. All authors have read and agreed to the published version of the manuscript.

Funding: This research received no external funding.

Institutional Review Board Statement: Not Applicable.

Informed Consent Statement: Not Applicable.

Data Availability Statement: The data presented in this study are available on request from the corresponding author.

Acknowledgments: This work was supported by the Volkswagen Group Polska Sp. z o.o.

Conflicts of Interest: Authors declare no conflict of interest.

Abbreviations

EV	Electric Vehicle
CAN	Controller Area Network
CC	Constant Current
CSS	Combined Charging System
GHG	Greenhouse Gas
GWP	Global Warming Potential
HEV	Hybrid Electric Vehicle
ICE	Internal Combustion Engines
LCA	Life Cycle Assessment
OBD	On Board Diagnostics System
PV	Photovoltaic
RES	Renewable Energy Source
RDC	Real Driving Conditions
RDE	Real Driving Emissions
SOC	State of Charge
TTW	Tank to Wheel
V2G	Vehicle to Grid
WTT	Well to Tank

References

1. Van Mierlo, J.; Maggetto, G.; Lataire, P. Which energy source for road transport in the future? A comparison of battery, hybrid and fuel cell vehicles. *Energy Convers. Manag.* **2006**, *47*, 2748–2760. [CrossRef]
2. Final Energy Consumption by Fuel Type and Sector. Available online: https://www.eea.europa.eu/ (accessed on 1 November 2020).
3. Leach, F.; Kalghatgi, G.; Stone, R.; Miles, P. The scope for improving the efficiency and environmental impact of internal combustion engines. *Transp. Eng.* **2020**, *1*, 100005. [CrossRef]
4. González Palencia, J.C.; Nguyen, V.T.; Araki, M.; Shiga, S. The Role of Powertrain Electrification in Achieving Deep Decarbonization in Road Freight Transport. *Energies* **2020**, *13*, 2459. [CrossRef]
5. Moreno, J.C.; Stenlaas, O.; Tunestal, P. Multi-Cylinder Adaptation of In-Cycle Predictive Combustion Models. *SAE Int. J. Adv. Curr. Pract. Mobility* **2021**, *3*, 299–311.
6. Balawender, K.; Ustrzycki, A.; Lejda, K.; Jakubowski, M.; Jaworski, A.; Kuszewski, H.; Siedlecka, S.; Zielińska, E. Modeling of Unburned Hydrocarbon Emission in a Di Diesel Engine Using Neural Networks. In *SAE Technical Papers, Proceedings of the SAE 2020 International Powertrains, Fuels and Lubricants Meeting, Krakow, Poland, 22–24 September 2020*; Society of Automotive Engineers: Warrendale, PA, USA, 2020.
7. Sitnik, L.J. Energy Demand Assessment for Long Term Operation of Vehicles. In *SAE Technical Papers, Proceedings of the SAE 2020 International Powertrains, Fuels and Lubricants Meeting, Krakow, Poland, 22–24 September 2020*; Society of Automotive Engineers: Warrendale, PA, USA, 2020.
8. Pielecha, J.; Merkisz, J.; Kurtyka, K.; Skobiej, K. Cold start emissions of passenger cars with gasoline and diesel engines in Real Driving Emissions tests. *Combust. Engines* **2019**, *179*, 160–168.
9. Del Pero, F.; Delogu, M.; Pierini, M. Life Cycle Assessment in the automotive sector: A comparative case study of Internal Combustion Engine (ICE) and electric car. *Procedia Struct. Integr.* **2018**, *12*, 521–537. [CrossRef]
10. Messagie, M.; Boureima, F.-S.; Coosemans, T.; Macharis, C.; Mierlo, J.V. A Range-Based Vehicle Life Cycle Assessment Incorporating Variability in the Environmental Assessment of Different Vehicle Technologies and Fuels. *Energies* **2014**, *7*, 1467–1482. [CrossRef]
11. Hawkins, T.R.; Singh, B.; Majeau-Bettez, G.; Strømman, A.H. Comparative Environmental Life Cycle Assessment of Conventional and Electric Vehicles. *J. Ind. Ecol.* **2012**, *17*, 53–64. [CrossRef]
12. Chau, K.; Wong, Y.; Chan, C. An overview of energy sources for electric vehicles. *Energy Convers. Manag.* **1999**, *40*, 1021–1039. [CrossRef]
13. Xu, L.; Yilmaz, H.Ü.; Wang, Z.; Poganietz, W.-R.; Jochem, P. Greenhouse gas emissions of electric vehicles in Europe considering different charging strategies. *Transp. Res. Part D Transp. Environ.* **2020**, *87*, 102534. [CrossRef]
14. Cox, B.; Bauer, C.; Beltran, A.M.; Van Vuuren, D.P.; Mutel, C.L. Life cycle environmental and cost comparison of current and future passenger cars under different energy scenarios. *Appl. Energy* **2020**, *269*, 115021. [CrossRef]
15. Burchart-Korol, D.; Jursova, S.; Folęga, P.; Pustejovska, P. Life cycle impact assessment of electric vehicle battery charging in European Union countries. *J. Clean. Prod.* **2020**, *257*, 120476. [CrossRef]
16. Yang, Y.; Lian, C.; Ma, C.; Zhang, Y. Research on Energy Storage Optimization for Large-Scale PV Power Stations under Given Long-Distance Delivery Mode. *Energies* **2019**, *13*, 27. [CrossRef]

17. Khuong, P.M.; McKenna, R.; Fichtner, W. A Cost-Effective and Transferable Methodology for Rooftop PV Potential Assessment in Developing Countries. *Energies* **2020**, *13*, 2501. [CrossRef]
18. Coffman, M.; Bernstein, P.; Wee, S. Integrating electric vehicles and residential solar PV. *Transp. Policy* **2017**, *53*, 30–38. [CrossRef]
19. Kobashi, T.; Yoshida, T.; Yamagata, Y.; Naito, K.; Pfenninger, S.; Say, K.; Takeda, Y.; Ahl, A.; Yarime, M.; Hara, K. On the potential of "Photovoltaics + Electric vehicles" for deep decarbonization of Kyoto's power systems: Techno-economic-social considerations. *Appl. Energy* **2020**, *275*, 115419. [CrossRef]
20. Novoa, L.; Brouwer, J. Dynamics of an integrated solar photovoltaic and battery storage nanogrid for electric vehicle charging. *J. Power Sources* **2018**, *399*, 166–178. [CrossRef]
21. Gudmunds, D.; Nyholm, E.; Taljegard, M.; Odenberger, M. Self-consumption and self-sufficiency for household solar producers when introducing an electric vehicle. *Renew. Energy* **2020**, *148*, 1200–1215. [CrossRef]
22. Falvo, M.C.; Graditi, G.; Siano, P. Electric Vehicles integration in demand response programs. In Proceedings of the 2014 International Symposium on Power Electronics, Electrical Drives, Automation and Motion, Ischia, Italy, 18–20 June 2014; pp. 548–553.
23. Wu, Y.; Zhang, J.; Ravey, A.; Chrenko, D.; Miraoui, A. Real-time energy management of photovoltaic-assisted electric vehicle charging station by markov decision process. *J. Power Sources* **2020**, *476*, 228504. [CrossRef]
24. Raza, W.; Ko, G.S.; Park, Y.C. Induction Heater Based Battery Thermal Management System for Electric Vehicles. *Energies* **2020**, *13*, 5711. [CrossRef]
25. Vaz, W.; Nandi, A.K.; Landers, R.G.; Koylu, U.O. Electric vehicle range prediction for constant speed trip using multi-objective optimization. *J. Power Sources* **2015**, *275*, 435–446. [CrossRef]
26. Rhode, S.; Van Vaerenbergh, S.; Pfriem, M. Power prediction for electric vehicles using online machine learning. *Eng. Appl. Artif. Intell.* **2020**, *87*, 103278. [CrossRef]
27. Liu, X.; Deng, X.; He, Y.; Zheng, X.; Zeng, G. A Dynamic State-of-Charge Estimation Method for Electric Vehicle Lithium-Ion Batteries. *Energies* **2019**, *13*, 121. [CrossRef]
28. Zhang, J.; Wang, Z.; Liu, P.; Zhang, Z. Energy consumption analysis and prediction of electric vehicles based on real-world driving data. *Appl. Energy* **2020**, *275*, 115408. [CrossRef]
29. Gong, J.; He, J.; Cheng, C.; King, M.; Yan, X.; He, Z.; Zhang, H. Road Test-Based Electric Bus Selection: A Case Study of the Nanjing Bus Company. *Energies* **2020**, *13*, 1253. [CrossRef]
30. Donkers, A.; Yang, D.; Viktorović, M. Influence of driving style, infrastructure, weather and traffic on electric vehicle performance. *Transp. Res. Part D Transp. Environ.* **2020**, *88*, 102569. [CrossRef]
31. Xie, Y.; Li, Y.; Zhao, Z.; Dong, H.; Wang, S.; Liu, J.; Guan, J.; Duan, X. Microsimulation of electric vehicle energy consumption and driving range. *Appl. Energy* **2020**, *267*, 115081. [CrossRef]
32. Guo, J.; Jiang, Y.; Yu, Y.; Liu, W. A novel energy consumption prediction model with combination of road information and driving style of BEVs. *Sustain. Energy Technol. Assess.* **2020**, *42*, 100826. [CrossRef]
33. Gis, W.; Merkisz, J. The development status of electric (BEV) and hydrogen (FCEV) passenger cars park in the world and new research possibilities of these cars in real traffic conditions. *Combust. Engines* **2019**, *178*, 144–149.
34. Cieślik, W.; Szwajca, F.; Golimowski, J. The possibility of energy consumption reduction using the ECO driving mode based on the RDC test. *Combust. Engines* **2020**, *182*, 59–69.
35. PV Market in Poland, Institute for Renewable Energy. Available online: https://ieo.pl/en/pv-report (accessed on 24 November 2020).
36. PV Market in Poland. 2020. Available online: https://ieo.pl/ (accessed on 24 November 2020).
37. Number of Electric Passenger Cars in Poland from 2019 to 2020, by Type of Vehicle. Available online: https://www.statista.com/ (accessed on 24 November 2020).
38. Gis, W.; Waśkiewicz, J.; Menes, M. Experts forecasts on the demand for energy carriers in motor vehicle transport in Poland up to year 2035. *Combust. Engines* **2019**, *178*, 162–165.
39. Chłopek, Z.; Lasocki, J.; Wójcik, P.; Badyda, A.J. Experimental investigation and comparison of energy consumption of electric and conventional vehicles due to the driving pattern. *Int. J. Green Energy* **2018**, *15*, 773–779. [CrossRef]
40. Wang, F.; Zhu, Y.; Yan, J. Performance of solar PV micro-grid systems: A comparison study. *Energy Procedia* **2018**, *145*, 570–575. [CrossRef]
41. Monna, S.; Juaidi, A.; Abdallah, R.; Itma, M. A Comparative Assessment for the Potential Energy Production from PV Installation on Residential Buildings. *Sustainability* **2020**, *12*, 10344. [CrossRef]
42. Fachrizal, R.; Shepero, M.; Van Der Meer, D.; Munkhammar, J.; Widén, J. Smart charging of electric vehicles considering photovoltaic power production and electricity consumption: A review. *eTransportation* **2020**, *4*, 100056. [CrossRef]
43. Mohammad, A.; Zamora, R.; Lie, T.T. Integration of Electric Vehicles in the Distribution Network: A Review of PV Based Electric Vehicle Modelling. *Energies* **2020**, *13*, 4541. [CrossRef]
44. Savio, D.A.; Juliet, V.A.; Chokkalingam, B.; Padmanaban, S.; Holm-Nielsen, J.B.; Blaabjerg, F. Photovoltaic Integrated Hybrid Microgrid Structured Electric Vehicle Charging Station and Its Energy Management Approach. *Energies* **2019**, *12*, 168. [CrossRef]
45. UNISON GROUP. Energy Self-Sufficiency. Available online: https://www.unison.co.nz/ (accessed on 1 November 2020).

46. Cieslik, W.; Zawartowski, J.; Fuc, P. The Impact of the Drive Mode of a Hybrid Drive System on the Share of Electric Mode in the RDC Test. In *SAE Technical Papers, Proceedings of the SAE 2020 International Powertrains, Fuels and Lubricants Meeting, Krakow, Poland, 22–24 September 2020*; Society of Automotive Engineers: Warrendale, PA, USA, 2020.
47. *The European Commision Regulation 2017/1154 Amending Regulation (EU) 2017/1151 Supplementing Regulation (EC) No 715/2007 of the European Parliament and of the Council on Type-Approval of Motor Vehicles with Respect to Emissions from Light Passenger and Commercial Vehicles (Euro 5 and Euro 6) and on Access to Vehicle Repair and Maintenance Information, Amending Directive 2007/46/EC of the European Parliament and of the Council, Commission Regulation (EC) No 692/2008 and Commission Regulation (EU) No 1230/2012 and repealing Regulation (EC) No 692/2008 and Directive 2007/46/EC of the European Parliament and of the Council as Regards Real-Driving Emissions from Light Passenger and Commercial Vehicles (Euro 6) (Text with EEA Relevance)*; Official Journal of the European Union: Brussels, Belgium, 7 June 2017.
48. Pielecha, J.; Skobiej, K. Evaluation of ecological extremes of vehicles in road emission tests. *Arch. Transp.* **2020**, *56*, 33–46. [CrossRef]
49. Jacobs-Crisioni, C.; Kompil, M.; Baranzelli, C.; Lavalle, C. *Indicators of Urban form and Sustainable Urban Transport. Introducing Simulation-Based Indicators for the LUISA Modelling Platform*; JRC Technical Reports; Joint Research Centre: Ispra, Italy, 2015; pp. 1–33.
50. Chiara, B.D.; Deflorio, F.; Pellicelli, M.; Castello, L.; Eid, M. Perspectives on Electrification for the Automotive Sector: A Critical Review of Average Daily Distances by Light-Duty Vehicles, Required Range, and Economic Outcomes. *Sustainability* **2019**, *11*, 5784. [CrossRef]
51. Solar Resource Map. Available online: https://solargis.com/ (accessed on 20 November 2020).
52. SMA Solar Technology. Available online: https://www.sma-solar.pl (accessed on 10 December 2020).
53. Karkosiński, D.; Pacholczyk, M.; Sienkiewicz, Ł. Experimental study of the use of electric car powered with stationary solar and electrochemical batteries in Northern Poland. *MATEC Web Conf.* **2018**, *180*, 0209. [CrossRef]
54. Sarniak, M. Analysis of energy efficiency of photovoltaic installation in central Poland. *E3S Web Conf.* **2018**, *46*, 00002. [CrossRef]
55. Pielecha, J.; Skobiej, K.; Kurtyka, K. Exhaust Emissions and Energy Consumption Analysis of Conventional, Hybrid, and Electric Vehicles in Real Driving Cycles. *Energies* **2020**, *13*, 6423. [CrossRef]
56. Iora, P.; Tribioli, L. Effect of Ambient Temperature on Electric Vehicles' Energy Consumption and Range: Model Definition and Sensitivity Analysis Based on Nissan Leaf Data. *World Electr. Veh. J.* **2019**, *10*, 2. [CrossRef]
57. Doyle, A.; Muneer, T. Energy consumption and modelling of the climate control system in the electric vehicle. *Energy Explor. Exploit.* **2018**, *37*, 519–543. [CrossRef]

Article

Description of Acid Battery Operating Parameters

Józef Pszczółkowski 

Faculty of Mechanical Engineering, Military University of Technology, 00-908 Warsaw, Poland;
jozef.pszczolkowski@wat.edu.pl

Abstract: In this paper, the operating principles of the acid battery and its features are discussed. The results of voltage tests containing the measurements conducted at the terminals of a loaded battery under constant load conditions, and dependent on time, are presented. The article depicts the principles of the development of electric models of acid batteries and their various descriptions. The principles for processing the results for the purpose of the determination and description of the battery model are characterized. The characteristics under stationary and non-stationary conditions are specified using glued functions and linear combinations of exponential functions, and the electrical parameters of the battery are determined as the components of the circuit, i.e., its electromotive force, resistance, and capacity. The dynamic characteristic of the battery in the form of transmittance was determined, using the Laplace transform. Possible uses of the crankshaft driving signals as diagnostic signals of the battery, electric starter, and internal combustion engine are also indicated.

Keywords: acid battery; battery operating parameters; testing and modeling

Citation: Pszczółkowski, J. Description of Acid Battery Operating Parameters. *Energies* **2021**, *14*, 7212. https://doi.org/10.3390/en14217212

Academic Editor: Wojciech Cieslik

Received: 12 October 2021
Accepted: 29 October 2021
Published: 2 November 2021

Publisher's Note: MDPI stays neutral with regard to jurisdictional claims in published maps and institutional affiliations.

1. Introduction

The lead–acid battery is a chemical source of electric energy in which current is generated as a result of chemical processes taking place on its electrodes in the presence of sulfuric acid. The factor that forces the course of the current generating processes is an electromotive force of the cell resulting from a difference in the electrodes constituting the cell normal potentials. The basic parameters characterizing the electrical and energy properties of the battery are: voltage, twenty amp hour (Ah) rate capacity, and the ability to start an engine (CCA—Cold Cranking Amps). CCA is a rating used to define the ability of the battery to start an engine in a cold temperature. The existing chemical models of the battery explain a mechanism of the generation of an electromotive force and a sum of its electrical and energetic capacities, e.g., electric capacity. However, the chemical models are not useful for analyzing electrical circuits where the acid battery is a component. When using an acid battery, it is not possible to determine the current electrical parameters of the circuit, current, and voltage. In such a circuit it is necessary to use electric battery models composed of the typical electrical circuit components: electromotive force, resistance, capacitance, inductance, and others [1]. Modeling of the batteries, including acid batteries, has become necessary and is carried out in a particularly intensive manner due to the increased demand for electricity in vehicles resulting from the arrival of electric and hybrid drives. The modeling and determination of the battery model parameters is considered a difficult, unclear, laborious, expensive, and ambiguous process [2].

A lead–acid battery consists of a negative electrode made of porous lead and a positive electrode consisting of lead dioxide. Both electrodes are immersed in electrolyte which is a solution of sulfuric acid and water. The overall reversible chemical reaction, which enables lead–acid batteries to store energy, is as follows:

$$PbO_2 + Pb + 2H_2SO_4 \Leftrightarrow 2PbSO_4 + 2H_2O$$

Discharging a battery causes the formation of lead sulphate at both the negative and positive electrodes. Sulphate from the sulfuric acid electrolyte surrounding the battery is used in the formation of this lead sulphate. When the battery is in the fully discharged state, its two electrodes are of the same material and there is no chemical potential or voltage between these two electrodes. Between the fully charged and discharged states, the lead–acid battery experiences a gradual reduction in voltage. A voltage level is used to indicate the state of charge of the battery. Thus, there is a dependence of the battery voltage on the battery state of charge. The battery is in equilibrium only in the state characterized by no load. The battery voltage and its capacity have specified values. The battery under load is not in equilibrium, and its voltage and battery capacity differ significantly from the equilibrium values. The difference between the voltage at equilibrium and that under a load, with a current flow, is termed the battery polarization.

The battery voltage value or its dependence on time versus battery operating condition parameters is the basic battery operating parameter or characteristic. The operating characteristics of the acid battery are the object of research and modeling for the implementation of many practical and theoretical objectives: the evaluation of the correlation of the starting parameters of the internal combustion engine [3]; the design of the internal combustion engine start-up systems [4]; the analysis of the dynamic properties of the battery in electric vehicles [5]; the possibility of using the characteristics in the process of diagnosing the internal combustion engine and its starting system [6]. In the case of the lead–acid battery model in electric or hybrid vehicles, the charging and discharging process is of great importance, i.e., a charging/discharging voltage and state of charge (SoC) [7]. Very often the model of the lead–acid battery for the Stop-Start Technology is a circuit model with two resistance–capacitance (RC) blocks [8]. The simulated battery operating parameters are the voltages, currents, and state of charge (SoC). The battery models for the different designs of the lead–acid-based batteries, i.e., batteries with gelled electrolyte and an Absorbent Glass Mat (AGM), differ from the common lead–acid batteries models in regards to the parameters of the battery model, although they are based on the same chemistry [9]. There are also different models of the lead–acid battery in terms of their ageing processes, i.e., deep discharge models which are combined with a sulfation model [10]. The ageing processes determine the battery state of health (SoH). The purpose of some works is to investigate factors which affect the failure of automotive batteries or battery durability. The main factors influencing the aging process of batteries are the battery temperature and the discharge current [11]. Statistical methods are used for analysis and prediction of battery degradation in electric vehicle use [12], including regression models for estimation of the battery state of health [13]. In regression models, charge/discharge cycle number, battery terminal voltage, and internal resistance are used as independent variables.

The existing methods of battery testing have been systematically developed, and new approaches are used to determine the characteristics of batteries, e.g., based on neural networks, genetic algorithms, or Kalman filters. These often concern the determination of model parameters, the battery state of charge, and energy management in energy storage systems using batteries. Genetic algorithms are used to optimize the energy system of electric vehicles because of the growing number of electricity consumers in the vehicle [14]. For the state of charge of batteries, and its dynamic determination, supervised chaos genetic algorithms have been used [15]. The use of the Kalman filter based on the *RC* model for estimation of model parameters and the state of charge of lead–acid batteries requires knowledge of the value of the process covariance and the measurement noise [16]. The prediction voltage and lifetime of a lead–acid battery may be determined using neural network methods [17]. Ref. [18] describes the design of a measurement system to conduct the electrical tests, and an estimation algorithm for automatic analyses and reporting proceedings for lead–acid started batteries. Determination of the state of charge (SoC) of a lead–acid battery was tested using the electrochemical impedance spectroscopy (EIS) method [19]. Lead–acid cells were explored during intermittent discharge and charge processes. More battery parameters were taken into account in the design and simulation of a model of a lead–acid battery [20]. These parameters were the SoC, battery voltage, and temperature of the battery in the charge and discharge state.

Previous research has also investigated the problems of the physical phenomena that determine the operation of the energy storage system, i.e., the lead–acid battery [21]. A relatively similar new modeling method for lead–acid batteries combined the physicochemical model with the equivalent circuit model [22]. Ref. [23] drew attention to the fact that the battery equivalent circuit model has two time constants. Because the load process duration is often short, the test data during the load period may not contain sufficient information for extracting these time constants. In contrast, the relaxation period may last hours, and thus may provide sufficient data for this purpose. The problems of the modern technology used in the battery production process were considered in [24]. Carbon materials are widely used as an additive to the negative active mass and allow the battery specific energy and active mass utilization to be increased. A constant problem concerning the use of battery energy relates to the starting of an automobile engine in low or negative temperature conditions. When using an autonomous means of engine pre-heating, it is necessary to optimally distribute the battery energy to the pre-start and start-up discharges [25].

The objective of this paper is to present the author's mathematical models of the acid battery for stationary and non-stationary dynamic operating conditions. The basis for the development of the models was the research results of voltage measurements at the terminals of the loaded batteries under constant load conditions, i.e., the dependence of the voltage on time. The battery tests were carried out on a test stand that was placed in a low-temperature chamber, which allowed the ambient and tested battery temperatures to be changed.

In the literature, the accumulator battery models are presented in graphic or mathematical form, and a mathematical description is not frequently used. Therefore, in this work the mathematical form of the model is particularly emphasized and explained. On the basis of test results, a linear model of the dependence of the battery terminals' voltage on its nominal electric capacity, loaded current, temperature, and battery state of charge (SoC) is elaborated. This multidimensional model was developed for the stationary operating conditions within the time period of several seconds following switching on the load. In this case, the principles of planning the experiment were applied [26]. The dynamic characteristics of the battery are also presented, i.e., its voltage at the dynamic state of operating just after switching on the load, and after switching it off.

The principles of processing the results for the purpose of the determination and description of the battery models are characterized. The characteristics under the stationary and non-stationary conditions are specified using glued functions and linear combinations of exponential functions, and the electrical parameters of the battery are determined as the components of the circuit, i.e., its electromotive force, resistance, and capacity. Possible uses of the crankshaft driving signals as diagnostic signals of the battery, electric starter, and internal combustion engine are also indicated.

2. Materials and Methods

The battery performance tests were carried out on a test stand that was placed in a low-temperature chamber, which allowed the ambient and the tested battery temperatures to be changed. The equipment of the test bench enabled the test implementation and the recording of the battery operation parameters to be controlled. The tested battery was loaded with a constant resistance value within approx. 10 s. The values of the current and voltage were recorded by means of a computer measuring system, including after switching off the load, to observe changes in the electromotive force of the battery polarization during this period. The examples of the recorded dependencies of the current and voltage at the terminals within the load test of the battery of 54 Ah capacity are shown in Figures 1 and 2.

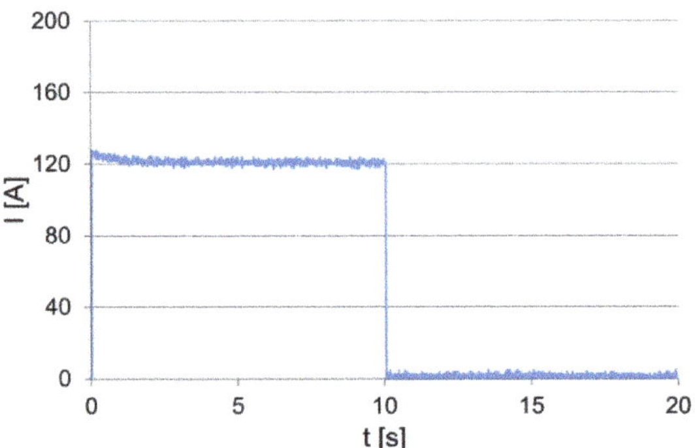

Figure 1. The current drawn from the battery.

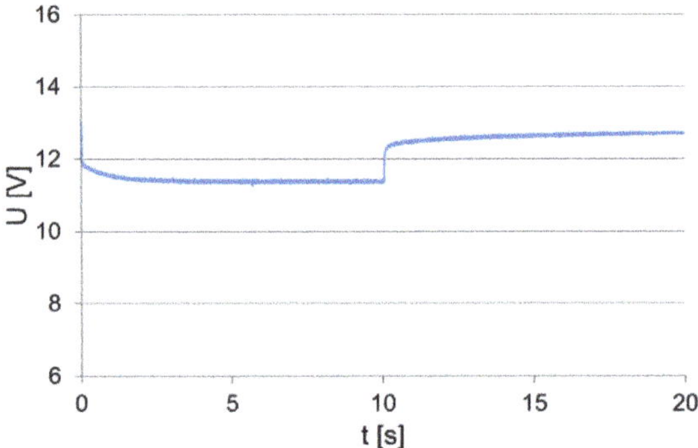

Figure 2. The voltage (to 10 s) at the loaded and unloaded battery terminals.

At the moment of switching on the load, characteristic and correlated, proportional changes in the intensity of the absorbed current and voltage at the battery terminals can be observed. During the initial discharge period, the voltage at the battery terminals and the current decrease approximately exponentially, and then their values stabilize. When the load is switched off, the voltage increases rapidly and then increases exponentially (Figure 2). This is due to an increase in its electromotive force caused by changes in the electrolyte concentration in the vicinity and in the inner layers of the active mass of the battery plates.

The recorded characteristics have, in addition to the visible and clear trend of changes, significant irregularity. This can cause difficulties in their further processing to determine and interpret electrical characteristics and the battery model. Therefore, the courses were subjected to pre-processing aimed at smoothing them. The causes of signal distortion were analyzed, and methods of their elimination were developed. The following sources of interference were identified:

- The own noise of the measuring system;
- Interference from the external electricity network;
- Quantization errors of the measuring system.

The various forms and principles of averaging were adopted as the methods of smoothing of the received signals. These can only be used in the case of a good recognition of the signal and an understanding of the nature of its changes, to ensure that useful signal components are not lost. The own noise of the measuring system, particularly high values, is usually represented as a single isolated deviation of its value from the set level. In principle, all distortions can be reduced using a method analogous to the moving average; the difference is that the moving average is a forecasting method in which the forecast value is assumed as the moving average of the preceding values. In this case, a calculated value of the mean was taken as the central data value. It is preferred that this uses an odd number of datapoints. Depending on the degree of a signal interference, the smoothing of the curve can be used several times.

As a characteristic of the battery load, the dependence of the voltage at its terminals at a constant value resistance load is considered. The analysis concerns the characteristics during the battery load period, as shown in Figure 3 as U_{load}. After the load is turned off, the voltage at the battery terminals increases abruptly, and then gradually stabilizes to the value in the no-load state.

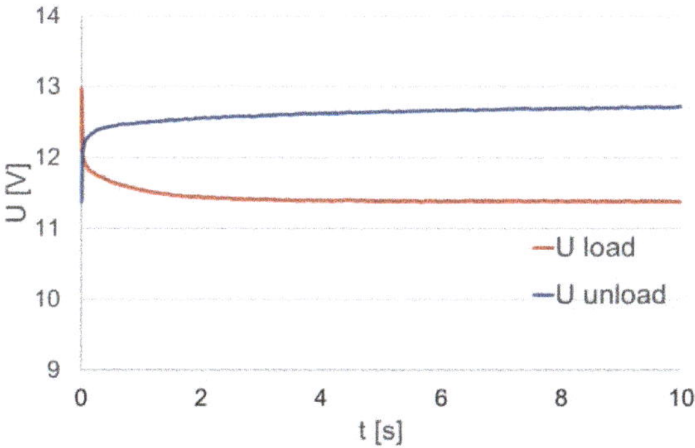

Figure 3. The voltage at the loaded (U_{load}) and unloaded (U_{unload}) battery terminals.

The different battery models are applied to the specific purposes and the different methods used to test the battery characteristics [27,28]. The simplest model of the battery presents it as an ideal voltage source, i.e., an electromotive force that does not exhibit even any internal resistance. The lead–acid battery is most often treated as a voltage source of electric current with a defined electromotive force and a variable internal resistance. In the electrical circuit, certain voltage changes at its terminals (at a constant value resistance load) can be justified by a change in its electromotive force or internal resistance. The changes in the electromotive force (or the internal resistance) are caused by the processes in the electrolyte around the electrodes or on their surface. When under the given discharge conditions, a constant value of electromotive force is accepted, and a classic equivalent electrical circuit of the battery can be presented, as shown in Figure 4.

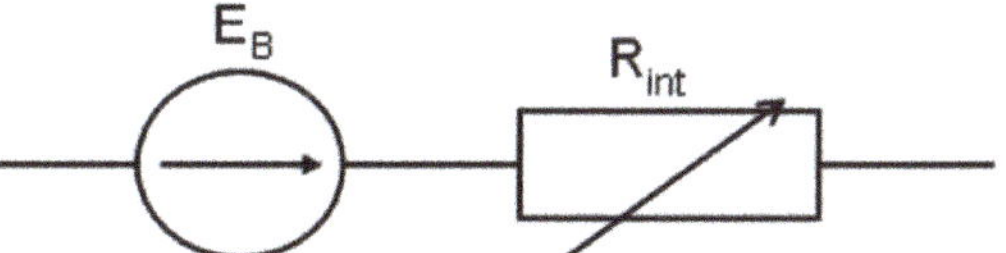

Figure 4. The classic equivalent electrical circuit of the battery.

When such a battery is loaded with the external resistance R or the constant current I, the voltage at the battery and receiver terminals is as follows (1):

$$U = RI = E_B - IR_{int}. \tag{1}$$

A complex battery model takes into account its electromotive force, internal resistance, inductance or capacitance, and other characteristics. The model presented in this article considers the dependencies of the characteristics of the battery on its rated capacity, temperature, and state of charge. However, the model should reflect the principle of the lead–acid battery. It also should be simple, fast, and effective to implement and use. The equations of the lead–acid model always contain constants that must be determined experimentally by laboratory tests. Any battery model can be validated by a simulation using, for example, the MATLAB/Simulink Software [29].

3. Results

3.1. Stationary Operating Characteristics of an Acid Battery

The operating voltage of the lead–acid battery depends on its rated capacity, current consumption, temperature, and state of charge (technical state). Previous research of one- and two-dimensional operating characteristics of acid batteries formulated a conclusion about the linear nature of the dependences of the battery operating voltage on the above-mentioned independent parameters [30]. In order to determine the multidimensional characteristics of the acid battery operation, an experiment was developed that enabled determination of the coefficients of a linear equation that describes the relationship between the voltage of the loaded battery and the factors influencing it. Testing the characteristics of the batteries was carried out on a test stand prepared and placed in a low-temperature chamber, which enabled the operating conditions of the battery to be changed. In the initial period of load, changes in the current and voltage values are visible, resulting from the dynamic nature of the tested battery operation (load switching on). Because the objective of the study was the determination of the characteristics under the steady-state load conditions, the values of the parameters describing the operating state of the battery were determined for the load duration time of approx. 10 s, i.e., after their stabilization.

Gaining an understanding of the properties of the research object and its behavior under the influence of extortion requires many, often costly and time-consuming experiments. The number of measurements performed depends on the complexity of the model, the number of independent variables affecting the research object, and the variables' values. For the purpose of limiting the number of measurements and, at the same time, obtaining as much information as possible, it is necessary to plan the experiments and then perform them according to the principles resulting from the adopted plan. In the experimental research, the most commonly adopted approach is a linear structure of the model, in addition to exponential or logarithmic structures that can be reduced to a linear form. The method of least squares is most often used to determine the coefficients of these models. The method of least squares is used to identify the linear models and the second-order polynomials, in addition to the power or logarithmic functions.

The research object is characterized by the independent (input) variables x_i, i.e., a set of parameters influencing its properties; and dependent variables y_i, i.e., output quantities (the result of the interaction of input and disturbing quantities). The disturbing quantities z_i are the result of an impact of random factors on the research object and the inaccuracy of measurement methods and means.

The research is carried out according to a prepared experiment plan, usually in accordance with the experiment table included in the plan. The determination of the inaccuracy of the measurement results is possible when the same experiment is repeated several times. The arithmetic mean can be used as a position measure and the standard deviation as a dispersion measure. The optimization of the model describing the real object mainly consists in finding the best of all possible limitations, and a model that describes the relationships between the studied variables.

As mentioned above, the experiment plan was developed with the assumption of a linear structure of the mathematical model that describes the relationship between the voltage of the loaded battery (dependent variable) and the physical quantities that influence the voltage (independent variables): battery nominal capacity—Q [Ah]; load current intensity—I [A]; temperature (of electrolyte)—T [°C] (in this case it is more advisable and convenient to use the Celsius temperature scale than the Kelvin temperature scale); and battery condition—k. A two-level, static, determined, and complete experiment was assumed. The plan assumes that the input factors, i.e., the independent variables, take two levels of values: the upper ones are marked as "+1" and the lower ones are marked as

"-1". Therefore, the number of tests in the planned experiment for the four independent variables is: $n = 2^4 = 16$.

For individual independent variables, the appropriate symbols x_i were adopted, and the levels of their variability were assumed. The levels of variability define the range within which they take the values $[x_{min} - x_{max}]$. To develop the model, it is necessary to code the input quantities. Coding consists of transforming the value of any input quantity into a coded (normalized) value which is within the range limited by the conventional levels of the input variables, and falling into the following set $[-1 \div +1]$. For this purpose, mathematical operations were performed, consisting of determining central values; that is, calculating arithmetic means for the individual variables and determining a unit of variation for the individual quantities, which is the unit value of the input factor change. These values were then coded. The units of the variables' variation were determined on the basis of the changes in the parameters of the operation of the engine starting system under the average engine starting conditions with the use of an electric starter.

A calculation of the units of variation consists of determining the unit value of the change in the independent variable. The unit of variation was determined on the basis of Equation (2). The value of x_{imax} and x_{imin} in Equation (2) corresponds to the maximum and minimum value, respectively, of the independent variable with the number, i.e., x_i in the adopted variation range.

$$\Delta x_i = \frac{x_{i\,max} - x_{i\,min}}{2}. \tag{2}$$

The central values are the arithmetic means of the maximum and minimum values of each individual independent variables Equation (3):

$$x_{io} = \frac{x_{i\,max} + x_{i\,min}}{2}. \tag{3}$$

Coding of independent variables results in transforming the values of the input quantities into dimensionless numbers contained in the following set $[-1; +1]$. Coding makes the experiment plan independent of the real values and the physical meaning of independent variables describing the research object, and replaces the independent variables with dimensionless values. Hence, the methods of planning the experiment become universal and independent of the physical importance of factors describing a given phenomenon, and can be used in various fields of research.

Thus, the coded value of any independent variable, according to Equation (4), is:

$$x_{ik} = \frac{x_i - x_{io}}{\Delta x_i}, \tag{4}$$

where the individual component of Equation (4) has the following meaning:

- x_{ik}—the coded value of the independent variable;
- x_i—the independent variable subjected to the coding;
- x_{io}—the central value, determined by the Equation (3);
- Δx_i—the unit of variation of the independent variable subjected to the coding.

The appropriate levels of the variability of the factors were adopted, for which the values were, respectively:

- The battery nominal capacity: $Q = 110$ and 170 Ah;
- The current: $I = 84$ and 224 A;
- The temperature: $T = 0$ and $+22\ °C$
- The battery condition (state of charge): $k = 0.7$ and 1.

The notional levels of the factor values are described as -1 for the lower value and $+1$ for the higher value. The first step of the experiment is to code the variables, which then assume conventional, dimensionless values. The central values of the independent variables were determined in the form of the arithmetic mean of the values assumed by these variables at the upper and lower levels according to Equation (3). Then, the units of variability of the factors considered in the experiment were calculated. After calculating the central moments and units of variability, the variables were coded in accordance with Equation (4). The results of the activities leading to the presentation of the variables in the coded form are recorded in Table 1.

Table 1. Variables' coding results.

Variable	Characteristic Value		
	Central Value	**Unit of Variation**	**Coded Variable**
Capacity: Q	140	30	$x_Q = \frac{Q-140}{30}$
Current: I	154	70	$x_I = \frac{I-154}{70}$
Temperature: T	11	11	$x_T = \frac{T-11}{11}$
SoC: k	0.85	0.15	$x_k = \frac{k-0.85}{0.15}$
Voltage: U	–	–	$y = U$

The dependent variable, i.e., the voltage at the terminals of the loaded battery, U [V], is also coded. Following the coding of the variables, the next step in the preparation of the experiment plan is the plan table arrangement, according to which the measurements are carried out. Table 2 presents the table for the planned experiment in question. The number of planned experiments results from the number of variables describing the research object and the number of levels of the values that these variables take.

Table 2. The matrix of the experiment.

Number	x_0	x_1	x_2	x_3	x_4	y_{mean}
1	+1	+1	+1	+1	+1	11.32
2	+1	+1	+1	+1	−1	11.03
3	+1	+1	+1	−1	+1	11.09
4	+1	+1	+1	−1	−1	10.63
5	+1	+1	−1	+1	+1	11.89
6	+1	+1	−1	+1	−1	11.78
7	+1	+1	−1	−1	+1	11.82
8	+1	+1	−1	−1	−1	11.34
9	+1	−1	+1	+1	+1	11.00
10	+1	−1	+1	+1	−1	10.78
11	+1	−1	+1	−1	+1	10.86
12	+1	−1	+1	−1	−1	10.41
13	+1	−1	−1	+1	+1	11.7
14	+1	−1	−1	+1	−1	11.61
15	+1	−1	−1	−1	+1	11.59
16	+1	−1	−1	−1	−1	11.21

The x_0 value is an intercept of the linear model describing the object, and the subsequent columns represent independent variables, respectively: battery rated capacity, load current, electrolyte temperature, and battery condition (SoC—state of charge). A single experiment from Table 2 defines the measurement system as a set of independent variable values. Only one value of each variable belongs to each set, and all independent variables describing the research object were simultaneously taken into account. In the created plan, the number of experiments was 16. It is also assumed that the individual experiments included in Table 2 should be performed in a random order. The measurements were made for the tests presented in the table.

The linear regression equation describing the relationships between the variables for the presented plan takes the form of Equation (5):

$$y = a_0 + a_1x_1 + a_2x_2 + a_3x_3 + a_4x_4. \tag{5}$$

The coefficients of the regression equation were determined using the following relationships Equations (6) and (7):

$$a_0 = \frac{1}{N} \sum_{i=1}^{N} x_{0i} \cdot y_{mean},$$

$$\tag{6}$$

$$a_{1 \div 4} = \frac{1}{N} \sum_{i=1}^{N} x_{1 \div 4i} \cdot y_{mean}.$$

$$\tag{7}$$

After confirming the adequacy of the model, the equations were decoded and written in the form of a linear function taking into account all independent variables. The sought-after linear model of the object describes the relationships between the variables Equation (8).

$$y = a_0 + a_i \cdot \frac{x_i - x_{io}}{\Delta x_i} + \cdots + a_n \cdot \frac{x_n - x_{no}}{\Delta x_n}.$$

$$\tag{8}$$

After carrying out the measurements according to the plan review table, the coefficients of the linear equation describing the relationships between the variables were determined according to Equations (6) and (7). After determining the coefficients, the following equation was obtained in a coded form Equation (9):

$$y = 11.25 + 0.109 x_Q - 0.364 x_I + 0.135 x_T + 0.155 x_k.$$

$$\tag{9}$$

This expression presents a linear mathematical model of the object, i.e., an acid battery, the coefficients of which, determined on the basis of the data from Table 2, indicate how much the value of the dependent variable (battery voltage) will change when the value of the coded independent variable changes by one.

The mathematical description of the research object, i.e., the lead–acid battery, obtained as a result of the tests, is presented below. For this purpose, Equation (9) was decoded in order to determine the coefficients describing the quantitative influence of the individual physical variables on the value of the loaded battery voltage. The decoded linear model of the research object is a quantitative model that describes the dependence of the loaded battery terminal voltage U, as the dependent variable, on the nominal (rated) capacity Q, the discharge current I, the ambient (electrolyte) temperature T, and the (technical) state of charge (SoC) k, as the independent variables. This is presented as Equation (10). This makes it possible to know the "degree of influence" of the individual independent variables on the dependent variable.

$$U = 10.53 + 0.0036Q - 0.0052I + 0.012T + 1.033k; [V].$$

$$\tag{10}$$

Thus, using the principles of experiment planning, a multidimensional model of the acid battery under the stationary operating conditions was obtained. This method significantly reduces the time needed to conduct experiments in order to achieve the intended research objective, especially when the objective is to develop a mathematical model with a known (assumed) form that describes the relationships between the factors.

3.2. Acid Battery Non-Stationary Operating Characteristics

In the electrical circuit of the acid battery under the non-stationary operating conditions, the dynamic characteristics of the battery are revealed, which can be represented by the variability of its internal resistance. On the basis of Equation (1), the internal resistance of the analyzed battery was determined (Figure 5), considering that the voltage at the terminals of the unloaded battery, i.e., its electromotive force, was equal to 12.97 V.

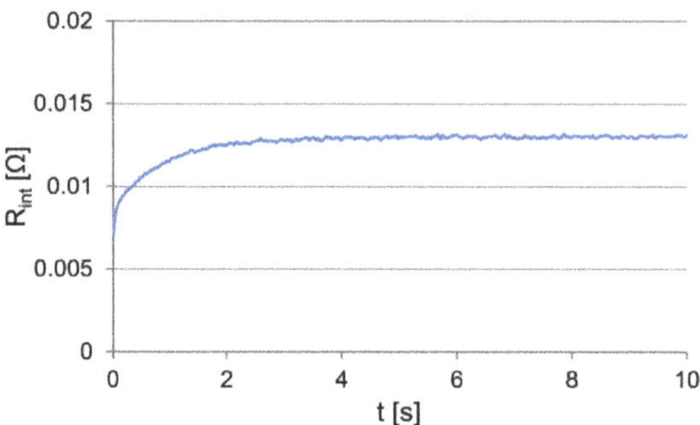

Figure 5. The changes in the internal resistance of the loaded battery.

The analytical form of the obtained dependencies (internal resistance and analogously of the voltage and current consumption) is convenient for the engineering calculations for predicting the features of an object. One of the significant problems in this case is the choice of the form of a regression function that is appropriate as an object or a process model. In the analyzed case, e.g., for the internal resistance, it is advisable to adopt the exponential function featuring the form Equation (11) because of the nature of the variability of the observed dependency. In addition:

- It is a function commonly used in science and technology;
- It is easy to interpret;
- It enables the development, through the interpretations, of a structural model of the object that undergoes an exponential response to a step extortion and its analytical description.

$$R_{int} = R_s - R_v \exp\left(-\frac{t}{\tau}\right), \tag{11}$$

where:

- R_s—battery resistance in the steady operating state;
- R_v—amplitude of the variable resistance component of the battery;
- τ—time constant of the change process of the internal resistance.

The obtained signal courses (Figures 1–3, and 5) indicate the need to isolate fixed and variable parts of the dependences. Clear determining the value of the specified course is difficult because, under exponential variability, this value is reached in infinity. In addition, especially at high current values, low temperature, and poor battery condition, the changes in the value of the analyzed signals can also be a result of the battery discharge, and thus a permanent change in its properties.

The variable part of the course, as presented in Figure 5, cannot be easily described using one exponential function. In this case, the description can be made using a glued function, i.e., a set of exponential functions defined in the different time intervals. The functions should meet the condition of continuity at the limits of the time intervals. The general form of the glued function, F, and the continuity condition can be written as in Equation (12):

$$F(t) = F_i(t); \quad t_{i-1} \leq t < t_i,$$
$$F_i(t_i) = F_{i+1}(t_i); \quad i = 1, \ldots, n-1. \tag{12}$$

In this case, another problem is the choice of the number and domain of each function, which are related to the description complexity and accuracy. As a criterion for the choice and assessment of these properties, the coefficient of determination R^2 for an individual function can be used. With regard to the analyzed dependencies in Figure 6, the fixed voltage values at the loaded and unloaded battery terminals were determined. Using the value of the coefficient of determination as a criterion, the time intervals of voltage stabilization at the terminals of the loaded and unloaded battery were determined, in addition to the variation intervals, for which the voltage change characteristic has an invariable value with respect to the time constant (Figure 6). In this manner, it was established that the voltage stabilization time of the loaded battery is approximately 4.5 s, and the stabilization of the voltage value after the load was turned off did not end within 20 s. It is important to note that the approximation of the value of the variable voltage of the loaded battery using the spline function requires two components, and for an unloaded battery, the required number of components is equal to three.

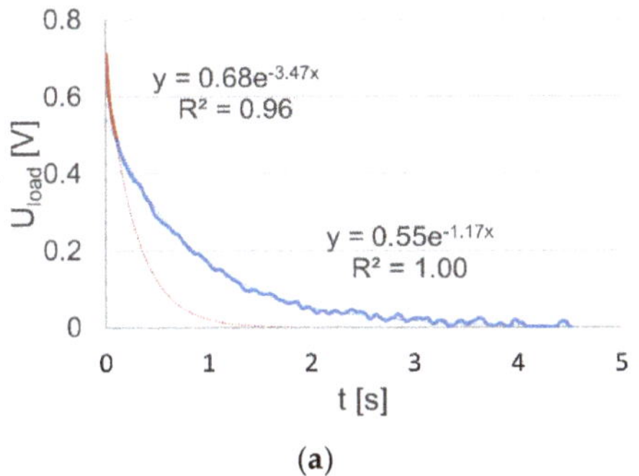

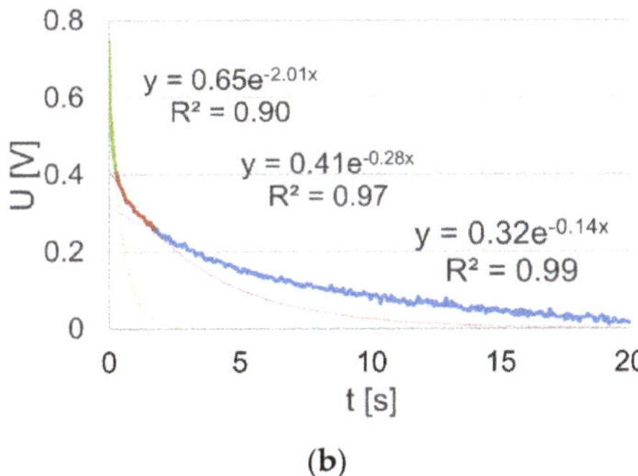

Figure 6. The variable component of the: (**a**) loaded battery voltage together with the approximating glued function lines (regression equations are written in the text (13)); (**b**) unloaded battery voltage together with the approximating glued function lines (regression equations are written in the text (14)).

The final form of the analytical description of the two dependencies is presented by Equations (13) and (14), and its illustration, together with the exponential analytic functions, is shown in Figure 6.

$$U_{\text{load}}(t) = \begin{cases} 0.68\exp(-3.47t)\ 0 \le t < 0.12; \\ 0.55\exp(1.17t)\ 0.12 \le t < 4.5. \end{cases} \tag{13}$$

$$U_{\text{unload}}(t) = \begin{cases} 0.65\exp(-2.01t)\ 0 \le t < 0.25; \\ 0.41\exp(-0.28t)\ 0.25 \le t < 1.7; \\ 0.32\exp(-0.14t)\ 1.7 \le t < 20. \end{cases} \tag{14}$$

The second possible means of describing the presented dependencies with the exponential functions is using their linear combination, i.e., a mixture of exponential functions. The mixture of functions, F_i, can be undertaken as follows Equation (15):

$$F(t) = \sum_{i=1}^{n} a_i F_i(t), \tag{15}$$

where ai represents the function weighting factors, which are also amplitudes of each individual function.

In the case of the analyzed battery, the description was made using a mixture of the voltage characteristic curve functions within the time interval from 0 to 4.5 s for a loaded battery. A stable component of the value of $U_s = 11.37$ was extracted. Hence, a very good correspondence of the description with the real dependency was obtained, by distinguishing the range of fast polarization voltage variations in the period up to 0.1 s. In this case, a description according to Equation (16) was obtained, and the separated intervals and their approximation functions are shown in Figure 7a. Similarly, the voltage dependencies at the battery terminals after switching off the load were described using a mixture of exponential functions (Figure 7b, Equation (17)). In addition, in the case of a mixture of functions, it was necessary to use the three components of the exponential function.

$$U_{\text{load}}(t) = 0.19 \exp(-30.88t) + 0.55 \exp(-1.17t). \tag{16}$$

$$U_{\text{unload}}(t) = 0.2548 \exp(-26.12t) + 0.21 \exp(-2.12t) + 0.32 \exp(-0.14t). \tag{17}$$

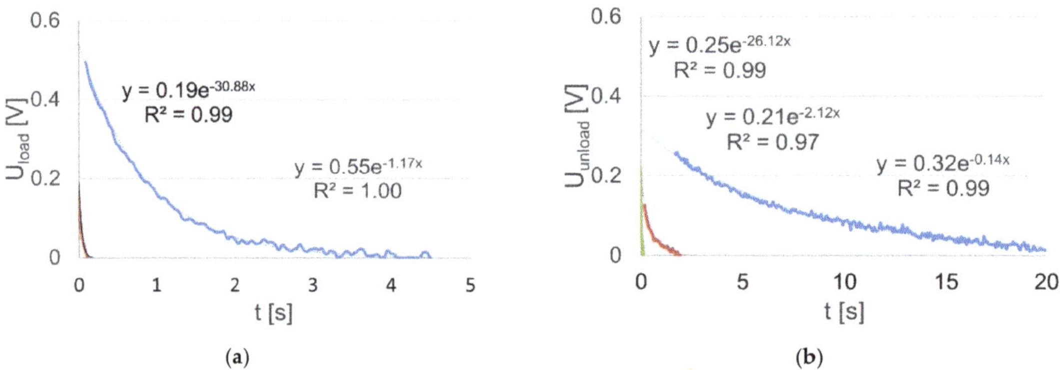

Figure 7. Determined voltage components at the terminals of the: (**a**) loaded battery within the time range up to 4.5 s (regression equations, mixture of exponential functions, are written in the text (16)); (**b**) unloaded battery within the time range up to 20 s (regression equations, mixture of exponential functions, are written in the text (17)).

Attention should be paid to the significant, more than 25-fold, differentiation in the time constants of both functions (for the loaded battery), which is equal to about 0.032 s for the fast-changing component, i.e., in the time interval up to 0.1 s and 0.86 s for the slow-changing component (they are equal to the inverse of coefficients specified in the function exponents).

3.3. Battery Structural Model

In the previous considerations, according to the electrical diagram given in Figure 4, the reason for the voltage change at the terminals of the loaded battery was recognized to be the change in its internal resistance. The primary reasons for the change are the changes in the electrolyte density around the electrodes and in the inner layers of the active mass of the battery plates. The change in the electrolyte density is also the reason for the changing potentials of the electrodes, i.e., the electromotive force of the battery. Therefore, the change in the voltage at the terminals can also be considered to be the change in the component of the electromotive force called the electromotive force of polarization. The polarity of each electrode (anode and cathode) can be distinguished, depending on the location of the polarization processes and the voltage drop in the electrolyte. In general, the electrical resistance of the battery is constituted by resistance, capacitance, and inductance components.

The electrical diagram of the acid battery shown in Figure 4 can be used to describe the operation of the battery under a constant current load or constant resistance. The variability of the internal resistance with time under dynamic load conditions makes it practically impossible to use this diagram to determine the response of the accumulator to the variable, dynamic force.

The description of the battery discharge characteristics (voltage at its terminals) using the exponential function enables the introduction of the electric components to the equivalent battery circuit, whose electrical properties generate responses in the form of exponential function. Such a component of vicarious battery diagrams may consist of a capacitor and a resistor through which the capacitor is charged or discharged. Therefore, it is possible to connect the RC circuits to a stationary source of the electromotive force of the battery E_B, as in Figure 8. Under variable load conditions, these circuits generate the electromotive force of the polarization components E_{pi} (Figure 8) according to Equations (16) and (17). A description of the battery discharge characteristics by means of a linear combination of two or more exponential functions indicates the possibility and need to also apply a larger number of RC circuits connected in series to the equivalent circuit of the battery.

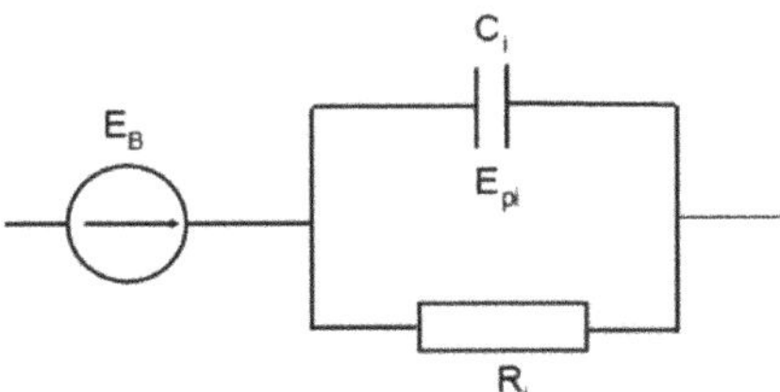

Figure 8. Electrical equivalent circuit for an acid battery.

Determination of the parameters of the circuit components is made possible on the basis of the test results of the battery discharge characteristics, as specified in Equations (16) and (17). It is known that the capacitor discharge characteristic is the exponential curve in the following form Equation (18):

$$U(t) = U_o \exp(-\frac{t}{\tau}) = U_o \exp(-\frac{t}{RC}). \tag{18}$$

This enables physical values to be assigned to the indicated components of the battery equivalent circuit. This is due to the fact that the time constant of the discharge process is $\tau = RC$.

4. Discussion

The characteristic of a device functioning (operating) is the dependence of its output signal feature value, i.e., device response, on the value of the input signal. Devices have static and dynamic characteristics. A device's static characteristic is most often the function $y = y(u)$, which represents the dependence of the value of the output signal feature y of the device on the input value, i.e., the input signal feature u under the steady-state conditions. An example of such a characteristic may be the dependence of the voltage at the battery terminals on the factors influencing its value, as presented above. The most frequently expected static characteristic is a linear one-dimensional or multi-dimensional characteristic, for example, as shown in Equation (10).

The dynamic characteristic of the device determines the transformation of the input signal $u(t)$ (extortion signal), which varies as a function of time, into the output signal $y(t)$; that is, the variable as a function of time constitutes a response of the system to this input. The dynamic characteristics of a device are most frequently described using transmittance. The operator transmittance, also referred to as a function of the device transition, is the ratio of the Laplace transform of the output signal $Y(s)$ to the Laplace transform of the input signal $U(s)$ under the zero initial conditions Equation (19):

$$G(s) = \frac{Y(s)}{U(s)}, \tag{19}$$

where the Laplace transformation is a transformation of the time function $f(t)$ into a complex function of the complex variable $F(s)$ Equation (20):

$$F(s) = L\{f(t)\} = \int_0^\infty f(t)e^{-st}dt. \tag{20}$$

On the basis of the transition function, using the inverse Laplace transform, it is possible to determine the signal that will be obtained at the output of the system for any input signal Equation (21)—or vice versa—to determine the driving signal, which should be given at the input of the device to obtain the desired response of the device.

$$y(t) = L^{-1}\{U(s)G(s)\}. \tag{21}$$

The transmittance is most often presented in the form of the amplitude characteristic $|A(f)|$ and phase characteristic $\Phi(f)$ of the device as a function of frequency f or angular frequency ω, i.e., amplitude and phase transfer in the event of a sinusoidal input. A lack of linearity of the amplitude and phase characteristics may cause dynamic deviations of the output signal. The amplitude and phase characteristics of the device are derived from the spectral transmittance.

The spectral transmittance is the ratio of the value of the complex Y response of the system caused by a sinusoidal extortion to the complex value of this extortion in the stationary state. Thus, spectral transmittance characterizes the response of the system to a sinusoidal extortion. The spectral transmittance can be obtained from the operator transmittance by substituting $s = j\omega$, thus obtaining the corresponding Fourier transform.

According to the aforementioned dependencies Equations (19) and (20), the battery operator transmittance was determined for the signal presented in Figure 7a and described using Equation (16). It was assumed that the input signal is the leap in the polarization electromotive force with a value equal to the sum of component amplitudes in Equation (16), given as Equation (22):

$$E_{p0} = E_{p10} + E_{p20}. \tag{22}$$

Therefore, in accordance with the above, the output signal Equation (16) can be written as Equation (23):

$$E_p = E_{p10}\left[1 - \exp\left(-\frac{t}{\tau_1}\right)\right] + E_{p20}\left[1 - \exp\left(-\frac{t}{\tau_2}\right)\right]. \tag{23}$$

Performing the Laplace transforms of the above-written expressions, the transformations of the input and output signal were obtained in the forms of Equations (24) and (25):

$$E_{p0}(s) = \frac{E_{p10} + E_{p20}}{s}. \tag{24}$$

$$E_p(s) = \frac{E_{p10}}{s(1 + s\tau_1)} + \frac{E_{p20}}{s(1 + s\tau_2)}. \tag{25}$$

As stated above, the battery operator transmittance can be obtained by dividing the Laplace transforms of the output signal and the input signal. Thus, after performing the appropriate transformations, the transmittance of the battery can be presented in the form Equation (26) as below:

$$F(s) = \frac{1}{E_{p10} + E_{P20}} \cdot \frac{E_{p10}(1 + s\tau_2) + E_{p20}(1 + s\tau_1)}{(1 + s\tau_1)(1 + s\tau_2)}. \tag{26}$$

Knowledge of the operating characteristics of the acid battery, both static and dynamic, in the form of its transmittance Equation (26), enables determination of the battery response, i.e., the voltage at its terminals, for any load value, i.e., the current drawn from the battery. In the expression describing the static characteristic Equation (10), the independent factors include the technical condition, i.e., the state of charge of the battery, k (it is advisable and necessary to determine similar relationships for the dynamic characteristics). The comparison of the battery voltage value determined on the basis of the characteristic for the reference battery, with the voltage value measured under the experimental conditions, may enable the (SoC) value k of the researched battery to be determined. In this case, value k becomes i.e., the battery status indicator, and can be used as a diagnostic parameter for the determination of the status of the battery. It is expected that the developed dependencies will be used in the diagnostic procedure of the battery, based on the measurement of the engine start signals, i.e., the driving of its crankshaft by the starting system. The proposed method will allow not only the diagnosis of the condition of the acid battery, but also the electric starter and the internal combustion engine in terms of the resistance to motion, taking into account the compression pressure of cylinder charges.

5. Conclusions

The acid battery is a functionally and structurally complex non-linear power source, whose features are dependent on many parameters. Its static characteristic, i.e., the loaded battery terminal voltage U in stationary operating conditions, depends on the nominal (rated) capacity Q, the discharge current I, the ambient (electrolyte) temperature T, and the (technical) state of charge (SoC) k as the independent variables. For the purpose of describing the dynamic characteristic of its operation, that is, the response to a rapid leap in current (i.e., a load with a constant current or resistance), it is convenient to use the exponential functions in the form of glued functions or a mixture of functions. Both methods of description correspond to two different electric models (equivalent circuits) of the acid battery in the form: (1) the electromotive force and the variable internal resistanc; (2) the stationary electromotive force and the (two or three) RC systems having different characteristics, resulting in changes in the electromotive force of the polarization of the battery in the circuit. The presence of two or three different components of the electromotive force of polarization indicates that the equivalent circuit of a lead–acid battery should include at least two RC circuits connected in series, with the significantly different parameters defined by means of time constants. In fact, the change in the time constant of the polarization electromotive force occurs continuously, from very small values to theoretically equal to infinity. The consideration of many independent parameters in the description of the battery and its structure requires long-term extensive experimental research.

A feature of modern machine exploitation is the constant, systematic increase in the role and meaning of technical diagnostics. The broad possibilities of its application result from the change in the properties of the exploitation objects, including motor vehicles and the development of methods and means of diagnosis using digital signal recording and processing techniques. In the diagnostics of internal combustion engines and their starting systems, the diagnostic parameters of the working and accompanying processes of the driving of the crankshaft can be used. These diagnostic parameters include: the current consumed by the starter, the voltage at the battery or starter terminals, and the speed of the crankshaft forced by the starter. The set of electric starter characteristics depends on the properties of the energy source, i.e., the acid battery.

The developed models, both for the stationary and non-stationary conditions, will be used in the proposed and currently developed diagnosis method for the internal combustion engine–electric starting system based on the engine start-up signals and its driving by the starting system. It will be possible to determine the state of charge of the battery on the basis of Equation (10). The dynamic components of the model in the form of transmittance Equation (26) will allow determination of the significance of the influence of the battery's dynamic characteristics on the process of driving the crankshaft of the engine, and the necessity to include this in the diagnostic test of the system.

As a result of the presented research and characteristics of the acid battery, and the analyses performed, conclusions important for the knowledge and understanding of the principles of operation of the acid battery can be formulated:

- Regression methods that enable general, symbolic, and formal descriptions of the test results play an important role in the study of the characteristics and parameters of the acid battery structure.
- The static characteristics of the battery as a function of the essential independent parameters, such as rated capacity, state of charge, current intensity, and electrolyte temperature, are linear under a wide range of usable operating conditions.

- Under dynamic operating conditions, the battery is a complex control object with time-varying characteristics and properties of an equivalent structural electrical circuit. This is expressed by the necessity to include two or even three RC circuits in its structure.
- The determined stationary and dynamic model enables determination of the influence on the operation of the starter and the parameters of driving the crankshaft of the internal combustion engine by the battery under specific operating conditions.
- Estimation of the time constants of dynamic voltage characteristics of the battery enables the determination of its operator and spectral transmittance, and evaluation of the impact of the dynamic characteristics, not only on the functioning of the battery, but also on the internal combustion engine starting system and its crankshaft drive characteristics.
- The dynamic characteristics of the load phase and voltage stabilization after switching off the load are significantly different, and the equivalent circuit has two or three RC circuits, respectively.
- Significant differences in the load phase and voltage stabilization characteristics indicate a significant differentiation in the causes of the appearance and disappearance of the imbalance of the loaded battery and after the load is turned off.
- The determination procedures, especially of the dynamic characteristics, are laborious and require precise separation of the variable, dynamic part from the recorded course. It is possible to achieve high accuracy in determining the characteristics, as assessed by the value of the determination coefficient, which can be as high as 0.99.
- The criterion for assessing the accuracy of determining the dynamic characteristics may be the value of the determination coefficient, both in the range of the value of the stable voltage and in the time intervals of the exponential function with a specific value of the time constant.

Funding: This research received no external funding.

Institutional Review Board Statement: Not applicable.

Informed Consent Statement: Not applicable.

Data Availability Statement: Exclude this statement.

Conflicts of Interest: The author declares no conflict of interest.

References

1. Ceraolo, M. New dynamical models of lead-acid batteries. *IEEE Trans. Power Syst.* **2000**, *15*, 1184–1190. [CrossRef]
2. Jackey, R.A. *A Simple, Effective Lead-Acid Battery Modeling Process for Electrical System Component Selection*; SAE International: Warrendale, PA, USA, 2007.
3. Droździel, P. Impact of selected operation conditions of a car combustion engine on its start-up parameters. *Eksploat. Niezawodn. Maint. Reliab.* **2003**, *4*, 22–30.
4. Enache, B.-A.; Constantinescu, L.-M.; Lefter, E. Modeling aspects of an electric starter system for an internal combustion engine. In Proceedings of the 2014 6th International Conference on Electronics, Computers and Artificial Intelligence (ECAI), Bucharest, Romania, 23–25 October 2014; pp. 39–42.
5. Kasprzyk, L. Modelling and analysis of dynamic states of the lead-acid batteries in electric vehicles. *Eksploat. Niezawodn. Maint. Reliab.* **2017**, *19*, 229–236. [CrossRef]
6. Pszczółkowski, J. Rozruch silnika tłokowego jako proces diagnostyczny. *Diagnostyka* **2002**, *27*, 48–53.
7. Chacón, H.E.A.; Banguero, E.; Correcher, A.; Pérez-Navarro, Á.; Morant, F. Modelling, Parameter Identification, and Experimental Validation of a Lead Acid Battery Bank Using Evolutionary Algorithms. *Energies* **2018**, *11*, 2361. [CrossRef]
8. Lee, S.; Cherry, J.; Safoutin, M.; McDonald, J. Modeling and Validation of 12V Lead-Acid Battery for Stop-Start Technology. *SAE Tech. Pap. Ser.* **2017**, *1*. [CrossRef]
9. Dost, P.; Sourkounis, C. Generalized Lead-Acid based Battery Model used for a Battery Management System. *Athens J. Technol. Eng.* **2016**, *3*, 255–270. [CrossRef]
10. Mohsin, M.; Picot, A.; Maussion, P. Lead-acid battery modelling in perspective of ageing: A review. In Proceedings of the 2019 IEEE 12th International Symposium on Diagnostics for Electrical Machines, Power Electronics and Drives (SDEMPED), Toulouse, France, 27–30 August 2019; pp. 425–431.
11. Park, C.H.; Yoon, J.H.; Choi, J.D. A Quantitative Study for Critical Factors of Automotive Battery Durability. In *Automotive Electronics Reliability*; SAE International: Warrendale, PA, USA, 2010; pp. 187–192.

12. Barré, A.; Suard, F.; Gérard, M.; Montaru, M.; Riu, D. Statistical analysis for understanding and predicting battery degradations in real-life electric vehicle use. *J. Power Sources* **2014**, *245*, 846–856. [CrossRef]
13. Tseng, K.-H.; Liang, J.-W.; Chang, W.; Huang, S.-C. Regression Models Using Fully Discharged Voltage and Internal Resistance for State of Health Estimation of Lithium-Ion Batteries. *Energies* **2015**, *8*, 2889–2907. [CrossRef]
14. Korotunov, S.; Tabunshchyk, G.; Okhmak, V. Genetic Algorithms as an Optimization Approach for Managing Electric Vehi-cles. Charging in the Smart Grid. Available online: http://ceur-ws.org/Vol-2608/paper15.pdf (accessed on 26 October 2021).
15. Shen, Y. Supervised chaos genetic algorithm based state of charge determination for LiFePO4 batteries in electric vehicles. In *Proceedings of the 2nd International Conference on Advances in Materials, Machinery, Electronics (AMME 2018), Xi'an, China, 20–21 January 2018*; AIP Publishing: College Park, MD, USA, 2018; Volume 1955, p. 040050.
16. Loukil, J.; Masmoudi, F.; Derbel, N. A real-time estimator for model parameters and state of charge of lead acid batteries in photovoltaic applications. *J. Energy Storage* **2021**, *34*, 102184. [CrossRef]
17. Wang, Z.-H.; Hendrick; Horng, G.-J.; Wu, H.-T.; Jong, G.-J. A prediction method for voltage and lifetime of lead-acid battery by using machine learning. *Energy Explor. Exploit.* **2019**, *38*, 310–329. [CrossRef]
18. Demirci, O.; Taskin, S. Development of measurement and analyses system to estimate test results for lead-acid starter batteries. *J. Energy Storage* **2021**, *34*, 102172. [CrossRef]
19. Křivík, P.; Vaculík, S.; Bača, P.; Kazelle, J. Determination of state of charge of lead-acid battery by EIS. *J. Energy Storage* **2019**, *21*, 581–585. [CrossRef]
20. Raji, S.; Kubba, Z.M. Design and Simulation of Lead-Acid Battery. *J. Al-Nahrain Univ. Sci.* **2020**, *23*, 39–44. [CrossRef]
21. Laadissi, M.; Filali, A.; Zazi, M.; Ballouti, A. Comparative Study of Lead Acid Battery Modelling. *ARPN J. Eng. Appl. Sci.* **2006**, *13*, 4448–4452.
22. Shi, M.; Yuan, J.; Dong, L.; Zhang, D.; Li, A.; Zhang, J. Combining physicochemical model with the equivalent circuit model for performance prediction and optimization of lead-acid batteries. *Electrochim. Acta* **2020**, *353*, 136567. [CrossRef]
23. Jiang, S. A Parameter Identification Method for a Battery Equivalent Circuit Model. In *SAE Technical Paper 1*; SAE International: Warrendale, PA, USA, 2011. [CrossRef]
24. Lach, J.; Wróbel, K.; Wróbel, J.; Podsadni, P.; Czerwiński, A. Applications of carbon in lead-acid batteries: A review. *J. Solid State Electrochem.* **2019**, *23*, 693–705. [CrossRef]
25. Zhuravsky, B.V.; Zanin, A.; Kvasov, I.N. Simulation of the electric starter system of the internal combustion engine start-up to study the impact on its operation of the pre-start battery discharge. *J. Phys. Conf. Ser.* **2020**, *1441*, 012030. [CrossRef]
26. Dyga, G.; Pszczółkowski, J. Plan eksperymentu identyfikacji modelu napięcia obciążonego akumulatora kwasowego. *Technika Transportu Szynowego.* **2015**, *12*, 1843–1849.
27. Chang, W.-Y. The State of Charge Estimating Methods for Battery: A Review. *ISRN Appl. Math.* **2013**, *2013*, 953792. [CrossRef]
28. Nikdel, M. Various battery models for various simulation studies and applications. *Renew. Sustain. Energy Rev.* **2014**, *32*, 477–485. [CrossRef]
29. Moubayed, N.; Kouta, J.; El-Ali, A.; Dernayka, H.; Outbib, R. Parameter identification of the lead-acid battery model. In Proceedings of the 2008 33rd IEEE Photovolatic Specialists Conference, San Diego, CA, USA, 11–16 May 2008; pp. 1–6. [CrossRef]
30. Pszczółkowski, J.; Dyga, G. Dwuwymiarowe liniowe zależności funkcyjne napięcia akumulatora kwasowego. *Logistyka* **2014**, *6*, 8984–8995.

Article

Analysis of Scenarios for the Insertion of Electric Vehicles in Conjunction with a Solar Carport in the City of Curitiba, Paraná—Brazil

Ana Carolina Kulik [1], Édwin Augusto Tonolo [1,*], Alberto Kisner Scortegagna [1], Jardel Eugênio da Silva [2] and Jair Urbanetz Junior [1]

[1] Department of Electrotechnics, Industrial Electrical Engineering—Electrotechnics, Federal University of Technology—Paraná (UTFPR), Avenue Sete de Setembro, Curitiba 80230-901, Paraná, Brazil; kulik.anacarolina@gmail.com (A.C.K.); albertok_s@hotmail.com (A.K.S.); urbanetz@utfpr.edu.br (J.U.J.)
[2] Grazziotin Engineering & Solar Energy, Street Manoel Furtado Neves, 895, São Mateus do Sul 83900-000, Paraná, Brazil; jardel.eugenio@hotmail.com
* Correspondence: edwintonolo@gmail.com; Tel.: +55-45-991087892

Citation: Kulik, A.C.; Tonolo, É.A.; Scortegagna, A.K.; da Silva, J.E.; Urbanetz Junior, J. Analysis of Scenarios for the Insertion of Electric Vehicles in Conjunction with a Solar Carport in the City of Curitiba, Paraná—Brazil. *Energies* **2021**, *14*, 5027. https://doi.org/10.3390/en14165027

Academic Editor: Wojciech Cieslik

Received: 28 July 2021
Accepted: 12 August 2021
Published: 16 August 2021

Abstract: The growing environmental impact and rising emission of greenhouse gases have accelerated the research toward renewable energy sources and electric vehicles since one of the main sources of pollution is the CO_2 emissions produced by conventional combustion vehicles. This article presents the analysis of the energy balance between a photovoltaic carport with 4.89 kWp installed capacity and an EV, model Renault Fluence ZE DYN, driven in real conditions. The driving tests were performed during the winter season in the city of Curitiba, the capital of the state of Paraná, Brazil, with approximately 1.7 million inhabitants and 1.1 million vehicles. During the test period, we attempt to reproduce the citizen's daily routes through the city, presenting an average consumption of 15.75 kWh/100 km. The carport PV module's energy generation and in-plane incident irradiation were acquired to calculate the performance ratio, making a comparison after cleaning maintenance possible. The solar carport system has 4.89 kWp and has generated an average of 465.37 kWh during its 24 months of operation. The analysis scenarios consist of replacing part of the city's combustion vehicle fleet with the EVs (the same as used in the study) and thus determining how many replicas of the presented photovoltaic systems might be needed, as well as the area required for the installations. In a simulation with 15% of the fleet's replacement, it would be necessary to generate 17,151.8 MWh, which requires the construction of 36,856 carports, covering an area of approximately 1,105,685 m^2. Finally, an economic comparison between an internal combustion vehicle and the EV determined that the expenditures involving electric energy to charge the batteries are 3.3 times lower than buying gasoline, assuming the same driving routines.

Keywords: vehicle charging stations; electric vehicles; energy consumption; energy generation; renewable energy sources

1. Introduction

The environmental degradation resulting from greenhouse gas emissions and the upcoming shortage of fossil fuels highlights the importance of generating energy with minimal environmental impacts, confirming the relevance of including different energy sources in the national energy matrix [1–3].

The transport sector is one of the biggest consumers of fossil fuels, contributing to the increase in greenhouse gas emissions and global warming [4]. In 2019, this sector produced 8.2 Gt of CO_2 (gigatons of carbon dioxide), representing 24% of direct emissions from fuel combustion [5,6]. Public and private transportation both have a crucial role in the development of a region or entire country. However, there are also commonly associated problems such as pollution, fossil fuel consumption, noise, accidents, resources use and waste, etc. [7]. Improvements aiming to decrease the dependence on these non-renewable fuels and to reduce the environmental impacts are considered vital to ensure energy security and population welfare [7,8].

Knowing the need to decrease pollution (mainly air) around the globe, we must reduce the emissions caused by the transport sector by shifting away from the traditional fossil fuel-based concept to an alternative system [8–10]. With numerous objectives to be achieved (and quickly), electric vehicles (EVs) are set to be the key to shift into electric mobility, considering that they have already been playing a significant role in recent years [6,8].

The electric motor is characterized by high efficiency, lack of emissions, and lower noise, compared to traditional combustion engines [11–13]. In general, EVs have not been successful in the past because of the limitations in battery technology in terms of high weight and price, short life, and long charging time [14]. Supported by recent advances and developments in technology, they are now on the market as strong competitors to traditional internal combustion vehicles [12,14].

EVs can be divided into battery electric vehicles (BEVs), plug-in hybrid electric vehicles (PHEVs), and fuel cell electric vehicles (FCEVs) [15]. BEVs are only powered by an electric motor and traction batteries [16]. FCEVs use fuel cells to deliver energy and power the electric motor [16]. Hybrid vehicles have two sources and are powered by an internal combustion engine and batteries [16,17]. They are not completely free of non-renewable fuels but cost less and can be used to travel long distances without stopping to recharge [16,17].

EVs are known for their low maintenance, high performance and efficiency, and zero air pollutant emissions, explaining their infiltration into the automobile market [15]. Along with increasing environmental concerns and energy-related issues, EVs have become one of the main subjects of research [16].

In the year 2020, there were 10 million electric vehicles moving around the globe, an increase of 43% compared to 2019, with BEVs accounting for nearly 66% of the sales. The sales forecast, in a stated policies scenario, declares that the EV stock in the year 2030 will be approximately 150 million [18].

The largest fleet can be found in China, representing 45% of the global electric car stock [18]. The Brazilian market is developing at a slow pace, and according to [19], approximately 44 thousand EVs have been sold, representing less than 0.1% of the total national fleet.

Public accessibility to chargers increased by 45%, considering both slow and fast chargers, reaching 1.3 million stations [18]. In [20], an explanation about the different kinds of plugs and charging levels is presented, pointing out the major characteristics and differences. However, [18] found that when the power is lower than 22 kW, it is considered a slow charge. Interurban transportation is one of the cases in which EVs may not be able to provide a solution yet, due to the large distances between locations. To guarantee safe interurban transportation, [21] proposed the deployment of fast-charging infrastructure on the highways.

Continuous population growth and urban expansion have led to an increase in demand for electricity [22]. Reduced carbon emissions are one of the main goals in urban planning and energy policies [7,23]. With the cost of electricity generated by solar power decreasing significantly and becoming competitive, the rapid development of photovoltaic (PV) infrastructure was achieved, along with a strong market. The design flexibility of PV systems allows the energy to be generated with a wide range of options, meeting the needs of distinct levels of consumers, from homes to large industries [23,24].

The global installed capacity of renewable energies grew by more than 250 GW (gigawatts) in 2020, led by photovoltaic solar energy, and it is estimated that approximately 29% of the electricity generated in 2020 came from renewable resources. For the sixth consecutive year, renewable energy system installations have surpassed those of fossil fuel and nuclear power capacity combined. More specifically, the photovoltaic solar energy segment showed an increase of approximately 22% in installed capacity, with 139 GW being added. The greatest demand for photovoltaic solar has occurred in China, the United States of America, Europe, and emerging markets around the world. The global installed capacity is approximately 760 GW, which includes both on-grid and off-grid solar generation capacity, compared to a total of less than 40 GW just ten years earlier [25].

Renewable energy sources for electricity generation, and the electrification of the transport sector, offer great potential for reducing the use of fossil fuels–one of the major causes of air pollution and health problems globally [12,26]. Thus, the benefits between photovoltaic energy production and EV charging are greater when the integration between charging stations and photovoltaic systems are considered, on what is a promising solution to the abovementioned environmental problems [3,14,24,27]. To make the best use of both technologies, the EVs must be charged during the day, using the energy generated by the PV system or other renewable energy sources [28].

The combination of EV charging stations and PV generation can be achieved through the construction of solar car parking, also named solar canopies or solar carports. These structures are built to park a car under, and PV modules are installed on their roofs to generate electricity. These systems are versatile and can be used in the most diverse places, such as markets, hotels, shopping malls, restaurants, public institutions, country houses, camping areas, shopping centers, business parks, sports centers, train stations, airports, etc., benefiting areas that in most cases are available to be reshaped [6,14,29]. Any of these places can provide convenient charging while the EVs are parked, supporting the development and use of renewable energy systems [22,23].

During the life cycle of a PV module, situations like natural degradation, possible components failure, various weather conditions, electrical stress, and others are faced that momentarily and/or permanently alter some characteristics [30]. Out of the main encountered situations, dust accumulation is one the most common, defined by the particles that cover the module surface, blocking the cells from receiving the energy from the sun, negatively impacting the PV energy generation [30,31]. During wet seasons, when it rains more frequently, there is less dirt accumulation compared to dry seasons, which can be accredited to the rain's natural cleaning [32,33]. Although it is not as effective as a deep cleaning made by a human or machine, it still exhibits positive effects.

The use of photovoltaic energy generation together with EV charging infrastructure still poses challenges, mainly because of the uncertain energy demand pattern of EVs, which is based on the drivers' behavior and preferences and the intermittence of the PV generation [9,34]. Ref. [24] presented a complete study of the vehicles' charging requirements based on the period of the year, distance traveled, and location of the route. Meanwhile, ref. [35] modeled the driving patterns and energy demand of the EV for the country of Austria. Studies in the field are essential for accurate planning of the investments aiming to promote electric mobility [36].

EVs represent a storage capacity, in a great new approach known as Vehicle-to-Grid (V2G), which gives the option of a bidirectional energy flow, positively affecting the vehicle and the grid; this can be interpreted as another functionality for intelligent energy networks (smart-grids) [29,34]. It can supply back-up electricity, shift the electricity load, and respond quickly to balance the grid, representing a new important power source [15]. The work presented by [15] centers the attention on a technical and economic analysis for the application of V2G techniques in the power grid.

In 2012, the National Electric Energy Agency—ANEEL, released Resolution 482 for the regulation of the distributed energy generation in Brazil, which was amended in 2015 [37]. It is classified as micro-generation when the PV installation is up to 75 kW (kilowatt) and mini generation for 75 kW to 5 MW (megawatt). It also gives the consumer the possibility to exchange the surplus energy through a free loan with the grid, reducing the energy bill and generating energy credits that can be consumed within 60 months [38]. The energy does not need to be generated at the time or in the month it is consumed [39].

The constant increase in energy demand, at least in Brazil, has been leading to an increase in the contribution of non-renewable energy sources in the production of Primary

Energy, as can be seen in [40]. Observing the graphics, it can be seen that from 1970 to 1999, the production of renewable sources was greater than the non-renewable ones, a scenario that was reversed after the year 2000. Although both cases are presenting a growing pattern, non-renewable energy production is growing at a higher rate. There is another challenge regarding the grid voltage stability [41]. Ref. [42] proposes a combination of different local resources to alleviate the potential grid problems. Charging EVs through renewable power generation systems must be optimized in order to reduce the cost of operation and the grid problems. Based on the Brazilian regulation, ref. [43] proposed a tariff model for public access points, reducing the costs for the consumer.

In our work, considering an EV, we intend to define how much electrical energy is consumed during daily and monthly driving routines, considering only working days during the winter season in the city of Curitiba, Brazil. Regarding the PV systems, the energy generation data and local irradiation information are acquired, making it possible to calculate the performance ratio. At first, it is discussed if there are improvements in the energy generation efficiency by performing a cleaning service in the modules. Examining the integration of both technologies, it is discussed if the installed PV system is capable of generating the electric vehicle required energy that was measured during the tests. Based on fleet substitution scenarios, the manuscript discusses how many carports may be necessary to recharge the EVs batteries in each condition. An economic analysis to compare the monthly fuel costs of an EV and an internal combustion vehicle is presented.

The scope of this study includes the energy generated by the carport during its 24 months of operation, covering two parking spaces. In addition, the electric vehicle's actual consumption will be displayed. Finally, the performance ratio of the PV system will be presented, and the maintenance of the modules will be discussed.

2. Materials and Methods

The aim of this study is the evaluation and analyses of the energy generated by a PV system installed on the parking lot of the university and the required energy for charging an EV, emulating the average driven distance by the citizen from the city of Curitiba, state of Paraná, Brazil. The analysis of energy generation from the PV system is verified from July 2019 (date when it was inaugurated) until June 2021, totalizing 24 months.

The average EV energy consumption is also presented. The following questions were evaluated: How much energy does an electric vehicle in the city consume for its daily activities? Does the 4.89 kWp (kilowatt-peak) carport photovoltaic system guarantee the energy demand necessary to power two electric vehicles?

The system performance ratio is also calculated, and after the cleaning results are presented, the system maintenance and meteorological factors that influence its operation are discussed.

This article presents an analysis of two stages of energy conversion. The first one studies the energy consumption of the electric vehicle Renault Fluence ZE DYN in real daily conditions of the urban environment, while the second one deals with the maintenance of energy generation of the solar carport. Figure 1 illustrates the proposed division.

2.1. Electric Vehicle

The electric vehicle used in the driving tests is a Renault Fluence ZE DYN, equipped with a 22 kWh (kilowatt-hour) lithium-ion battery bank and an engine with a maximum power of 95 hp (horsepower), corresponding to 70 kW. The main characteristics of the EV are summarized in Table 1.

Recollecting the information about the EVs market in Brazil, it still represents a small portion of the vehicle fleet, mainly due to their high prices. Under these circumstances, the presented car is the only data source available for this type of test in the university. The EV was recently purchased.

(**a**) (**b**)

Figure 1. Analyses proposed for this study: (**a**) Carport maintenance and energy generation; (**b**) Renault Fluence ZE DYN energy consumption.

Table 1. Electric vehicle technical information [44].

Electric Motor/Car		Battery	
Parameter	**Value**	**Parameter**	**Value**
Max. voltage	400 V	Battery type	LI-ion
Max. output power	70 kW	Battery total capacity	22 kWh
Max. speed	135 km/h	Charge Port	AC-Type 2

2.2. Carport

The generated energy from the carport is acquired from the online website of the installed inverter. The system is working connected to the grid since July 2019, and all the generated energy that is not used, the surplus, is used as energy credit in the deduction of the university's monthly energy bill.

The entire PV system of the carport was built by the donation from some companies that have business in the energy sector. From the academic point of view, it aims at the integration of photovoltaic energy generation with EV battery recharging. In addition to the energy generation and energy cost savings, it is an important source of information for research in the field.

The PV system contains 15 polycrystalline solar modules, arranged in 2 strings, one containing 6 modules from a manufacturer with 335 Wp each, while the other contains 9 modules from another manufacturer with 320 Wp each, totaling 4.89 kWp. The installation also includes a single-phase 5 kW inverter with integrated monitoring.

The carport was assembled aligned with the parking lot's geographical position, distant from shadows of trees and light posts, with a north orientation, a 10° inclination, and azimuth deviation of 22° to the west, allowing two vehicles to be parked simultaneously. Table 2 summarizes the main parameters of the PV system, and Table 3 presents the inverter electrical characteristics.

Table 2. Electrical parameters of the PV system strings [29].

	Modules Number	P_{mp} Modules	P_{mp} String	V_{oc} String	V_{mp} String	I_{mp} String	I_{sc} String
String 1	6	335 W	2.01 kWp	274.8 V	224.4 V	8.96 A	9.54 A
String 2	9	320 W	2.88 kWp	417.6 V	336.6 V	8.56 A	9.05 A

Table 3. Electrical parameters of the PV system inverter [29].

	Parameters
MPPT	2
MPPT Individual Maximum Current DC	12 A
Maximum Input Voltage DC	1000 V
Voltage Range MPPT	240–800 V
Rated Output Power AC	5000 W
Maximum Output Current AC	21.7 A
Connection Network Voltage	1-NPE 220/230 V
Network Frequency	50/60 Hz

For the electrical security of the PV system, there is a connection board in the DC (direct current) side, containing 4 inputs and 2 output pairs, supporting the connection of both strings into the MPPT (maximum power point tracking) channels of the inverter. It also contains, for every string, a set of SPDs (surge protection devices) model Dehn Type II for 1000 V (volts), a particular model for PV applications, and a disconnection switch of 1000 V and 25 A (amps). Regarding the AC (alternating current) side, there is a connection board with a circuit breaker and an SPD.

It was decided to build a wall in front of the carport, measuring 3.17 m × 1.63 m (meters), so that the fundamental equipment could be protected from the rainfall, but also exhibited to the community and other students from the university, contributing to the awareness of technologies and its benefits. In this wall, the inverter, both connection boards, a pair of standard AC outlets, and the vehicular charger were fixed.

The two standard AC outlets were installed for charging the EV in Level 1 Charge Mode or to use general electrical devices. The one and only vehicular charger so far, model ProEV1, was donated and installed by Egnex company and is illustrated in Figure 2. There is enough space for two more charges.

Figure 2. EGNEX ProEV1 charger.

This charger automatically communicates with the EV and starts charging it when there is an established connection with the vehicle, in accordance with current technical standards. To charge the EV, using the AC grid, the user can configure the equipment according to the local electrical installation, from 1 to 7 kWh in a single-phase and from 2 to 22 kWh in a three-phase network. The charger is available with a Type 1 or Type 2 connector, which is the most accessible connector on the national market [45]. Figure 3 displays the carport's internal and external aspects, including the wall and the fixed equipment.

(a)   (b)

Figure 3. Current carport installations: (**a**) External; (**b**) Internal.

Each parking space has an area of approximately 15 m^2 (square meter). Knowing that two cars can be parked under the structure for simultaneous charging, the total area covered by the carport is 30 m^2.

3. Results and Discussion

3.1. Vehicle Energy Consumption

In order to measure the energy consumption of the Renault Fluence ZE DYN, it was necessary to purchase and connect a specific scanner, "Konnwei OBDING BT 3.0", model KW902, to the EV's on-board computer. Using an Android smartphone to install the Fluence ZE Spy application, the device and the software establish communication via Bluetooth, exhibiting the required information. Figure 4 shows the scanner, and the application used to display the electrical consumption of the EV.

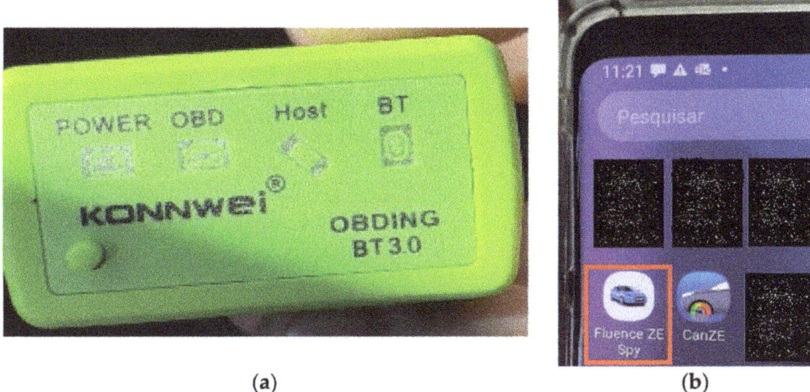

(a)   (b)

Figure 4. Required gadget and software: (**a**) Scanner Konnwei OBDING BT 3.0, model KW902; (**b**) Fluence ZE Spy application.

After installing the scanner hardware and the application to display the required results, the energy consumption of the EV was measured in the urban environment in

June 2021, the winter season in Brazil. We executed two measurements with the same driving distances, with the objective of obtaining a reliable result. The driving tests were performed in the central area of the city, around 11 a.m., facing a light traffic jam. Eco mode was enabled.

The total driving distance was 42.9 km (kilometer), and both attempts ended up with approximately the same outcome, with 15.75 kWh/100 km (kilowatt-hour per a hundred kilometer) for the first trip and 15.55 kWh/100 km for the second drive, both presented in Figure 5.

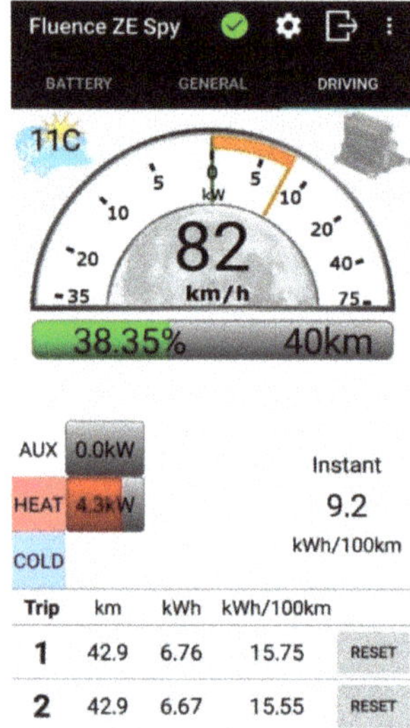

Figure 5. Energy consumption measurements for the Renault Fluence ZE DYN in the urban environment during the driving tests.

To perform the evaluations proposed in this article, the highest obtained result from the two-driving tests has been used; therefore, in the following scenario calculations, it will be considered that the EV energy consumption, considering the urban traffic of the city of Curitiba, is 15.75 kWh/100 km.

3.2. Carport Energy Generation and Performance Ratio

Based on the mass memory of the inverter, Table 4 presents the monthly energy generation of the PV system from July 2019 to June 2021, totaling 24 months of analysis. Summing the data from all operational months, the PV system has already generated 11,168.77 kWh. The monthly average energy generation is 465.37 kWh, and it is the value that will be used in the scenario analyses, taking into consideration the local renewable energy regulation.

One way to evaluate the generation of a photovoltaic system is by analyzing the merit indices. These are important indicators that show whether the energy is being generated by the systems in an optimized way and enable the comparison between other photovoltaic installations or other energy sources [46–48].

Among the merit indices, only the performance ratio (PR) will be analyzed here, which is the relation between the generated energy (kWh), the reference irradiance (kW/m^2, kilowatt per square meter), the incident irradiation in the plane of the photovoltaic panel (kWh/m^2, kilowatt-hour per square meter), and the installed PV power (kWp), according to Equation (1).

$$PR = \frac{Energy\ Generation \times Reference\ irradiation}{Irradiation \times PV\ Power} \tag{1}$$

Table 4. Monthly Energy Generation of the Carport (kWh).

Month	2019	2020	2021
January	-	488.72	400.08
February	-	432.54	433.34
March	-	571.15	412.02
April	-	556.07	380.35
May	-	501.02	413.40
June	-	299.53	360.00
July	485.01	436.63	-
August	461.21	496.96	-
September	422.06	488.39	-
October	582.13	516.33	-
November	521.90	543.31	-
December	541.93	424.69	-
Annual	3014.24	5755.34	2399.19

It is impossible to achieve an efficiency of 100%, as photovoltaic installations present losses that are typical, and among them, panel degradation, temperature, dirt, internal network failures, cabling, inverter, transformer, and system availability can be highlighted. For the dimensioning of photovoltaic systems, it is common to adopt a PR between 70% and 80% [47].

Due to the dirt conditions that can be identified by visual inspection and the calculated low-performance ratio indicator, a cleaning service was executed on the photovoltaic modules on 1 June 2021 in order to verify its impacts on the generation of electricity. A comparison between the electricity generated in June with the previous months is performed. Figure 6 illustrates the conditions of the modules before and after cleaning them.

(a)

(b)

Figure 6. Comparison of the PV modules conditions: (**a**) before washing; (**b**) after washing.

The irradiation data were acquired from a solarimetric station called EPESOL—Solar Energy Research Station, located in the university's campus, the same site where the carport system is installed. However, the installation of the carport was prior to the solarimetric station, which has been in operation since March 2020. The previous irradiation values were obtained from the INMET (National Institute of Meteorology). Originally, these measurements are acquired for the horizontal plane and later converted to the plane of the carport, according to the inclination and azimuthal deviation, through software, and are presented in Table 5.

Table 5. Monthly average irradiation in the plane of the carport ($kWh/m^2/day$).

Month	2019	2020	2021
January	-	5.16	4.37
February	-	4.57	6.24
March	-	6.04	4.75
April	-	5.44	4.26
May	-	4.45	3.82
June	-	2.87	3.01
July	3.69	3.99	-
August	3.98	4.43	-
September	4.00	4.69	-
October	5.49	5.24	-
November	5.23	6.06	-
December	5.30	5.18	-

With the acquired data and Equation (1), it is possible to calculate the monthly performance ratio for the entire period of operation of the system.

In Table 6, the results of the performance ratio for the 24 months of the carport's operation are presented. For the periods when the calculated outcome is below 70%, it can be understood as a warning signal, which may indicate that the PV system is experiencing long periods of shading, the modules are dirty, or there are electrical problems.

Table 6. Performance ratio (%).

Month	2019	2020	2021
January	-	62.48%	60.39%
February	-	66.74%	50.72%
March	-	62.38%	57.22%
April	-	69.68%	60.86%
May	-	74.27%	71.39%
June	-	71.14%	81.53%
July	86.71%	72.19%	-
August	76.44%	74.00%	-
September	71.93%	70.98%	-
October	69.95%	65.00%	-
November	68.02%	61.11%	-
December	67.45%	54.08%	-
Annual	73.42%	67.01%	63.69%

The performance ratio measures the efficiency of the photovoltaic panel in a given location, deducting the losses, measures the onsite quality, and compare PV installations in other locations. Considering the 24 months studied, an average performance ratio of 68.04% was reached.

As mentioned earlier, scheduled cleaning of the PV modules was performed with the objective of exploring its effect on the performance of the system. According to the results presented in Table 6, it is possible to verify that there was an improvement of approximately 10%, comparing June 2021 with May 2021, which can be assigned to the executed maintenance.

Analyzing the abrupt variation between the months of April and May 2021 (Table 6), it is possible to indicate two climatic factors that had a positive influence on the improvement of efficiency. Rainfall rates collected from the same solarimetric stations indicate that the month of April 2021 was extremely dry, recording only 7 mm (millimeters), in contrast to the month of May 2021, where 115.8 mm were registered. The rainfall collaborates with the self-cleaning of the photovoltaic modules, eliminating superficial layers of deposited dirt.

Another determining factor for the improvement in the performance was the decrease in the temperature, which, according to the datasheet of the PV modules, results in an increase in the PV power. Comparing May 2021 with April 2021, the daily maximum temperature dropped by more than 1 °C (degree Celsius), and the daily average temperature decreased by more than 2 °C. It may seem that the temperature variation is not very significant, but the two climatic factors combined are strong contributors to an increase in photovoltaic energy generation.

3.3. Vehicle Energy Consumption and Carport Energy Generation

According to [49], at the end of 2020, there were 1,099,979 vehicles registered in the city of Curitiba, capital of the State of Paraná, whose population is 1,751,907 inhabitants. To make the calculations simpler, it was considered that the fleet is composed of 1,100,000 vehicles. According to [50], the population of the city travels, per day, an average of 13.7 km to get to work, totaling 27.4 km considering round trips. Adding 10% as a margin for supermarkets, gas stations, restaurants, or other emergencies, an average distance of 30 km per day is used, resulting in 660 km in a month. Only the working days of each month are included, which will be set at 22, considering that on weekends generally there is no associated driving pattern. Table 7 presents a summary with the main values.

Table 7. Main parameters.

EV Consumption [kWh/100 km]	Mean Generation [kWh]	Distance Per Month [Km]
15.75	465.37	660

After defining the variables, it was calculated that under these conditions, the electric vehicle would consume 103.95 kWh of electrical energy per month. Therefore, with the electricity generated by the carport, it would be possible to charge the battery of approximately 4 vehicles like the one used in the example.

Expanding the analysis scenarios, the simulations do replace a percentage of the city's vehicle fleet with EVs, based on the results obtained with the Renault Fluence ZE DYN, and compare with the electricity generation data from the presented carport. For each case that the number of EVs is increased, the amount of required charging energy, the number of identical carports, and the necessary available area for the installation of the PV system also increase.

The work presented by [51] analyses the impacts in the electric power system for the city of Curitiba in the case of a substitution of 15%, 30%, or 50% of the internal combustion vehicle fleet with EVs.

The scenario analyses follow the same replacement rates and define the average electric power generation of the carport as the basis for comparisons. The first two suppositions can be considered as a condition for 15 to 30 years in the future. The main obtained results are shown in Table 8.

Table 8. Comparison of the results.

15% Fleet			30% Fleet			50% Fleet		
165,000 EVs			330,000 EVs			550,000 EVs		
Charging energy [MWh]	Carport units	Area [m²]	Charging energy [MWh]	Carport units	Area [m²]	Charging energy [MWh]	Carport units	Area [m²]
17,151.8	36,856	1,105,685	34,303.5	73,712	2,211,369	57,172.5	122,854	3,685,616

3.4. Economic Analysis

According to [52], the price of photovoltaic systems in 2021 increased compared to 2020. In the study, it is stated that the low-power residential installations are costing approximately USD 0.95 per Wp. Reminding that the installed PV system has 4.89 kWp; as a result, the considered cost is USD 4645.00.

As stated previously, the metallic structure of the carport was donated to the university. However, the cost to purchase the product is about USD 2000.00. Another USD 200.00 is required for the electrical devices, like the SPD, cabling, connection board, disconnection switch, circuit breaker, and the construction expenses like the foundation and walls. Finally, the total expenditure of the carport is summed as USD 6845.00.

The energy tariff charged by the local power distribution concessionaire is USD 0.18 per kWh. The PV system average energy generation data from the past two years is reported as 465.37 kWh, resulting in an average avoided cost of USD 83.76 per month. The calculations are based on the average energy generation considering the renewable energy regulation of the country.

Performing a simple payback calculation for the carport and indexing the inflation and increasing in the energy tariff in the variation of the dollar exchange rate results in 81.7 months, approximately six years and nine months.

According to the technical data presented in [53], the same vehicle, a Renault Fluence, powered by a combustion engine and gasoline, presents a final consumption (in the city) of 9 L/100 km.

Considering the same usage scenario of 660 km/month, the car would consume 59.4 L to complete the entire drive. Taking into account that the average price of gasoline in the city of Curitiba is USD 1.05 per liter, the vehicle would generate a fuel cost of USD 62.37 per month.

If the same electric vehicle, used on an identical route, was recharged without the photovoltaic system, using only the energy made available by the local utility, the consumption of 103.95 kWh/month would generate an average extra electricity cost of USD 18.71 per month.

In the last comparison, when the EV is recharged by the electricity generated in the carport, after the payback period of the PV system, it no longer has a recharge cost.

4. Conclusions

The selected EV is a Renault Fluence ZE DYN, whose driving consumption tests in the city resulted in an average of 15.75 kWh/100 km. Investigating the responsible agencies, the number of vehicles registered in the city is determined, making it possible to simulate scenarios of the replacement of internal combustion vehicles by EVs.

The article presents the complete electric energy generation records of a PV system constructed as a carport to cover two parking spaces of the parking lot of the Federal University of Technology—Paraná, Brazil, with 4.89 kWp of installed power. Along with the electricity generation, the irradiation data was collected from a solarimetric station installed in the same area of the university, making it possible to calculate the system performance ratio, both for monthly and annual analyses.

Calculated performance ratio values were lower than what is commonly accepted as normal, and with high dirt accumulation, all the PV modules were cleaned, aiming for an overall performance improvement of the PV installation. The task had a positive effect, increasing the performance ratio by more than 10% when compared to the previous month, demonstrating that the PV systems demand periodic maintenance, which is simple and can improve efficiency.

Investigating the performance ratio of the other months, it was detected that after a month with very low rainfall records, the rain performed a natural cleaning on the photovoltaic modules, which, combined with the lower temperature due to the winter season, there was an increase in the PV system performance ratio. The rain has the capacity to perform superficial cleaning and reduce the rate of deposition of dirt on the modules.

Defining the daily drive based on the distance to work, returning home, and a margin for extra activities, while limiting it to weekdays, it was calculated that the EV energy consumption was 103.95 kWh per month, and the average energy generated by the carport was enough to charge the battery of approximately 4 EVs. The study does not define a specific charging routine and also does not analyze the impact on the electric power system. The energy balance is linked to the resolution of the National Electric Energy Agency, where the credits from the months with high PV generation can be used within 60 months.

The economic analyses compared the monthly expenses of an internal combustion vehicle and the EV for the same driving scenario. The EV presented approximately 3.3 times less expenditure. Purchasing and maintenance costs were not compared.

The EV used in the driving test was recently purchased and is the only source of energy consumption available at the moment. Future research should use larger driving samples during different periods of the year and hours of the day.

Author Contributions: Conceptualization, A.C.K., A.K.S., J.E.d.S. and J.U.J.; Data curation, J.E.d.S. and J.U.J.; Formal analysis, É.A.T.; Investigation, É.A.T., J.E.d.S. and J.U.J.; Methodology, A.C.K. and J.U.J.; Project administration, A.C.K.; Resources, J.E.d.S.; Software, É.A.T.; Supervision, J.U.J.; Visualization, A.K.S.; Writing—original draft, A.C.K. and A.K.S.; Writing—review & editing, É.A.T. and A.K.S. All authors have read and agreed to the published version of the manuscript.

Funding: This study was financed in part by the Coordenação de Aperfeiçoamento de Pessoal de Nível Superior—Brasil (CAPES)—Finance Code 001.

Conflicts of Interest: The authors declare no conflict of interest.

References

1. Garcia, G.; Nogueira, E.F.; Betini, R.C. Solar energy for residential use and its contribution to the energy matrix of the state of Parana. *Braz. Arch. Biol. Technol.* **2018**, *61*, e18000510. [CrossRef]
2. Kulik, A.C.; Silva, J.G.; Gabriel, J.D.; Tonolo, E.; Urbanetz, J., Jr. Integration of a pilot PV parking lot and an electric vehicle in a university campus located in Curitiba: A study case. *Braz. Arch. Biol. Technol.* **2021**, *64*. [CrossRef]
3. Bhatti, A.R.; Salam, Z.; Aziz, M.J.B.A.; Yee, K.P.; Ashique, R.H. Electric vehicles charging using photovoltaic: Status and technological review. *Renew. Sustain. Energy Rev.* **2016**, *54*, 34–47. [CrossRef]
4. International Energy Agency. Tracking Transport. 2020. Available online: https://www.iea.org/reports/tracking-transport-2020 (accessed on 30 May 2021).
5. Branco, N.C.; Affonso, C.M. Probabilistic approach to integrate photovoltaic generation into PEVS charging stations considering technical, economic and environmental aspects. *Energies* **2020**, *13*, 5086. [CrossRef]
6. Miceli, R.; Viola, F. Designing a sustainable university recharge area for electric vehicles: Technical and economic analysis. *Energies* **2017**, *10*, 1604. [CrossRef]
7. Calise, F.; Cappiello, F.L.; Carteni, A.; D'Accadia, M.D.; Vicidomini, M. A novel paradigm for a sustainable mobility based on electric vehicles, photovoltaic panels and electric energy storage systems: Case studies for Naples and Salerno (Italy). *Renew. Sustain. Energy Rev.* **2019**, *111*, 97–114. [CrossRef]
8. Biresselioglu, M.E.; Kaplan, M.D.; Yilmaz, B.K. Electric mobility in Europe: A comprehensive review of motivators and barriers in decision making processes. *Transp. Res. Part A* **2018**, *109*, 1–13. [CrossRef]
9. Fachrizal, R.; Shepero, M.; Meer, D.V.D.; Munkhammar, J.; Widén, J. Smart charging of electric vehicles considering photovoltaic power production and electricity consumption: A review. *eTransportation* **2020**, *4*, 2020.
10. Straub, F.; Streppel, S.; Göhlich, D. Methodology for estimating the spatial and temporal power demand of private electric vehicles for an entire urban region using open data. *Energies* **2021**, *14*, 2081. [CrossRef]
11. Badea, G.; Felseghi, R.A.; Varlam, M.; Filote, C.; Culcer, M.; Iliescu, M.; Raboaca, M.S. Design and simulation of Romanian solar energy charging station for electric vehicles. *Energies* **2019**, *12*, 74. [CrossRef]
12. Faddel, S.; Al-Awami, A.T.; Mohammed, O.A. Charge control and operation of electric vehicles in power grids: A review. *Energies* **2018**, *11*, 701. [CrossRef]
13. Zweistra, M.; Janssen, S.; Geerts, F. Large scale smart charging of electric vehicles in practice. *Energies* **2020**, *13*, 298. [CrossRef]
14. Nour, M.; Chaves-Ávila, J.P.; Magdy, G.; Sánchez-Miralles, A. review of positive and negative impacts of electric vehicles charging on electric power systems. *Energies* **2020**, *13*, 4675. [CrossRef]
15. Huda, M.; Koji, T.; Aziz, M. Techno economic analysis of vehicle to grid (V2G) integration as distributed energy resources in Indonesia Power System. *Energies* **2020**, *13*, 1162. [CrossRef]
16. Sun, X.; Li, Z.; Wang, X.; Li, C. Technology development of electric vehicles: A review. *Energies* **2020**, *13*, 90. [CrossRef]
17. Wróblewski, P.; Kupiec, J.; Drozdz, W.; Lewici, W.; Jaworski, J. The economic aspect of using different plug-in hybrid driving techniques in urban conditions. *Energies* **2021**, *14*, 3543. [CrossRef]

18. International Energy Agency. Global EV Outlook 2021. Available online: https://iea.blob.core.windows.net/assets/ed5f4484-f5 56--4110--8c5c-4ede8bcba637/GlobalEVOutlook2021.pdf (accessed on 6 May 2021).
19. Automotive Fleet. Available online: https://www.automotive-fleet.com/10139574/state-of-the-fleet-market-in-brazil (accessed on 4 August 2021).
20. Khaligh, A.; D'Antonio, M. Global trends in high-power on-board chargers for electric vehicles. *IEEE Trans. Veh. Techol.* **2019**, *68*, 3306–3324. [CrossRef]
21. Santos, A.C.; Palacio, C.; Diez, D.B.; Alejandro, O.M. Planning minimum interurban fast charging infrastructure for electric vehicles: Methodology and application to Spain. *Energies* **2014**, *7*, 1207–1229. [CrossRef]
22. Alghamdi, A.S.; Bahaj, A.B.; Wu, Y. Assessment of large scale photovoltaic power generation from carport canopies. *Energies* **2017**, *10*, 686. [CrossRef]
23. Erickson, L.; Ma, S. Solar-powered charging networks for electric vehicles. *Energies* **2021**, *14*, 966. [CrossRef]
24. Cieslik, W.; Szwajca, F.; Golimowski, W.; Berger, A. Experimental analysis of residential photovoltaic (PV) and electric vehicle (EV) systems in terms of annual energy utilization. *Energies* **2021**, *14*, 1085. [CrossRef]
25. REN21. Renewables 2021—Global Status Report. Available online: https://www.ren21.net/wp-content/uploads/2019/05/ GSR2021_Full_Report.pdf (accessed on 20 July 2021).
26. Canizes, B.; Soares, J.; Costa, A.; Pinto, T.; Lezama, F.; Novais, P.; Vale, Z. Electric vehicles'user charging behaviour simulator for a smart city. *Energies* **2019**, *12*, 1470. [CrossRef]
27. Shepero, M.; Munkhammar, J.; Widén, J.; Bishop, J.D.K.; Boström, T. Modeling of photovoltaic power generation and electric vehicles charging on city-scale: A review. *Renew. Sust. Energy Rev.* **2018**, *89*, 61–71. [CrossRef]
28. Buresh, K.M.; Apperley, M.D.; Booysen, M.J. Three shades of green: Perspectives on at-work charging of electric vehicles using photovoltaic carports. *Energy Sustain. Dev.* **2020**, *57*, 132–140. [CrossRef]
29. Pereira, J.S. Investigation of the Energy Potential of Photovoltaic Solar Parks to Supply Local Energy and Electric Vehicles. Master's Thesis, Energy Systems UTFPR, Federal University of Technology, Paraná, Curitiba, Brazil, 2019; 112p. (In Portuguese)
30. Iftikhar, H.; Sarquis, E.; Branco, P.J.C. Why can simple operation and maintenance (O&M) practices in large-scale grid-connected pv power plants play a key role in improving its energy output? *Energies* **2021**, *14*, 3798.
31. Montesinos, J.A.; Martínez, F.R.; Polo, J.; Chivelet, N.M.; Battles, F.J. Economic effect of dust particles on photovoltaic plant production. *Energies* **2020**, *13*, 6376. [CrossRef]
32. Conceição, R.; Vázquez, I.; Fialho, L.; García, D. Soiling and rainfall effect on PV technology in rural Southern Europe. *Renew. Energy* **2020**, *156*, 743–747. [CrossRef]
33. Bosman, L.B.; Salas, W.D.L.; Hutzel, W.; Soto, E.A. PV system predictive maintenance: Challenges, current approaches, and opportunities. *Energies* **2020**, *13*, 1398. [CrossRef]
34. Hoarau, Q.; Perez, Y. Interactions between electric mobility and photovoltaic generation: A review. *Renew. Sustain. Energy Rev.* **2018**, *94*, 510–522. [CrossRef]
35. Hiesl, A.; Ramsebner, J.; Haas, R. Modelling stochastic electricity demand of electric vehicles based on traffic surveys—The case of Austria. *Energies* **2021**, *14*, 1577. [CrossRef]
36. Alcover, E.A.; Mas, B.; Moll, V.M.; Rosselló, J.L.; Roca, M.; Canals, V. Energetic and economic analysis of the electric vehicles charge impacts on public parking lots. In Proceedings of the 18th International Conference on Renewable Energies and Power Quality, Granada, Spain, 1–2 April 2020.
37. Distributed Photovoltaic Generation in Brazil: Technological Innovation, Scenario Methodology and Regulatory Frameworks. Available online: https://repositorio.cepal.org/bitstream/handle/11362/44928/4/S1900906_en.pdf (accessed on 7 August 2021).
38. Ribeiro, M.C.P.; Nadal, C.P.; Rocha Junior, W.F.; Fragoso, R.M.S.; Lindino, C.A. Institutional and legal framework of the Brazilian energy market: Biomass as a sustainable alternative for Brazilian agribusiness. *Sustainability* **2020**, *12*, 1554. [CrossRef]
39. Distributed Generation. Available online: https://www.aneel.gov.br/geracao-distribuida (accessed on 7 August 2021).
40. Primary Energy Production. Available online: http://shinyepe.brazilsouth.cloudapp.azure.com:3838/ben/ (accessed on 8 August 2021).
41. Kongjeen, Y.; Bhumkittipich, K. Impact of plug-in electric vehicles integrated into power distribution system based on voltage-dependent power flow analysis. *Energies* **2018**, *11*, 1571. [CrossRef]
42. Ilieva, I.; Bremdal, B. Utilizing local flexibility resources to mitigate grid challenges at electric vehicle charging stations. *Energies* **2021**, *14*, 3506. [CrossRef]
43. Dos Santos, P.D.; De Souza, A.C.Z.; Bonatto, B.D.; Mendes, T.P.; Neto, J.A.S.; Botan, A.C.B. Analysis of solar and wind energy installations at electric vehicle charging stations in a region in Brazil and their impact on pricing using an optimized sale price model. *Int. J. Energy Res.* **2021**, *45*, 6745–6764. [CrossRef]
44. Technical Data Renault Fluence ZE. Available online: https://imprensa.renault.com.br/upload/produto/ficha-tecnica/2862d5 2ef67b80d0b5d8cd09778e0728.pdf (accessed on 30 May 2021).
45. *Operation Manual Car Charger ProEV1*; EGNEX: Curitiba, Brazil, 2021; Volume 18, pp. 1–34. (In Portuguese)
46. Urbanetz, J., Jr.; Tiepolo, G.M.; Casagrande, E.F., Jr.; Tonin, F.S.; Mariano, J.D.A. Distributed Photovoltaic Generation: The Case of UTFPR's Photovoltaics Systems in Curitiba. In Proceedings of the X Brazilian Energy Planning Congress, Gramado, Brazil, 26 September 2016. (In Portuguese)

47. Khalid, A.M.; Mitra, I.; Warmuth, W.; Schacht, V. Performance ratio—crucial parameter for grid connected PV plants. *Renew. Sustain. Energy Rev.* **2016**, *65*, 1139–1158. [CrossRef]
48. Martínez, S.M.; Carretón, M.C.; Escribano, A.H.; Lázaro, E.G. Performance evaluation of large solar photovoltaic power plants in Spain. *Energy Conver. Manag.* **2019**, *183*, 515–528. [CrossRef]
49. IBGE—Brazilian Statistics Institute. Car Fleet. Available online: https://cidades.ibge.gov.br/brasil/pr/curitiba/pesquisa/22/28120?ano=2020 (accessed on 7 July 2021).
50. Bem Paraná. Available online: https://www.bemparana.com.br/noticia/curitibano-gasta-33-minutos-para-chegar-ao-trabalho-#.YOcfr-hKjIU (accessed on 8 July 2021).
51. Rossetto, M.B.; Lourenço, E.M. Analysis of the Impacts of the Electrification of the Vehicle Fleet in the Electric Power System in Curitiba. *Braz. Arch. Biol. Technol.* **2018**, *61*, 1–11.
52. Fluence 1.6 16V Fuel Consumption. Available online: https://motoreu.com/renault-fluence-1.6--16v-mpg-fuel-consumption-technical-specifications-5753 (accessed on 4 August 2021).
53. Strategic Study for Distributed Generation. Available online: https://greener.greener.com.br/estudo-gd-2s2020 (accessed on 6 August 2021).

Article

CO$_2$ Emissions of Electric Scooters Used in Shared Mobility Systems

Andrzej Kubik 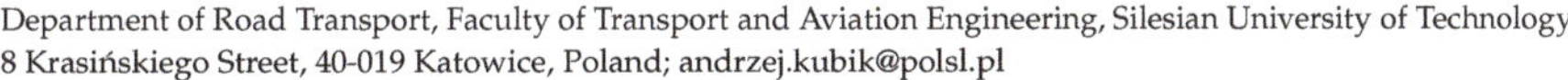

Department of Road Transport, Faculty of Transport and Aviation Engineering, Silesian University of Technology, 8 Krasińskiego Street, 40-019 Katowice, Poland; andrzej.kubik@polsl.pl

Abstract: The development of the electric mobility market in cities is becoming more and more important every year. With this development, more and more electric scooters are appearing in cities. Currently, the restrictions that result from the upcoming trends are reducing the number of vehicles powered by combustion engines in favor of vehicles equipped with electric motors. Considering the number of electric vehicles, the dominant type is an electric scooter. The aim of this article is to determine the CO$_2$ that is emitted into the atmosphere by using this type of vehicle. The main suppliers of this type of vehicle in cities are shared mobility systems. To recognize the research gap, consisting of the lack of CO$_2$ emissions of an electric scooter type vehicle, studies were carried out on the energy consumption of an electric scooter and CO$_2$ emissions, which were calculated based on the CO$_2$ emission value needed to produce a given energy value kWh. The plan of the research performed was developed on the basis of the D-optimal plan of the experiment, thanks to which the results could be saved in the form of mathematical models based on formulas.

Keywords: electric scooter; shared mobility systems; CO$_2$ emission; transportation engineering

Citation: Kubik, A. CO$_2$ Emissions of Electric Scooters Used in Shared Mobility Systems. *Energies* **2022**, *15*, 8188. https://doi.org/10.3390/en15218188

Academic Editor: Thanikanti Sudhakar Babu

Received: 8 September 2022
Accepted: 31 October 2022
Published: 2 November 2022

Publisher's Note: MDPI stays neutral with regard to jurisdictional claims in published maps and institutional affiliations.

1. Introduction

Climate change happening in the world forces man to become more and more interested in harmful gases produced by humanity. One of the harmful greenhouse gases, whose emissions in 2010 amounted to up to 76%, is CO$_2$ gas [1]. Noticing the harmful effects of CO$_2$ emissions into the atmosphere, humanity strives to reduce CO$_2$ emissions in all branches of the economy [2]. One of the solutions to reduce CO$_2$ emissions is to introduce policies that aim to reduce the use of gas emission sources—vehicles powered by internal combustion engines. Policies introduced in many countries around the world are coordinating the use of electric vehicles to mix in urban agglomerations [1,3].

To reduce CO$_2$ emissions, the Scandinavian countries have introduced a tax for owners of vehicles with an internal combustion engine that is 180% higher than the value of the tax on an electric vehicle [4]. In addition, owners of electric vehicles can count on free charging and parking of their vehicle in the city. The Chinese government is also paying a lot of attention to the development, research, and implementation of electrical vehicles. In addition, China is among the leading countries to demand an end to the use of internal combustion vehicles in favor of electric vehicles.

Another major contributing factor to CO$_2$ production is the planning of fossil fuels, from which many countries in the world produce electricity [5].

Therefore, it can be concluded that the elimination of vehicles with internal combustion engines, in favour of electric vehicles, could result in a drastic reduction in CO$_2$ emissions in cities since electric vehicles are advertised and presented as zero emission vehicles [6]. The question must therefore be asked: is the use of an electric vehicle a zero-emission vehicle? The research and results presented in this article indicate the answer to the question asked above.

With the development of electromobility and the emergence of electric vehicles, the availability of electric scooter vehicles in cities has increased exponentially. Kwangho et al. [7] found that the electric scooter works well as a last mile vehicle.

An electric scooter is a small device that allows you to move in an urban agglomeration, which is equipped with a low-power electric motor (usually about 300 W) and a battery that allows you to drive up to several dozen kilometers (depending on the scooter model) [8]. Due to the fact that electric scooters are available at every step in cities [9], and due to the possibility of their short-term rental from shared mobility systems, a research gap was noticed. What is the CO_2 emission of an electric scooter? What is the difference between the emission of an electric scooter and a motor vehicle with an internal combustion engine? It is worth emphasizing that the number of scooter vehicles moving around the city can be from several to several times higher than motor vehicles with electric drive [10]. For example, in Poland in 2021, there were about 8000 vehicles with an electric motor, whereas there were over 500% more available electric scooters [11]. Therefore, an electric scooter is one of the possible ways to increase the level of environmental performance of transport while having a positive impact on the economic aspects and quality of life of residents [12,13]. The introduction of this type of vehicle has also forced a number of actions that led to changes in the culture of movement in terms of orientation towards electric vehicles, which are referred to as electromobility [14]. The entire article is divided into four main chapters. The first is an introduction. Another is the description of the methodology and materials used in the research. The next chapter presents the results. The article ends with a chapter where the research results were discussed and related to the research of other scientists.

2. Materials and Methods

The subject used to conduct research is an electric scooter used in shared mobility systems, which is available in many urban agglomerations. The scooter was equipped with an electric motor with a power of 300 W and a battery capacity of 475 Wh. Figure 1 show an electric scooter and examples of values recorded during the tests.

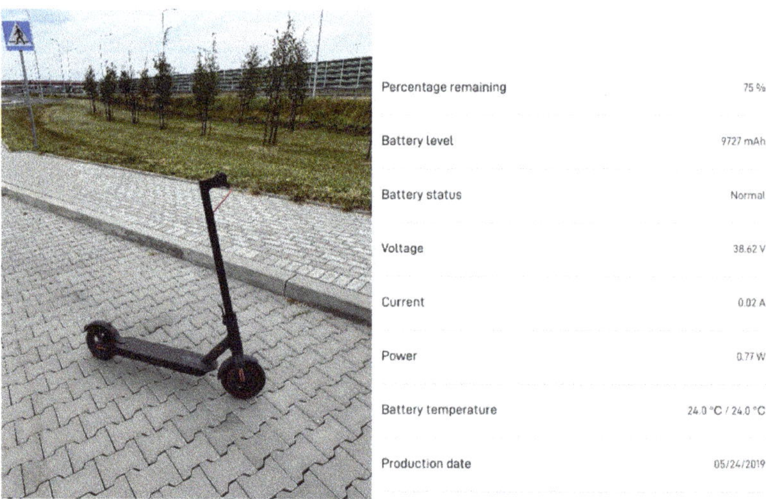

Figure 1. Electric scooter and sample values recorded during the tests.

The research plan assumed that the rides made with an electric scooter would correspond to the routes and driving style of the users of these systems. Then, determining the value of electricity consumption for the tests carried out, calculations were made regarding CO_2 emissions, which is the result of the consumption of electricity necessary to perform such a trip. The diagram of the research is shown in Figure 2.

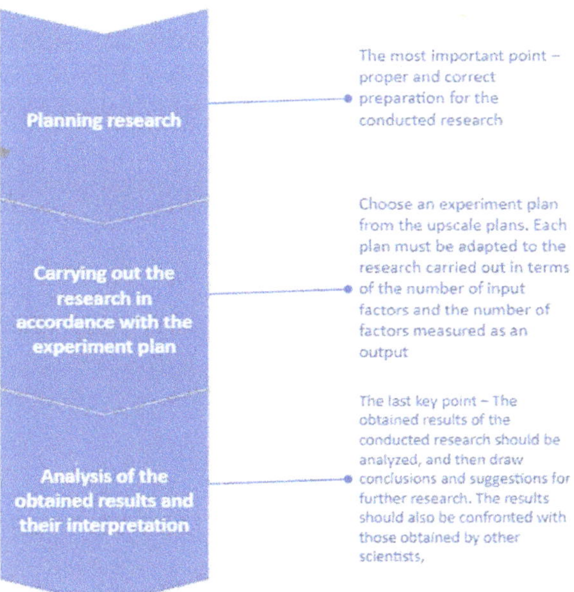

Figure 2. The diagram of the research implementation plan.

The first step was theoretical preparation for the assumed research. The next step (the most important) was to properly plan the research. On the basis of the literature, the factors influencing the value of energy consumption of an electric scooter have been determined. After selecting the experiment factors, the plan was selected, thanks to which it is possible to determine individual passes that are necessary for the mathematical modeling of the studied phenomenon [15,16]. Experimental studies were carried out according to a statistically established plan of a polyselective experiment. The third step was to determine CO_2 emissions based on electricity consumption. The last step of the research was the analysis of the obtained results and the presentation of appropriate conclusions.

Planning an experiment according to the D-optimal plan—Hartley's plan—is one of the methods of describing phenomena. Hartley's plan belongs to static, determined polyselection plans for three input values in which three different values are used for each input quantity The basic principle of creating polyselection plans is the deliberate selection of a combination of values of input quantities (within the previously assumed range) in such a way that it is possible to obtain the required scientific information under limited conditions [16]. The development of the plan consists of determining the dependence of the input quantities and their location relative to the base point—middle, zero. The experimental plans developed for the three input factors are based on a hypercube for which the coefficient $\alpha = 1$. The entry factors that were selected for the experiment plan are: x_1—the type of surface on which the scooter moves; x_2—the distance covered; and x_3—the scooter riding mode. Selected factors x_1, x_2, and x_3 are factors that, in particular, affect the possibility of differentiating the energy consumption of the scooter—results from the kinematics of moving the vehicle and attempts to simulate different driving styles of users. Factor x_3—the type of riding mode—limited the speed of movement of the scooter. In ECO mode, the maximum speed is 12 km/h; in NORMAL mode, the maximum speed is 20 km/h; and in TURBO mode, the maximum speed is 25 km/h. The overall form of the experiment plan on a standardized scale is restated in Table 1.

Table 1. Experiment plan for standardized scale factors.

Route No	X_1	X_2	X_3
1	−1	−1	1
2	1	−1	−1
3	−1	1	−1
4	1	1	1
5	−1	0	0
6	1	0	0
7	0	−1	0
8	0	1	0
9	0	0	−1
10	0	0	1
11	0	0	0

Thanks to the use of the general form of the experiment plan, it was possible to prepare a plan of experiments described in the makings on a real scale. Input factors have been introduced in the form of Formulas (1)–(3):

$$x_1 = \begin{cases} -1 \; for \; Asphalt/concrete \; road \\ 0 \; for \; Mixed \; road \; (50\% \; Asphalt/concrete, \; 50\% \; Road \; cobble - stones) \\ 1 \; for \; Road \; cobblestones \end{cases} \quad (1)$$

$$x_2 = \begin{cases} -1 \; for \; distance - 1km \\ 0 \; for \; distance - 3km \\ 1 \; for \; distance - 5km \end{cases} \quad (2)$$

$$x_3 = \begin{cases} -1 \; for \; eco \; mode \\ 0 \; for \; normal \; mode \\ 1 \; for \; turbo \; mode \end{cases} \quad (3)$$

Values based on Formulas (1)–(3): −1, 0, and 1 represent the variability of the individual factors x_1, x_2, and x_3. Based on Formula (4), it is possible to reconstruct the mathematical model of the phenomenon under study.

$$f(x_i, \ldots, x_j) = k_0 + \sum k_i(x_i) + \sum k_{ii}(x_i) + \sum k_{ij}(x_i)(x_j) \quad (4)$$

Where:
$f(x_i, \ldots, x_j)$—results,
$x_i, \ldots, x_j$—input factor in the normalized scale,
k_0, k_i, k_{ij}—regression coefficients.
Supplemented by a general table with real-scale factors, Table 2 is shown.

Table 2. Experiment plan for real-scale factors.

Route No	Type of Surface -X_1	Distance -X_2	Driving Type Mode -X_3
1	Asphalt/concrete road	1	turbo
2	Road cobblestones	1	eco
3	Asphalt/concrete road	5	eco
4	Road cobblestones	5	turbo
5	Asphalt/concrete road	3	normal
6	Road cobblestones	3	normal
7	Mixed road (50% Asphalt/concrete, 50% Road cobblestones)	1	normal
8	Mixed road (50% Asphalt/concrete, 50% Road cobblestones)	5	normal
9	Mixed road (50% Asphalt/concrete, 50% Road cobblestones)	3	eco
10	Mixed road (50% Asphalt/concrete, 50% Road cobblestones)	3	turbo
11	Mixed road (50% Asphalt/concrete, 50% Road cobblestones)	3	normal

3. Results

The results of the conducted research, which were carried out in accordance with the previously planned experiment plan, pre-set in Table 2, are presented in Table 3. Each run was repeated five times.

Table 3. The results of the conducted research.

Route No	Energy Consumption [kWh/100 km]	CO_2 Emissions [kg/100 km]
1	1.796	1.668
2	3.755	3.489
3	1.324	1.230
4	2.158	2.005
5	1.654	1.536
6	3.297	3.063
7	2.749	2.554
8	1.762	1.637
9	2.657	2.468
10	2.506	2.328
11	2.529	2.349

Thanks to the obtained results presented in Table 3 and the use of Formula (4), it was possible to determine the dependence of the impact of the selected factors x_1, x_2, and x_3 on the CO_2 emission value expressed in kg/100 km unit. Figures 3–11 predict the dependence of the type of surface, distance travelled, and driving style on CO_2 emissions, although some of the values shown in the diagrams are the result of prediction according to Formula (4). The biggest CO_2 emissions will be emitted by moving an electric scooter on a paving stone over a distance of 1 km. The lowest emissions were recorded for a scooter that moves on an asphalt road while covering a 5 km route. It is worth noting that the lowest consumption is almost three times lower than the value of the maximum CO_2 emissions that have been achieved.

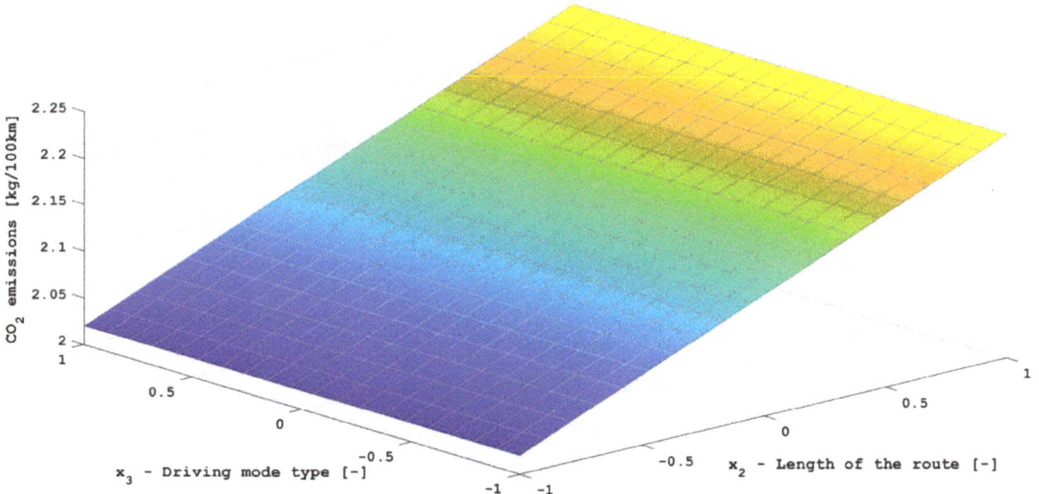

$$x_1 = -1 \wedge f(x_2, x_3) = 1.99 + 0.11x_2 + 0.86x_3 - 0.27x_2x_3$$

Figure 3. CO_2 emission value depending on driving style and length for asphalt road driving.

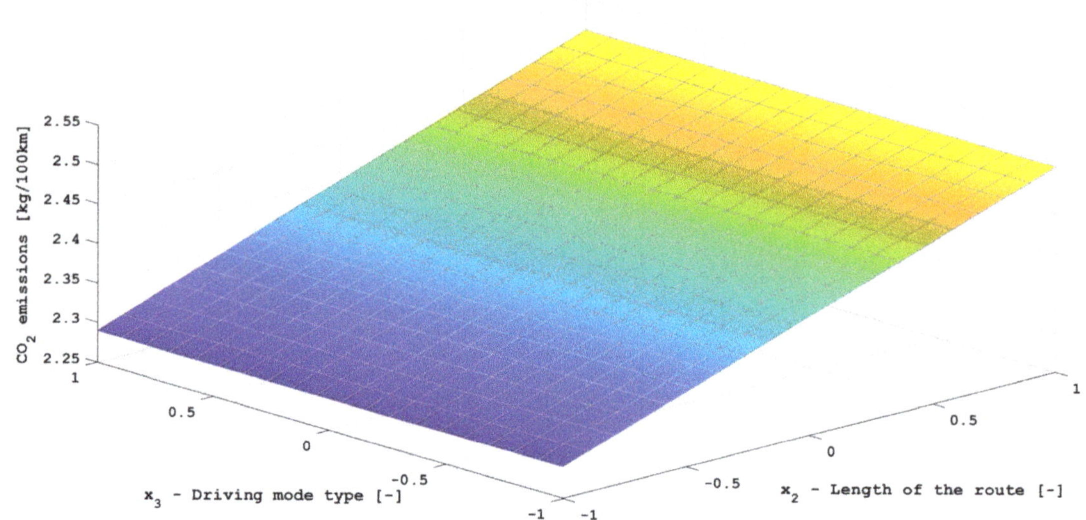

Figure 4. CO_2 emission value depending on driving style and length for driving on paving stones.

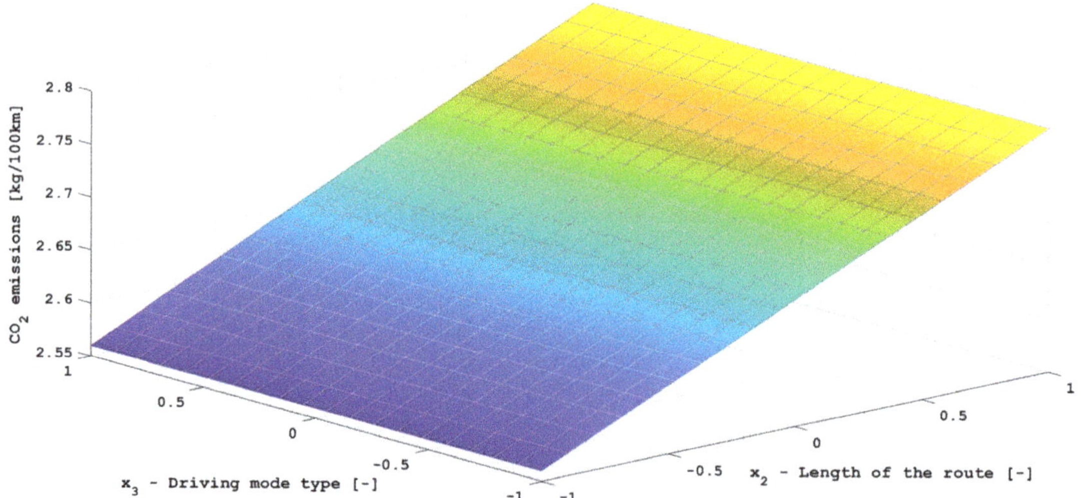

$$x_1 = 1 \wedge f(x_2, x_3) = 2.53 + 0.11x_2 + 0.86x_3 - 0.27x_2x_3$$

Figure 5. CO_2 emissions depending on driving style and length for mixed road driving.

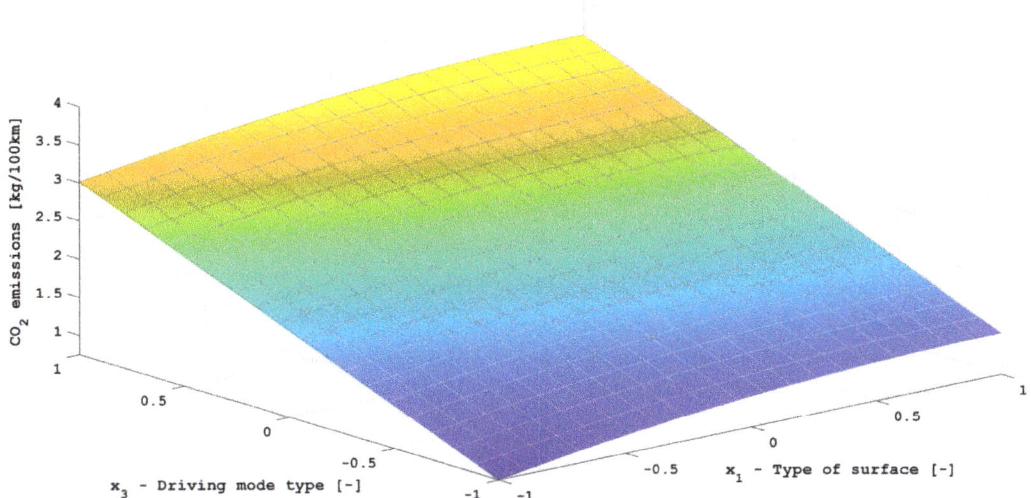

$$x_2 = -1 \ \wedge \ f(x_1, x_3) = 2.29 + 0.27x_1 + 0.86x_3 - 0.14x_1^2 + 0.27x_3$$

Figure 6. CO_2 emissions depending on driving style and type of surface when covering a distance of 1 km.

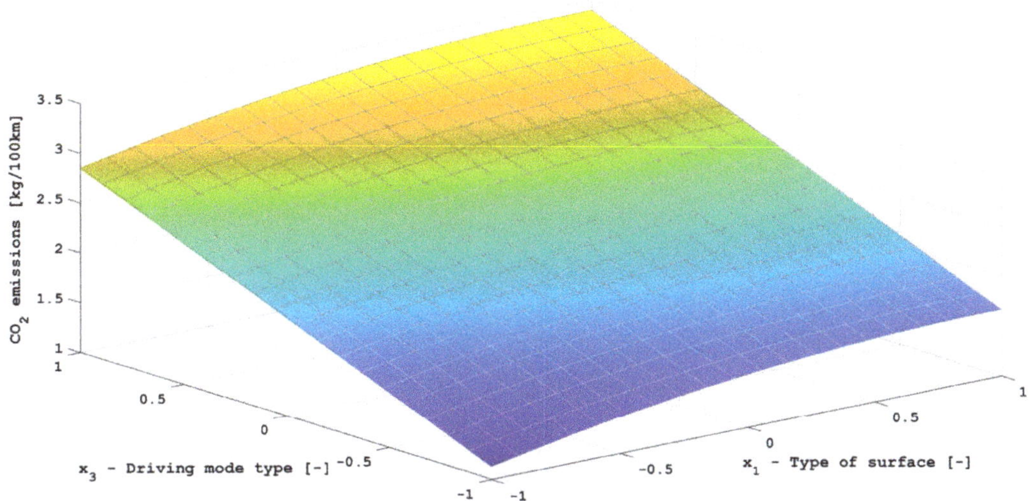

$$x_2 = 0 \ \wedge \ f(x_1, x_3) = 2.40 + 0.27x_1 + 0.86x_3 - 0.14x_1^2$$

Figure 7. CO_2 emissions depending on driving style and type of surface when covering a distance of 5 km.

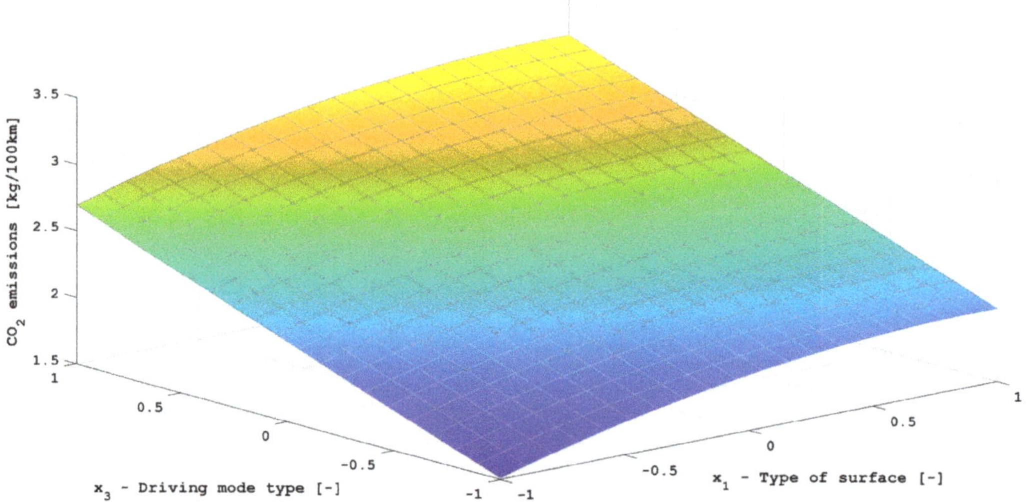

$$x_2 = 1 \ \wedge \ f(x_1, x_3) = 2.51 + 0.27x_1 + 0.86x_3 - 0.14x_1^2 - 0.27x_3$$

Figure 8. CO$_2$ emissions depending on driving style and type of surface when covering a distance of 10 km.

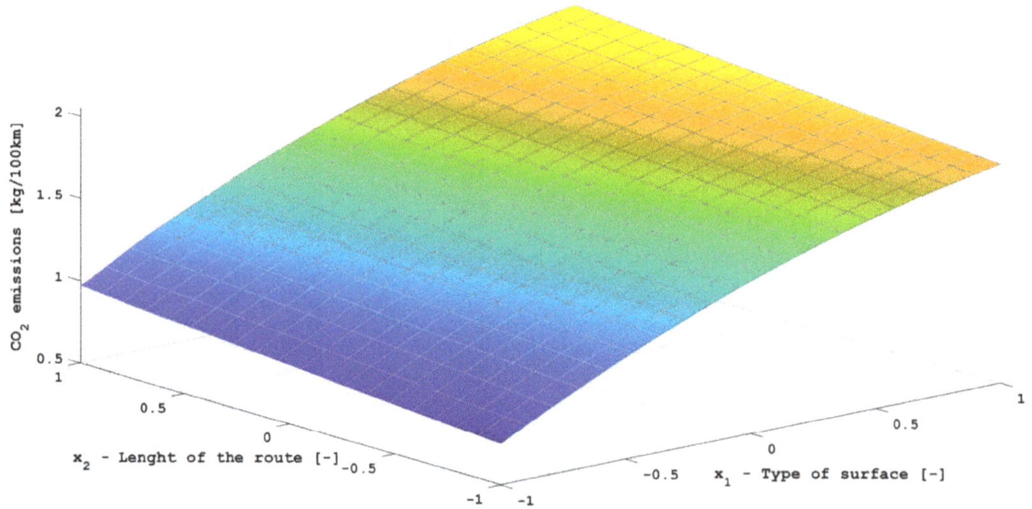

$$x_3 = -1 \ \wedge \ f(x_1, x_2) = 1.54 + 0.54x_1 + 0.11x_2 - 0.14x_1^2$$

Figure 9. CO$_2$ emission value depending on the length of the route covered and the type of surface for steering in ECO mode.

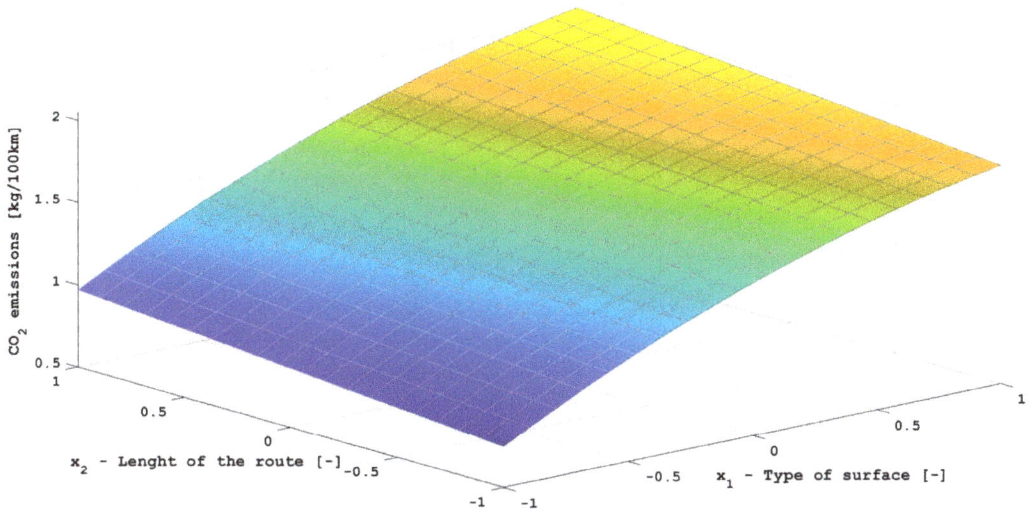

$$x_3 = -1 \ \wedge \ f(x_1, x_2) = 1.54 + 0.54x_1 + 0.11x_2 - 0.14x_1^2$$

Figure 10. CO_2 emission value depending on the length of the route covered and the type of surface for steering in NORMAL mode.

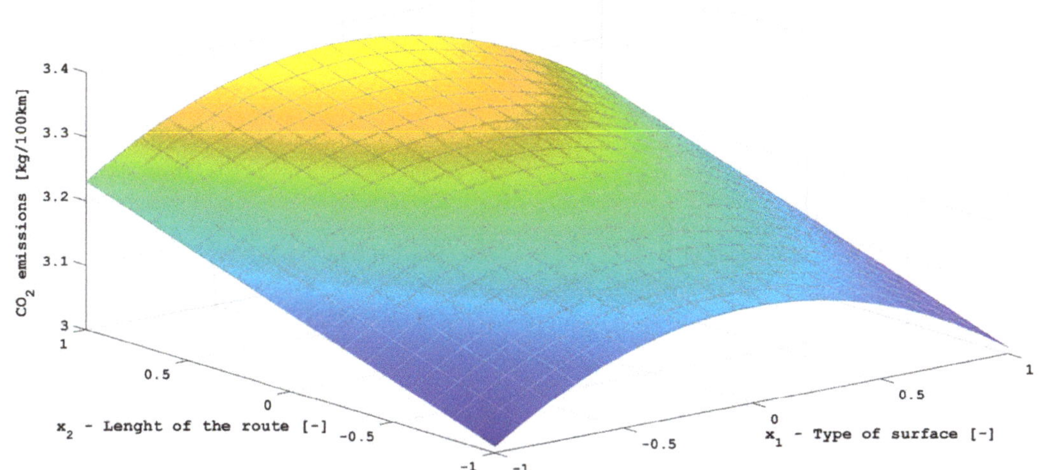

$$x_3 = 1 \ \wedge \ f(x_1, x_2) = 3.26 + 0.11x_2 - 0.14x_1^2$$

Figure 11. CO_2 emission value depending on the length of the route covered and the type of surface for TURBO mode.

By analyzing the graphs presented in Figures 3–5, it can be concluded that, regardless of the occurrence of the surface, the greatest impact on the emission value will be generated by the distance at which the scooter moves. It is also worth emphasizing that the type of surface affects the range of achieved emission values.

Figures 6–8 show the CO_2 emission values when travelling different distances with an electric scooter. It is worth noting that the impact of the driving mode generates a linear increase in emission values, while the asphalt surface type generates the lowest CO_2 emission values.

Figures 9–11 show the CO_2 emission values when moving a scooter in different riding modes. The greatest curiosity is the mode of moving the scooter in turbo mode. The extreme of the function is achieved for moving on the road paving stones. For other surfaces, the CO_2 emission value is at the same level and increases with increasing distance.

4. Discussion and Conclusions

The results obtained during the research suggest quite interesting and valuable conclusions. The research is a continuation of research that has already been carried out on electric scooters used in urban shared mobility systems. Studies have shown that factors, such as the type of pavement, can influence the value of energy consumption by linear dependence. Additionally, the type of surface that is transversed can cause a twofold increase in CO_2 emissions. Another factor, the length of the route, is characterized by the greatest variability of the final results, where electricity consumption and CO_2 emissions can be increased by up to three times. The least variability is characterized by scooter riding modes. Choosing the right mode will limit the user's CO_2 emissions according to the presented results, the variability of which is small. The objectives of the study presented in this article have been proven and executed. The results obtained in these studies indicated the dependence of the length of movement of the scooter, the type of surface, and the mode of riding the electric scooter. This fills a research gap among other authors who have conducted research on this subject. Other authors, such as Wang et al., conducted a study of motor vehicles with an internal combustion engine where they focused on the effect of ambient temperature on CO_2 emissions. The results indicate that a vehicle in low temperaments (-10 °C) emits more than two times more CO_2 than a vehicle used at 40 °C. It is worth noting that the results achieved by a vehicle with an internal combustion engine (gasoline) are about 300 g/km, which gives 30 kg CO_2/100 km. An electric scooter achieves a result 10 times smaller [17].

Buberger et al., in their article, touched on the total CO_2 emissions resulting from all stages of the vehicle's life. They found that vehicles running on renewable fuels (e.g., compressed biogas) have a similar impact on climate change as electric vehicles. Moreover, emissions of hybrid and electric vehicles are up to 89% lower compared to vehicles with an internal combustion engine. The total CO_{2-} emissions of a vehicle with an internal combustion engine that burns 7l/100 km is about 49,500 kg. For comparison, an electric vehicle emits about 5500 kg of CO_2 [18].

Reducing CO_2 emissions is one of the key actions to improve the quality of climate cities. The emergence of alternative vehicles to replace a motor vehicle has made users more and more willing to choose electric scooters as the primary means of transport. When comparing the replacement of one combustion vehicle—or several—with a dozen or so scooters, it is worth considering whether the emission of a dozen or so electric scooters is an acceptable result. Furthermore, the number of electric scooters in cities is growing at an amazing pace. Currently, it is estimated that there are about 360,000 electric scooters in 30 cities in Europe alone [19]. Estimating that electric scooters cover a distance of up to 100 km per week, all electric scooters can emit up to approx. 58,320,000 kg of CO_2 per year.

Sovacool, in his article, touched on what values of CO_2 emissions are emitted into the atmosphere to produce 1 kWh of energy. Currently, the most advantageous forms of industrial energy production are nuclear power plants, which emit 1.4 g/1 kWh, whereas coal-fired power plant emissions are about 790 g/1 kWh [20]. Furthermore, it should be noted that a nuclear power plant does not directly emit greenhouse gas emissions, and the total CO_2 emissions result from the life cycle (as a result of the construction and operation of the power plant, the extraction and grinding of uranium, and the decommissioning of the power plant).

To sum up, the research carried out in this article touches on a very interesting and modern form of mobility in cities, which is currently one of the fastest growing branches of vehicle sharing in cities. Moving around with the different power modes of a scooter can more than double CO_2 emissions. Cyclically, there are more and more operators providing scooter sharing services in cities, which means that the number of electric scooters in cities is increasing year by year. In the current era, it is necessary to consider, first of all, what source of energy the batteries of electric scooters are charged from. Studies have also shown that the highest CO_2 emissions result from the use of energy from coal-fired power plants. Due to the increasing number of scooters in cities, one should consider whether cities should not have energy from renewable sources or, for example, solar energy. Of course, the conducted research also has limitations. Primarily, apart from the measurements made, the other results are the result of prediction, which results from the applied plan of the experiment. Another limitation is the impact of wind, which can increase aerodynamic drag and energy consumption. Studies with the influence of different wind speeds will be carried out in the future. The next planned research will complement the existing research in order to learn about the impact of other factors affecting CO_2 emissions, such as electricity consumption or the dressage to which the user of this vehicle is exposed.

Funding: This research received no external funding.

Institutional Review Board Statement: Not applicable.

Informed Consent Statement: Not applicable.

Data Availability Statement: Not applicable.

Conflicts of Interest: The author declares no conflict of interest.

References

1. The Emissions Reduction Obligation Quota Policy is Reinforced with Increasing the Share of Renewables in Vehicle Fuels. Available online: https://www.regeringen.se/pressmeddelanden/2020/09/branslebytet-forstarks-med-hogre-inblandning-av-fornybart-i-drivmedel/ (accessed on 7 August 2022).
2. European Union. Regulation (EU) 2019/631 of the European Parliament and of the Council of 17 April 2019 setting CO2 emission performance standards for new passenger cars and for new light commercial vehicles, and repealing regulations (EC) No 443/2009 and (EU) No 510/2011. Available online: https://eur-lex.europa.eu/legal-content/PL/TXT/PDF/?uri=CELEX:32019R0631&from=EN (accessed on 7 August 2022).
3. Morfeldt, J.; Davidsson Kurland, S.; Johansson, D.J.A. Carbon Footprint Impacts of Banning Cars with Internal Combustion Engines. *Transp. Res. Part D Transp. Environ.* **2021**, *95*, 102807. [CrossRef]
4. Fridstrøm, L. The Norwegian Vehicle Electrification Policy and Its Implicit Price of Carbon. *Sustainability* **2021**, *13*, 1346. [CrossRef]
5. Energy and the Environment Explained. Available online: https://www.eia.gov/energyexplained/energy-and-the-environment/where-greenhouse-gases-come-from.php (accessed on 7 August 2022).
6. Towoju, O.A.; Ishola, F.A. A Case for the Internal Combustion Engine Powered Vehicle. *Energy Rep.* **2020**, *6*, 315–321. [CrossRef]
7. Baek, K.; Lee, H.; Chung, J.-H.; Kim, J. Electric Scooter Sharing: How Do People Value It as a Last-Mile Transportation Mode? *Transp. Res. Part D Transp. Environ.* **2021**, *90*, 102642. [CrossRef]
8. Kubik, A. Impact of the Use of Electric Scooters from Shared Mobility Systems on the Users. *Smart Cities* **2022**, *5*, 1079–1091. [CrossRef]
9. Dias, G.; Arsenio, E.; Ribeiro, P. The Role of Shared E-Scooter Systems in Urban Sustainability and Resilience during the Covid-19 Mobility Restrictions. *Sustainability* **2021**, *13*, 7084. [CrossRef]
10. Oeschger, G.; Carroll, P.; Caulfield, B. Micromobility and public transport integration: The current state of knowledge. *Transp. Res. Part D Transp. Environ.* **2020**, *89*, 102628. [CrossRef]
11. The Number of Electric Cars in Poland is Growing! Available online: https://globenergia.pl/rosnie-liczba-samochodow-elektrycznych-w-polsce-efekt-mojego-elektryka/ (accessed on 7 August 2022).
12. Pietrzak, K.; Pietrzak, O. Environmental Effects of Electromobility in a Sustainable Urban Public Transport. *Sustainability* **2020**, *12*, 1052. [CrossRef]

13. The Future of Urban Mobility. Towards Networked, Multimodal Cities of 2050. Available online: https://www.adlittle.com/sites/default/files/viewpoints/adl_the_future_of_urban_mobility_report.pdf (accessed on 7 August 2022).
14. Ingeborgrud, L.; Ryghaug, M. User perceptions of EVs and the role of EVs in the transition to low-carbon mobility. *ECEEE Summer Study Proc.* **2017**, *325*, 893–900.
15. Turoń, K.; Kubik, A.; Chen, F. Operational Aspects of Electric Vehicles from Car-Sharing Systems. *Energies* **2019**, *12*, 4614. [CrossRef]
16. Polański, Z. *Współczesne Metody Badań Doświadczalnych*; Wiedza Powrzechna: Warsaw, Poland, 1978.
17. Wang, Y.; Zhao, H.; Yin, H.; Yang, Z.; Hao, L.; Tan, J.; Wang, X.; Zhang, M.; Li, J.; Lyu, L.; et al. Quantitative Study of Vehicle CO2 Emission at Various Temperatures and Road Loads. *Fuel* **2022**, *320*, 123911. [CrossRef]
18. Buberger, J.; Kersten, A.; Kuder, M.; Eckerle, R.; Weyh, T.; Thiringer, T. Total CO2-Equivalent Life-Cycle Emissions from Commercially Available Passenger Cars. *Renew. Sustain. Energy Rev.* **2022**, *159*, 112158. [CrossRef]
19. Li, A.; Zhao, P.; Liu, X.; Mansourian, A.; Axhausen, K.W.; Qu, X. Comprehensive Comparison of E-Scooter Sharing Mobility: Evidence from 30 European Cities. *Transp. Res. Part D Transp. Environ.* **2022**, *105*, 103229. [CrossRef]
20. Sovacool, B.K. Valuing the Greenhouse Gas Emissions from Nuclear Power: A Critical Survey. *Energy Policy* **2008**, *36*, 2950–2963. [CrossRef]

Article

Comparative Assessment of Supervisory Control Algorithms for a Plug-In Hybrid Electric Vehicle

Nikolaos Aletras, Stylianos Doulgeris, Zissis Samaras and Leonidas Ntziachristos *

Mechanical Engineering Department, Aristotle University of Thessaloniki, GR 54124 Thessaloniki, Greece
* Correspondence: leon@auth.gr; Tel.: +30-2310996003

Highlights:

What are the main findings?

- Energy consumption reduction up to 24% with the use of ECMS algorithm
- Method for EMS algorithms comparison under the same required energy
- Back engineering extracted commercial algorithm based on experimental data

What is the implication of the main finding?

- ECMS adaptability advantage can be utilized under different driving conditions

Abstract: The study examines alternative on-board energy management system (EMS) supervisory control algorithms for plug-in hybrid electric vehicles. The optimum fuel consumption was sought between an equivalent consumption minimization strategy (ECMS) algorithm and a back-engineered commercial rule-based (RB) one, under different operating conditions. The RB algorithm was first validated with experimental data. A method to assess different algorithms under identical states of charge variations, vehicle distance travelled, and wheel power demand criteria is first demonstrated. Implementing this method to evaluate the two algorithms leads to fuel consumption corrections of up to 8%, compared to applying no correction. We argue that such a correction should always be used in relevant studies. Overall, results show that the ECMS algorithm leads to lower fuel consumption than the RB one in most driving conditions. The difference maximizes at low average speeds (<40 km/h), where the RB leads to more frequent low load engine operation. The two algorithms lead to fuel consumption differences of 3.4% over the WLTC, while the maximum difference of 24.2% was observed for a driving cycle with low average speed (18.4 km/h). Further to fuel consumption performance optimization, the ECMS algorithm also appears superior in terms of adaptability to different driving cycles.

Keywords: fuel consumption optimization; energy management system; hybrid vehicle control

Citation: Aletras, N.; Doulgeris, S.; Samaras, Z.; Ntziachristos, L. Comparative Assessment of Supervisory Control Algorithms for a Plug-In Hybrid Electric Vehicle. *Energies* **2023**, *16*, 1497. https://doi.org/10.3390/en16031497

Academic Editor: Wojciech Cieslik

Received: 29 December 2022
Revised: 16 January 2023
Accepted: 30 January 2023
Published: 2 February 2023

1. Introduction

Global warming due to increasing emissions of greenhouse gases (GHG) appears today as the main environmental pressure [1]. Transport is one of the key sources of manmade carbon dioxide (CO_2) emissions [1–3]. This has led authorities around the world to set targets and take measures to reduce these emissions [3,4]. The European Union (EU) has set a target of reducing CO_2 levels from new passenger cars by 37.5% by 2030, compared to 2021 [5]. Therefore, solutions such as electrified vehicles are promoted by the automotive industry to meet these targets [6].

Hybrid Electric Vehicles (HEVs) and Plug-in Hybrid Electric Vehicles (PHEVs) are currently the most widespread options for electrified vehicles in the market. HEVs and PHEVs have two independent energy sources, namely the battery and the fuel tank. However, only PHEVs can be charged directly by grid power and can cover substantial ranges (e.g., the latest models appear to have an electrical range of 100 km or 65 miles [7]) with electric power alone [6]. The share and mix between battery and engine power are constantly being decided during operation by an on-board energy management system (EMS) [1,6,8,9]. The EMS performance has a significant impact on fuel consumption (FC), which is directly linked to CO_2 emissions. Therefore, EMS supervisory algorithms can be optimized to further decrease CO_2 emissions from PHEVs.

Wu et al. [10] showed that there are a variety of principles for EMS algorithms. The main categories can be distinguished into rule-based (RB), optimization-based (OB), and learning-based (LB) ones. RB algorithms rely on a fixed set of rules, without a priori knowledge of driving conditions. OB algorithms are further split to offline or online ones. In online OB algorithms, such as the Equivalent Consumption Minimization Strategy (ECMS) [10–13], an instantaneous optimization is conducted based on current vehicle operation. In offline OB algorithms, a cost function is optimized for the complete driving cycle. Dynamic Programming (DP) [14,15] and Pontryagin's Minimum Principle (PMP) [10,16] are some of the common offline OB algorithms. LB algorithms are capable of instantaneously and in real time controlling and learning the optimal power split operations. Reinforcement Learning [17,18] and Artificial Neural Networks are some principles that are used in LB algorithms implementation [10].

There are only limited works in the literature on how different algorithm categories compare to each other under different operation conditions. Actually, Torreglosa et al. [19] mentioned that the optimization algorithms presented in the literature are seldom compared against commercial RB strategies. In their study, they compared different EMSs for HEVs with RB strategies using FASTSim, an open-source tool that includes validated HEV RB models. That analysis showed that optimum EMSs may provide fuel consumption benefits of 5% to 10%, compared to commercial RB EMSs. Wu et al. [20] proposed an optimization-based strategy that appeared to reduce the fuel consumption of a 2010 Toyota Prius hybrid by 3.5–6%, compared to an RB algorithm that was earlier published by Kim et al. [21]. Hwang et al. [22] applied particle swarm optimization to improve the fuel economy of a power split hybrid, and showed up to 9.4% improvement compared to an RB algorithm. This limited previous work showed that there are further margins to improve fuel consumption over commercially applied algorithms.

In assessing the performance of different algorithms, one needs to make sure that the exact operation profile is followed over computer simulations or real-world experimentation with the various EMS approaches. Fuel consumption differences of only a few percentage units, such as those expected when varying the EMS, can be observed only due to slight deviations of the original speed profile in consecutive simulations, e.g., due to underpowering accelerations. Moreover, it needs to be ensured that fuel consumption improvement is assessed under the same state of charge levels (SOC) to avoid part of the difference only being due to variation in the battery depletion levels, e.g., over consecutive simulations. Although such conditions may be self-evident, these are seldom if at all demonstrated in published EMS algorithm comparison studies.

The article focuses on the comparative assessment of commercial RB and ECMS based algorithms for a plug-in hybrid vehicle powertrain. The purpose is to examine if further FC reduction can be achieved by introducing an enhanced EMS over a commercial one. A method is first presented to compare fuel consumption over identical battery state of charge (SOC) levels, vehicle distance traveled and wheel power demand. The proposed method suggests novel corrections for the fuel energy consumption of the compared EMS algorithms. More specifically, the method introduces correction terms for the deviations in the final SOC values, propulsion energy and benefits from regenerative braking between the compared EMS algorithms. We propose that such a method needs to be used in all similar studies of fuel consumption comparison.

2. Methods

2.1. Back-Engineered EMS Algorithm

A vehicle simulator of a parallel P2 PHEV [23] has been built in the AVL Cruise simulation platform. Its overall performance has been validated with actual experimental data collected by tests on an actual vehicle in the chassis dynamometer of Aristotle University. The vehicle's technical specifications are listed in Table 1.

Table 1. Parallel P2 PHEV Technical Specifications.

Component	Specifications
Vehicle test mass	1700 kg
Fuel-type, displacement, engine power	Gasoline, 1560 cm^3, 77.2 kW
Battery, type	8.9 kWh Li-Ion Polymer
Electric motor	44.5 kW
Gearbox	6-speed dual-clutch automatic

Table 2 shows the tests conducted in the lab on the particular vehicle to understand the performance of its stock EMS algorithm. A Worldwide harmonized Light vehicles Test Cycle (WLTC) [24] and an ERMES cycle [25] have been used for the tests. The different cycles are distinguished into cold and hot start ones, depending on whether the start engine coolant temperature was lower than 35 °C or higher than 70 °C, respectively. A single case with intermediate start temperature is identified as warm start in Table 2.

Table 2. Driving cycles and specifications use for experimental validation of the back-engineered algorithm.

Cycle	ICE Condition	Initial SOC	Vehicle Mode	Short Name
ERMES	Cold start	35.7%	Charge depleting/sustain mode	ERMES CDCS
ERMES	Hot start	11.8%	Charge sustain mode	ERMES CS
WLTC	Hot start	20.4%	Charge sustain mode	WLTC CS HOT2
WLTC	Cold start	12.9%	Charge sustain mode	WLTC CS COLD
WLTC	Hot start	13.7%	Charge sustain mode	WLTC CS HOT1
WLTC	Cold start	71.4%	Charge depleting mode	WLTC CD
WLTC	Warm start	28.2%	Charge depleting/sustain mode	WLTC CDCS

These experiments have been used to back-engineer the rules of the heuristic controller on-board the commercial vehicle. In this paper, the RB algorithm is a specialization of the general methodology described by Doulgeris et al. [26], while the specific controller algorithm is described in detail by Doulgeris et al. [27].

Figure 1 shows the flowchart of the RB algorithm. The engine switches on if the power demand, vehicle speed or acceleration, and SOC level are above specific thresholds. The power demand threshold for engine start depends on the SOC level. The engine always shuts off when the power demand becomes negative.

Figure 2 shows the engine power output decided by the algorithm curve, depending on the gear engaged (x-axis) and the vehicle speed (parameter). If the engine meets the criteria for switch-on according to Figure 1, then the engine power output is determined by Figure 2 depending on current vehicle speed and gear.

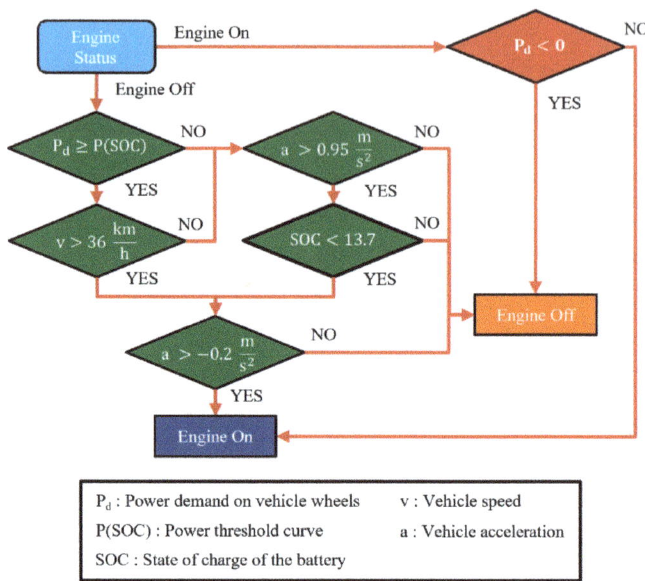

Figure 1. Stock rule-based algorithm extracted from experimental evidence.

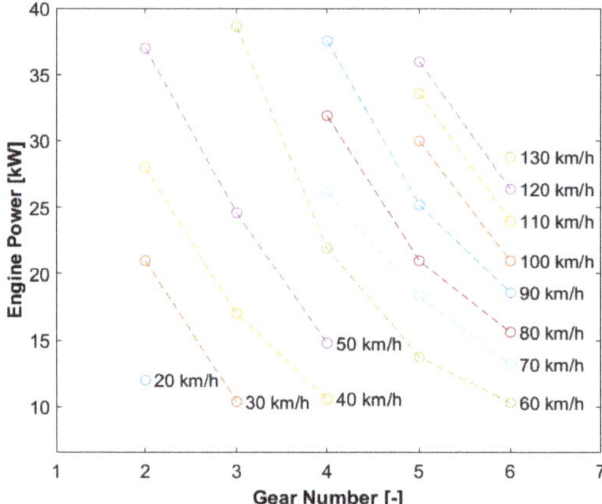

Figure 2. Engine power output decided by the RB algorithm depending on gear and vehicle speed.

2.2. Alternative Algorithm Description

An Equivalent Consumption Minimization Strategy (ECMS) algorithm has been developed in the current study, as an alternative to the back-engineered RB one. Both of the algorithms—ECMS and the back-engineered one—are applied to the same vehicle simulator platform of parallel P2 PHEV that has been built in the AVL Cruise simulator platform. The ECMS algorithm aims at optimizing a predefined FC cost function for given operation conditions. The general cost function for fuel consumption optimization of a hybrid vehicle is given in Equation (1), where J is a performance index that needs to be minimized. The integral term represents the total fuel consumption over a complete driving profile, as it integrates the instantaneous FC ($\dot{m}_{fuel}$) from an initial (t_0) to a terminal time stamp (t_f). FC depends on the normalized engine load function u that ranges between 0 (engine shut off) to 1 (operation at full power—$P_{e,max}$), according to Equation (2). The first term in Equation (1) is used as a soft constraint for the value of the SOC at the end of the cycle (SOC_f). With the use of the φ function, SOC deviations from the final target value (SOC_{target}) are being penalized. The choice of the SOC_{target} value depends on the examined study. Usually, in PHEVs applications, the charge depletion is permitted because the battery can be charged from the electric power grid. As a result, for PHEVs applications, the SOC_{target} can be lower than the initial value of SOC [8,9].

$$J = \varphi\left(SOC_f, SOC_{target}\right) + \int_{t_0}^{t_f} \dot{m}_{fuel}(u(t), t)d \tag{1}$$

$$u = \frac{P_e}{P_{e,max}} \tag{2}$$

The optimization of Equation (1) leads to the determination of the u for every second of the driving cycle, which in turn gives the instantaneous engine power output by means of Equation (2).

The ECMS optimization is subject to the conditions of the set of Equations (3)–(6). Equation (3) presents the power balance for the powertrain of a P2 parallel hybrid vehicle [28]. The sum of the demanded power at wheels ($P_{req,wheels}$) and the mechanical power losses ($P_{mech,losses}$) must be equal to the mechanical power output from the main power units (electric motor power—P_{em} and engine power—P_e). If the vehicle velocity is known, then the $P_{req,wheels}$ and $P_{mech,losses}$ can be determined by a vehicle power-based model. We have set up a vehicle model in AVL Cruise for this purpose. So, with the use of Equation (3), the power output of the electric motor is determined.

$$P_{req,wheels} + P_{mech,losses} = P_e + P_{em} \tag{3}$$

$$P_e \leq P_{e,max} \tag{4}$$

$$P_{em,min} \leq P_{em} \leq P_{em,max} \tag{5}$$

$$SOC_{min} \leq SOC \leq SOC_{max} \tag{6}$$

Equations (4)–(6) correspond to the physical constraints of the powertrain components. Equation (4) suggests that the engine cannot overcome its full load curve, represented by the maximum engine power output ($P_{e,max}$). The electric motor is also limited by its full load curve, depending on whether it works as a motor ($P_{em} > 0$) or generator ($P_{em} < 0$) (Equation (5)), with corresponding limits given by $P_{em,max}$ and $P_{em,min}$. Finally, the battery SOC cannot exceed a range of maximum (SOC_{max}) and minimum (SOC_{min}) levels for the purpose of maintaining battery life (Equation (6)).

Optimizing Equation (1) within the set of Conditions (3)–(6) is only possible when the operation mission is known a priori. In the real-world, a priori knowledge of the driving profile application is known only over in-lab tests and not for on-road driving. For on-road operation, the ECMS will have to be locally optimized according to the present driving conditions. Such local optimization is achieved by means of Equation (7), where the integral term of Equation (1) has been eliminated. In Equation (7), the engine fuel rate ($\dot{m}_{fuel}$) and the battery power flow expressed in terms of an equivalent fuel rate ($\dot{m}_{eq}$) (Equation (8)—where Q_{LHV} is the fuel's lower heating value) result in an equivalent total fuel mass rate ($\dot{m}_{tot}$) by means of the equivalence factor s (Equation (9)). The latter comprises the constant term s_0 and a penalization term $p(SOC)$ that depends on SOC. The s_0 term is used as the main weighting factor of the $\dot{m}_{eq}$ inside the cost function. The p penalizes deviations of the current SOC values from the target. The usage of the p term is similar to the one of the φ term in Equation (1). The difference is that the penalization is made for the instantaneous values of SOC instead of the SOC value at the end of the cycle, because the optimization is only carried out locally. The battery power (P_{batt}) in Equation (8) can be positive for power outflux from the battery (Equation (10a)) and negative when the

EM acts as a generator that charges the battery (Equation (10b)), with η_{batt} and η_{em} being the battery and electric motor efficiencies, respectively.

$$\dot{m}_{tot} = \dot{m}_{fuel} + s \times \dot{m}_{eq} \tag{7}$$

$$\dot{m}_{eq} = \frac{P_{batt}}{Q_{LHV}} \tag{8}$$

$$s = s_0 + p(SOC) \tag{9}$$

$$P_{batt} = \left(\frac{P_{em}}{\eta_{batt} \times \eta_{em}} \, |P_{em} \geq 0 \right) \tag{10a}$$

$$P_{batt} = (P_{em} \times \eta_{batt} \times \eta_{em} \, |P_{em} < 0) \tag{10b}$$

The algorithm basically decides on the engine operation variable u(t) in Equation (2) that leads to the lowest total equivalent fuel mass ($\dot{m}_{tot}$). This procedure is repeated in every second of the complete mission profile. The algorithm takes into account two conditions regarding the potential battery charge or discharge. The first one is that a potential battery charge will lead to an SOC surplus, which can be utilized in the future. The second condition is that a present battery discharge generates a requirement for a future battery charge in order to retain the battery SOC within certain limits. An optimal solution can be guaranteed if the s term in Equation (7) is adapted appropriately (Equation (9)). In this way, although the s-by-s optimization cannot achieve as good a performance as the global optimal solution, it still produces a practical optimization solution that can be integrated in EMS without knowledge of the forthcoming driving profile.

Three alternative expressions for p have been examined in the current work (Equations (11a)–(11c)). In Type A expression, p is proportional to the difference of current SOC over a constant reference value SOC_{ref} (Equation (12a)). Therefore, this expression tries to keep SOC at a value close to the reference one over the complete driving profile—and it is tuned by a proportional term (kp). In Type B, the SOC_{ref} value varies with travelled distance D (Equation (12b) [29]). The algorithm also tries to keep the current SOC close to the SOC_{ref} value, as in Type A. More specifically, in Type B, the SOC_{ref} value starts with an initial value (SOC_i) and then the SOC_{ref} decreases proportionally with the D until it reaches the target SOC value (SOC_{target}). In Type B, the total driving distance (D_{final}) must be either known or estimated. Some research articles mention that this type of linear SOC trajectory with distance seems to be close to the global optimal solution [30,31]. In Type C expression, a specific SOC window is used for determining p [8] (Equation (11c)). More specifically, the *p* value dependents on SOC, a target value for the SOC (SOC_{target}), selected maximum and minimum SOC values and a selected superscript for the penalization function (a). With this expression, SOC is retained above a certain level in order to ensure the battery physical constraints ($SOC > SOC_{min}$) proactively with the adaptation of the equivalence factor. Moreover, this expression constrains battery charging during charge sustain operation until a rational level (e.g., $SOC_{max} = 18\%$).

Attention is required in selecting the parameters for each expression to achieve feasible solutions. For example, in our effort for parameters tuning, we spotted that some parameter combinations led to extremely low SOC levels or even that the vehicle could not follow the speed profile. So, after a trial-and-error basis in order to achieve feasible solutions, the setup of the algorithm parameters is presented in Table 3.

$$p(SOC) = kp \times (SOC - SOC_{ref}) \tag{11a}$$

$$p(SOC) = kp \times (SOC - SOC_{ref}(D)) \tag{11b}$$

$$p(SOC) = kp \times \left(1 - \left(\frac{(SOC - SOC_{target})}{0.5(SOC_{max} - SOC_{min})} \right)^{a} \right) \tag{11c}$$

$$SOC_{ref} = SOC_{target} \tag{12a}$$

$$SOC_{ref}(D) = SOC_i - \left[\left(SOC_i - SOC_{target}\right) \times \left(\frac{D}{D_{final}} \right) \right] \tag{12b}$$

Table 3. Parameters selected for the three ECMS versions.

Version	s_0	kp	SOC_{target}	SOC_{max}	SOC_{min}	a
ECMS Type A	3.5	−0.5	15	-	-	-
ECMS Type B	3.5	−0.5	14	-	-	-
ECMS Type C	3	−1.5	13	18	8	3

The ECMS algorithm flowchart is illustrated in Figure 3. ECMS requires power demand, vehicle velocity, gear number and current SOC as input data. The first step is to select the numerical values for the physical limitations according Equations (4)–(6). After that, the algorithm calculates the equivalent fuel rate $\dot{m}_{tot}$ for the different candidate operating points by adjusting the equivalent consumption of the electrical motor according to the SOC, as described in Equations (9) and (11a)–(12b). Finally, the algorithm selects the case with the minimum fuel mass, which is then translated into specific torque outputs of the ICE and the electric motor.

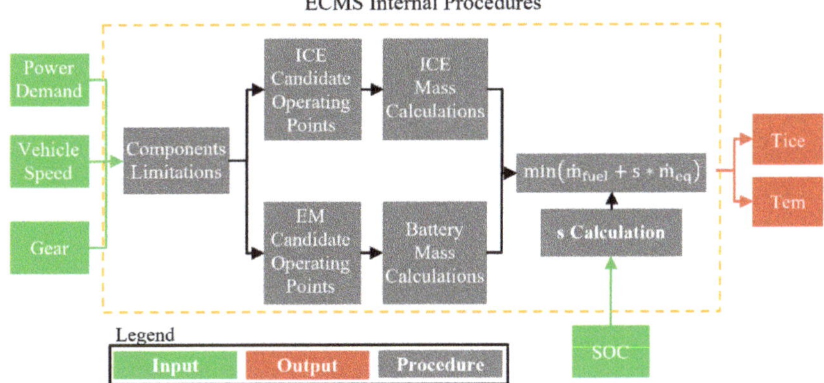

Figure 3. ECMS Procedure Overview.

2.3. Corrections for the Assessment of Different Algorithms

The assessment of different EMS algorithms needs to be carried out on a fair basis. Each EMS algorithm leads to different decisions for engine and motor engagement (Equation (2)) that may slightly affect the speed of the vehicle due to power availability and gear change interference. The different speed profiles will, in turn, result in a slightly different demanded power profile for each algorithm. These differences could have been avoided by backward modeling, because this approach guarantees that the vehicle exactly follows the target speed. However, in our approach, the forward modeling is chosen because forward simulators are based on physical causality. With these simulators, online control strategies can be developed [8]. Moreover, the SOC difference between trip start and end may differ between various algorithms. Nevertheless, when assessing the impacts of different algorithms on FC, one needs to make sure that the distance, demanded energy, and SOC differences are identical in the various simulations.

A method to adjust for such potential differences is, therefore, introduced. Assuming a total energy consumption over a theoretical accurate driving profile in each simulation, variations of this profile will lead to energy differences, not because of EMS performance but because of distance and SOC variations in each simulation. A corrected fuel-equivalent energy consumption (CE) can, therefore, be estimated from the simulated one (SE) by correcting for deviations in the SOC (ΔE_{SOC}), propulsion energy (ΔE_{PROP}) and contribution of regenerative braking (ΔE_{REG}) in Equation (13). ΔE_{SOC} correction is needed to make sure that all simulations result in identical final SOC. The ΔE_{PROP} term corrects for slight differences in the driving profile (speed, acceleration and distance) of the simulations of a given driving sequence. Finally, ΔE_{REG} adjusts the energy consumption when the simulated regenerative braking energy benefits are different from the ones calculated in the theoretically accurate driving profile. In that way, the energy differences due to driving profile variations at the braking phases can be corrected. The corrected energy correction of Equation (13) should be implemented in all relevant works where different optimization algorithms are being compared.

$$CE = SE + \Delta E_{SOC} + \Delta E_{PROP} - \Delta E_{REG} \tag{13}$$

Equation (14) describes ΔE_{SOC} as the fuel energy delivery that covers the difference between the simulated and reference depleted energies from the battery (E_{bat}). In the denominator of Equation (14), the average product of the individual components' efficiencies has been considered for the time moments that the battery is charged from the ICE, with η_e being the ICE efficiency. The calculation for the reference value of the depleted battery energy is presented in Equation (15). The value is calculated as a difference from a final SOC level, which in our case has been selected to be 14%, with C_{bat} and $\overline{V}_{bat}$ being the battery capacity and average battery voltage, respectively.

$$\Delta E_{SOC} = \frac{E_{bat} - E_{bat,SOCf}}{\left((\eta_{batt} \times \eta_{em} \times \eta_e) \,|P_{em} < 0 \,\wedge\, P_e > 0 \right)} \tag{14}$$

$$E_{bat,SOC_f} = (SOC_i - SOC_f) \times C_{bat} \times \overline{V}_{bat} \tag{15}$$

The ΔE_{PROP}—Equation (16)—is the fuel energy that should be supplied to equalize the simulated energy demand at gearbox (E_{GB}) with the one calculated from the theoretical speed profile ($E_{GB,th}$). The ICE efficiency should be the average one during positive power demand at gearbox inlet (P_{GB}). $E_{GB,th}$—Equation (17)—is the energy demand for vehicle motion for positive tractive force at the wheels (F_{th}), with v_{th} and $\overline{\eta}_{tr}$ being the force, theoretical vehicle speed and average transmission efficiency from gearbox inlet to vehicle wheels, respectively. $F_{tr,th}$ consists of a polynomial function of vehicle speed (which corresponds to road loads) and the term for vehicle acceleration (Equation (18)—F_2, F_1, F_0 are coast down test coefficients and M_v is the vehicle mass).

$$\Delta E_{PROP} = \left(\frac{E_{GB,th} - E_{GB}}{\overline{\eta}_e} \,|P_{GB} > 0 \right) \tag{16}$$

$$E_{GB,th} = \left(\frac{\sum F_{tr,th} \times v_{th}}{\overline{\eta}_{tr}} \,|F_{tr,th} \geq 0 \right) \tag{17}$$

$$F_{tr,th} = F_2 \times v_{th}^2 + F_1 \times v_{th} + F_0 + M_v \times \frac{dv_{th}}{dt} \tag{18}$$

ΔE_{REG}—Equation (19)—expresses the potential fuel energy that can be saved if the simulated speed profile was identical to the theoretical one ($E_{Bat, th}$) during decelerations. The simulated battery energy influx is calculated for the time instances that the power demand at gearbox inlet is negative. To convert the battery energy influx difference to fuel energy, the same average product of efficiencies as the one in Equation (14) has been used. The $E_{B-Bat,th}$—Equation (20)—is the potential energy of braking that can be recuperated. In this calculation, the negative energy influx from the wheels ($E_{B,th}$) is calculated based on Equation (21). In the current study, the share of the total braking energy that can be recuperated (b_{REG}) is assumed to be 40% of the total braking energy, while the rest is consumed at the mechanical brakes. The losses from the gearbox inlet up to the battery have been taken into account by the average product of the EM and battery efficiencies during the time events that P_{GB} is negative.

$$\Delta E_{REG} = (E_{B-Bat,th} - E_{Bat} \, | P_{GB} < 0) \times \left(\overline{(\eta_{batt} \times \eta_{em} \times \eta_e)} \, | P_{em} < 0 \wedge P_e > 0 \right) \quad (19)$$

$$E_{B-Bat,th} = \left(E_{B,th} \times b_{REG} \times \overline{\eta}_{tr} \times \overline{(\eta_{batt} \times \eta_{em})} \, | P_{GB} < 0 \right) \quad (20)$$

$$E_{B,th} = \left(\sum (F_{tr,th} \times v_{th}) \, | F_{tr,th} < 0 \right) \quad (21)$$

2.4. Drive Cycles Used in the Simulations

The performance of the Rule Based (RB) and the different types of ECMS Based (EB) algorithms are compared in different driving cycles, according to Table 4. Firstly, they are compared in WLTC driving cycle. After that, various driving cycles have been used in order to examine different conditions and identify the best algorithm in each case. For WLTC, ERMES and 10-15 Mode [32] cycles, the individual segments have been examined as well. Table 4 displays a summary of the characteristics of the different driving cycles that have been used. All simulations have been conducted with the same initial SOC of 20%, which allows a comparison of the algorithms during the most challenging charge sustain mode.

Table 4. Simulated Driving Cycles and characteristics.

Cycle	Repetitions	Average Speed [km/h]	RPA [m/s^2]	Total Trip Length [km]
10 Mode	6	16.0	0.198	12.0
WLTC Low	6	18.4	0.217	18.4
10-15 Mode	5	22.9	0.172	20.8
JC08 [33]	2	26.9	0.184	20.6
ERMES Urban	6	32.1	0.188	29.7
WLTC Medium	6	40.0	0.209	28.2
WLTCx2	2	46.1	0.160	46.1
WLTC High	6	56.0	0.137	42.6
ERMES	2	66.0	0.106	48.0
ERMES Extra Urban	6	69.9	0.120	44.1
WLTC Extra High	6	90.7	0.131	49.0
ERMES Motor	6	96.0	0.086	75.0

3. Results

3.1. Validation of the Rule Based Algorithm

Figures 4 and 5 present examples for model accuracy in SOC and fuel consumption, respectively, for four of the seven conducted tests (Table 2). The initial SOC was the actual one at the beginning of each test. The simulated SOC profile satisfactorily follows the measured one during the course of each cycle. This can be reflected by the high correlation coefficient values for SOC (R_{SOC}), which are higher than 69%. There is only one exception, for WLTC CS HOT1, where our back-engineered RB results in higher battery depletion in the 1000–1200 s range compared to the measured one. Apart from this, the simulated SOC follows the measured trend with a rather constant offset.

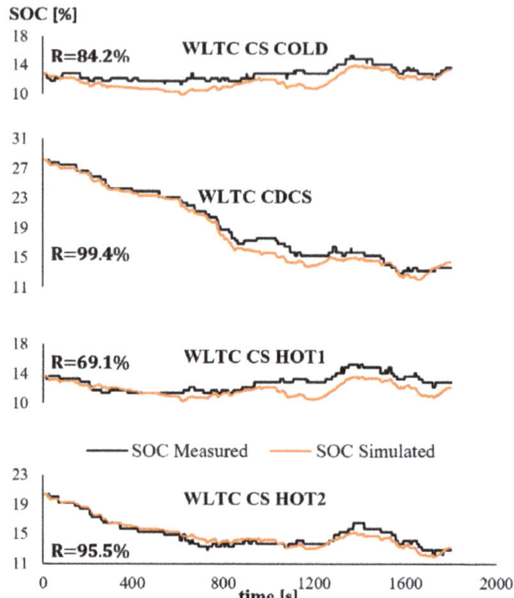

Figure 4. SOC measurement vs. simulation—examples.

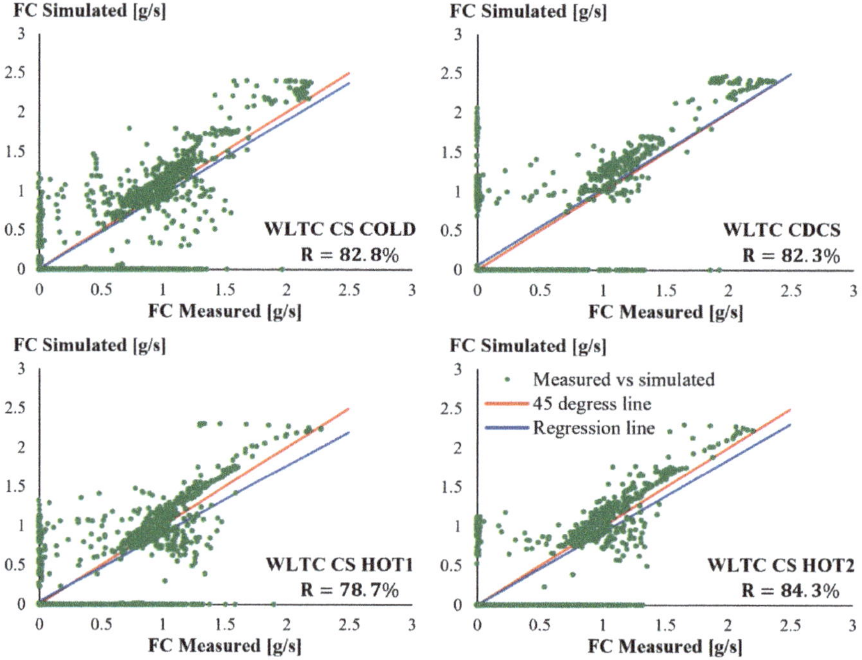

Figure 5. FC measurement vs. simulation—examples.

Table 5 compares the total simulated and measured FC and final SOC (SOC_f) values for the seven tests conducted. The absolute error in the simulation of total FC is lower than 7% for five of the seven tests. In the WLTC CD and WLTC CDCS tests, the FC estimation error is higher, but over very low FC values that overall lead to satisfactory deviations (WLTC CD: 0.5 g/km; WLTC CDCS: 4.2 g/km). The correlation coefficient between instantaneous measured and simulated fuel consumption (R_{fuel}) is always higher than 70% except in WLTC CD (57.7%). In WLTC CD, the initial SOC is rather high (71.4%) so the engine engagement is limited leading to few points where the instantaneous fuel consumption is non-zero, and this leads to a drop of the regression coefficient. With regard to battery SOC, the simulated final SOC values are quite close to the measured ones for WLTC. For the ERMES cases, the simulated SOC varies more than the measured one in WLTC. Additionally, for the ERMES CS the R_{SOC} value is negative. This means that the simulated battery charging events do not follow the measured ones for that test. It is worth mentioning that the parameters tuning for engine start/shut-down events (Figure 1) were based on the WLTC tests experimental data. As the ERMES cycle has different characteristics, we expected that the model could not have the same accuracy as in the WLTC tests. For all the other test cases, the R_{SOC} is higher than 69% which implies satisfactory model performance.

Table 5. Total FC and ΔSOC Comparison.

Magnitude	Cycle						
	WLTC CS COLD	WLTC CDCS	WLTC CS HOT1	WLTC CS HOT2	WLTC CD	ERMES CDCS	ERMES CS
FC_{meas} $\left[\frac{g}{km}\right]$	40.9	24.0	36.2	32.0	1.5	22.7	39.2
FC_{sim} $\left[\frac{g}{km}\right]$	39.8	28.2	35.3	30.4	2.0	24.0	41.0
$\frac{(FC_{meas}-FC_{sim})}{FC_{meas}}$ [%]	−2.8	17.4	−2.6	−5.0	27.9	6.1	4.8
R_{FC} [%]	82.8	82.3	78.7	84.3	57.7	89.5	76.3
SOC_i [%]	12.9	28.2	13.7	20.4	71.4	35.7	11.8
$SOC_{f,meas}$ [%]	13.7	13.7	12.9	12.9	28.2	11.8	12.2
$SOC_{f,sim}$ [%]	13.5	14.5	12.2	13.2	27.5	16.4	13.2
$SOC_{f,meas} - SOC_{f,sim}$ [%]	0.2	−0.8	0.7	−0.3	0.7	−4.6	−1.0
R_{SOC} [%]	84.2	99.4	69.1	95.5	99.9	99.7	−66.0

3.2. Comparative Assessment of RB and EB Algorithms in WLTC

Table 6 summarizes the simulation results for the three ECMS types (EBA, EBB, EBC) and the RB algorithm over WLTC. The FC values show correspondence to the simulation output (SFC) and the corrected ones (CFC), according to Equations (13)–(21). The table values clearly show that if no correction was introduced then one would come up with totally wrong conclusions regarding the relative performance of the different algorithms. For example, EBC leads to the highest FC difference over the RB (−3.4%), which is actually lower than the magnitude of the correction (−5.6%). Had the correction not been applied, the EBC would have actually resulted in +4.9% higher FC than the RB, i.e., this would have led to an entirely opposite conclusion than what is actually reached using the corrected values. This can be explained because the ΔE_{SOC} correction turns out negative for the EB and positive for the RB cases. The negative value for ΔE_{SOC} means that the final SOC is higher compared to the reference one in EB, and vice versa for RB. Additionally, Table 6 shows that ΔE_{PRO} has a higher impact on EB cases. This indicates that the simulated speed profile in the RB simulation better follows the theoretical one than in EB simulations.

The net energy flows between the different powertrain components during WLTC are presented in Figure 6. The shown energy flows stand for the positive propulsion instances. The electrical energy from the battery (2.89 MJ) and the energy demand at the gearbox inlet (13.2 MJ) have been adjusted to the exact same values in order to ensure that the comparison of the different algorithms is on a fair basis. Furthermore, the shown fuel energy consumption is the corrected one, which has been calculated from Equation (13).

Table 6. Consumption and mean efficiency values of the main powertrain components for WLTC.

Magnitude	Algorithm			
	RB	**EBA**	**EBB**	**EBC**
SFC [g]	698.5	750.0	736.1	732.6
$\Delta E_{SOC}/Q_{LHV}$ [g]	17.0	-78.0	-62.1	-66.8
$\Delta E_{PROP}/Q_{LHV}$ [g]	0.7	19.8	25.5	25.7
$\Delta E_{REG}/Q_{LHV}$ [g]	0.6	0.2	0.1	0.1
CFC [g]	715.6	691.6	699.4	691.4
(CFC-SFC)/SFC [%]	2.4	-7.8	-5.0	-5.6
$(CFC_{EBX}\text{-}CFC_{RB})/CFC_{RB}$ [%]	-	-3.3	-2.3	-3.4
Propulsion Efficiency [%]	39.5	40.7	40.3	40.8
ICE Mean Efficiency [%]	37.9	37.6	37.1	37.8
EM Mean Efficiency [%]	82.7	80.8	80.3	80.7

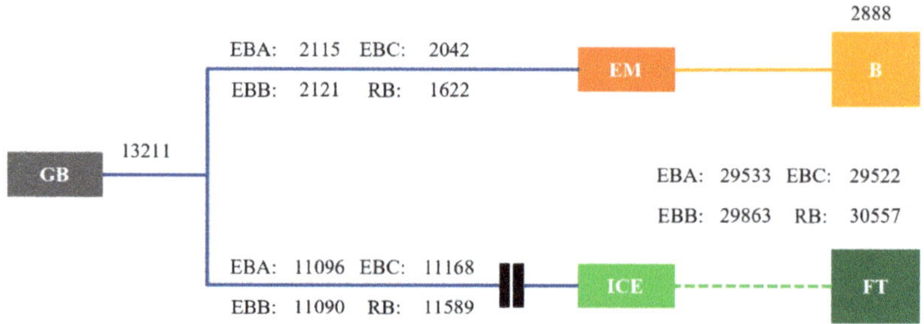

Figure 6. Net energy flows (kJ) for positive propulsion instances in WLTC for ECMS-Based Type A (EBA) algorithm, ECMS-Based Type B (EBB) algorithm, ECMS-Based Type C (EBC) algorithm and Rule-Based (RB) algorithm. ICE: Internal Combustion Engine, EM: Electric Motor, B: Battery and GB: Gearbox, FT: Fuel tank. Negative numbers correspond to energy loss.

The electric motor may function either as a propulsion device or as a generator for battery charging, depending on power flux direction. Assuming positive power flux from the battery to the gearbox, the energy flow through the EM ranges from 2121 kJ with the EBB to 1622 kJ with the RB algorithm. The difference in net energy flows is explained by the decisions of the different algorithms.

The lowest net EM energy flow is balanced by the maximum total energy outflux from the FT (30.6 MJ) with RB compared to EB. The simulation results in Table 6 show that the average ICE efficiency is quite close for all cases. Additionally, the demanded mechanical output energy from the ICE is higher in the RB case (11.59 MJ) compared to the EB (11.1–11.2 MJ). In RB, the electric motor contributes more for propulsion compared to the EB cases and results in higher battery discharge. This requires higher power demand from the ICE during hybrid mode to recharge the battery. That higher power demand for battery charging could have been saved had the EMS controller decided to directly command ICE propulsion with the assistance of the electric motor.

The above analysis showed that the description of the energy flows is very useful in order to understand the FC differences between RB and ECMS algorithms. For analysis convenience, in this paper a quantifiable metric for energy flow comparisons has been used, namely the average propulsion efficiency ($\bar{\eta}_{PROP}$). The $\bar{\eta}_{PROP}$ is defined as the ratio of the demanded energy at the gearbox (E_{GB}) over the sum of the two net energy flows from the energy sources during the positive propulsion phase (Equation (22)). The provided fuel energy is the corrected one, but only during the propulsion phases. Therefore, the correction terms of ΔE_{SOC} and ΔE_{PROP} have been added to the simulated fuel energy. Additionally, the reference battery depleted energy has been considered in the denominator. Table 6 shows that the RB algorithm results in the lowest propulsion efficiency and that leads to the highest fuel consumption.

$$\overline{\eta}_{PROP} = \frac{E_{GB}}{E_{bat,SOC_f=14\%} + SE + \Delta E_{SOC} + \Delta E_{PROP}} \tag{22}$$

As a benchmark, the results of Table 6 that have been obtained with localized optimization are compared to the global optimum for the known profile of WLTC. To do so, we have properly parameterized the hybrid electric vehicle model of Sundstrom et al. [34], which uses a DP a solution as an EMS algorithm. The parameterization has been achieved using exactly the same values for the individual components (ICE, EM, battery, Gearbox, axles, vehicle resistances and weight) with the ones used in the AVL Cruise simulated model. The DP model delivered 640.9 g as global optimum CFC, which is 7.3% lower than the one from the best-performing EBC (640.9 g vs. 691.4 g).

3.3. Comparison of RB and EB Algorithms for Other Cycles

Figure 7 illustrates the relative CFC difference between EB and RB algorithms for the different cycles. In most cases, EB achieves lower fuel consumption compared to RB. The highest FC reduction is observed for WLTC Low, which is 22.1–24.2% depending on the EB type, while the highest EB increase over RB is 2.7% over the ERMES Extra Urban. EB outperforms RB in all cycles of low speed. Evidently, the restrictions of RB on engine switch on criteria (Figure 1) and power output (Figure 2) have a cost on FC reached, while the adaptability of ECMS (Figure 3) allows for a much better result to be obtained.

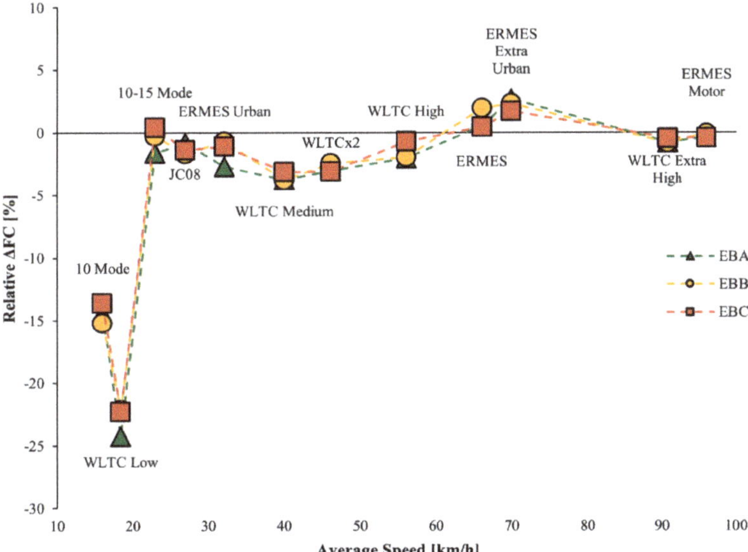

Figure 7. Fuel consumption change in ECMS algorithm types vs. Rule Based algorithm case for different cycles. EBA: ECMS Based Type A, EBB: ECMS Based Type B, EBC: ECMS Based Type C.

Table 7 better explains how EBC and RB perform for a low speed (WLTC Low) and a high speed (ERMES Extra Urban) cycle. In WLTC Low, EBC results in a higher propulsion efficiency by eight percentage units compared to the RB, and this leads to 22% FC reduction (RB: 582 g and EBC: 453 g). The improvement in propulsion is mainly linked to the mean ICE efficiency in this case. As Figure 2 shows, RB selects to operate the engine at lower load for vehicle speeds 0–30 km/h, which results in low engine efficiency. In reality, this selection may have to do with the need to retain low noise, vibration and harshness (NVH) levels in the vehicle cabin during low-speed driving. On the other hand, ECMS algorithm decisions for the engine load are dependent on the optimization of a cost function—Equation (7). As a result of its optimization-based functionality, ECMS leads to higher ICE Mean efficiency by 3.7 percentage units compared to RB in WLTC Low.

Table 7. Consumption and mean efficiency values of the main powertrain components for WLTC Low and ERMES Extra Urban in the case of Rule Based (RB) and ECMS Based—C (EBC) algorithms.

Cycle	Magnitude	Algorithm Type	
		Rule Based (RB)	**ECMS Based—C (EBC)**
WLTC Low	Fuel Consumption [g]	582	453
	Propulsion Efficiency [%]	32.5	40.5
	ICE Mean Efficiency [%]	33.3	37.0
	EM Mean Efficiency [%]	78.8	76.9
ERMES Extra Urban	Fuel Consumption [g]	1341	1363
	Propulsion Efficiency [%]	38.0	37.4
	ICE Mean Efficiency [%]	37.2	34.8
	EM Mean Efficiency [%]	86.7	84.3

The FC with ECMS appears higher than RB only in the ERMES cycle, and in particular in ERMES Extra Urban. In order to explain this, we can look the propulsion efficiency values at Table 7. The RB algorithm has a higher propulsion efficiency by 0.6 percentage units compared to EBC. The ERMES has a much higher power requirement and stronger accelerations than WLTC, which was used to tune the parameters of the ECMS algorithm (Equations (11a)–(12b) and Table 3). A better tuning for high power cycles could lead to lower FC compared to RB.

4. Discussion and Conclusions

This study makes an assessment and performance analysis for two types of energy management algorithms, including a back-engineered stock RB algorithm and an ECMS one.

A novel methodology to assess the different algorithms on a fair basis has been developed, introducing corrections for distance travelled, final SOC level and energy propulsion differences between alternative simulations. Our analysis showed that the impact of such corrections on fuel consumption can exceed 5%. Hou et al. [35] also considered SOC corrections when comparing different EMS algorithms, and found the impact of these corrections to be no more than 1% in fuel consumption. However, they did not correct for propulsion energy demand and regenerative braking. Our analysis shows that all three corrections can have a measurable result on fuel consumption values when different EMS algorithms are compared, and these should not be neglected in relevant studies.

The energy flow analysis also showed that the EMS affects both the efficiency of each powertrain component and the net energy flow between components. Therefore, it is the combination of these two magnitudes that affects overall total fuel consumption. As a result, the efficiencies of the individual components can be quite close for the two different EMS algorithms, but the FC can differ for the same energy demand.

Another important finding is that different driving conditions affect the magnitude of FC reduction that can be achieved with an alternative algorithm. Regarding the WLTC cycle, it has been found that the potential FC reduction with the use of ECMS algorithm is 3.4% over RB. For benchmarking purposes, a DP global optimization algorithm has been used for comparison. This requires a priori knowledge of the driving profile to find the optimum solution. The DP led to a 7.3% lower FC compared to the one from ECMS. Sun et al. [13] demonstrated that, on average, the FC achieved with DP can be 6.7% lower compared to an ECMS algorithm. ECMS algorithms, therefore, seem to offer good adaptability to driving conditions and sufficiently good final fuel consumption, which is not very distant from the global optimum.

With the use of ECMS, we found that fuel consumption in driving cycles with low average speeds (lower than 20 km/h) can be improved by up to 24.2% compared to a stock RB algorithm. The magnitude of improvement achieved was actually similar to the ones reported by Geng et al. [36], who proposed a combination of DP and ECMS to reduce FC by up to 19.9% in NEDC compared to RB. In another study, Hao et al. [37] used an adaptive ECMS to decrease the FC by 8–15% compared to an RB algorithm on a mild parallel hybrid vehicle. In our case, this large difference was partly because the RB forced the engine to operate at low loads when switched on in low average speeds. Although this may have been mandated in order to retain low NVH in the cabin at low speeds or due to pollutants emission control restriction, one needs to admit that an improvement margin of more than 20% is a significant incentive to invest more in powertrain efficiency control, especially in urban conditions.

This paper examines the energy optimization of the hybrid powertrain for propulsion. However, an EMS algorithm may consider additional parameters, such as consumption of auxiliaries for cabin comfort and the thermal management of the emission control devices. These parameters are not addressed in the current work, as they add complexity and extra investigation is required to achieve balance between real-time implementation and optimality. However, they can be addressed in a future work by extending the cost function expression to cover these terms and by extending the list of conditions that have been considered to the implemented the algorithms. Such conditions can also take into account NVH requirements that can be added as cost penalizing terms in relevant optimization.

In general, ECMS algorithms are simple to enforce in an ECU as they do not require a full set of guidance for all foreseeable conditions and could always be used as back-up, fail-safe algorithms. For example, in driving situations that have not been thoroughly considered when setting up an RB algorithm and which cause higher FC than what would be technically possible, an ECMS could step in instead. Moreover, an ECMS could be identical for different vehicle model variants, without needing to set exact limits for each version of the vehicle when, e.g., engine or motor sizes change while powertrain architecture stays the same. An ECMS algorithm can, therefore, have a more universal usage due to its adaptability features.

Supplementary Materials: The following supporting information can be downloaded at: https://www.mdpi.com/article/10.3390/en16031497/s1, Figure S1: Examined Cycles.

Author Contributions: Conceptualization, N.A., S.D. and L.N.; Methodology, N.A., S.D. and L.N.; Software, N.A. and S.D.; Writing—Original draft preparation, N.A.; Visualization, N.A.; Writing—Review and Editing, L.N.; Supervision, L.N. and Z.S. All authors have read and agreed to the published version of the manuscript.

Funding: The research work was supported by the Hellenic Foundation for Research and Innovation (HFRI) under the HFRI Ph.D. Fellowship grant (Fellowship Numbers: 607 and 6653).

Data Availability Statement: The data presented in this study are available in this article and the Supplementary Materials.

Acknowledgments: The authors would like to acknowledge support of this work by the personnel of the Laboratory of Applied Thermodynamics for performing the experimental campaign of the vehicle. The research work was supported by the Hellenic Foundation for Research and Innovation (HFRI) under the HFRI Ph.D. Fellowship grant (Fellowship Numbers: 607 and 6653).

Conflicts of Interest: The authors declare no conflict of interest. The funders had no role in the design of the study; in the collection, analyses, or interpretation of data; in the writing of the manuscript, or in the decision to publish the results.

Abbreviations

B	Battery
CO_2	Carbon dioxide
CFC	Corrected fuel consumption
DP	Dynamic programming
EBA	ECMS based type A
EBB	ECMS based type B
EBC	ECMS based type C
ECMS	Equivalent consumption minimization strategy
EM	Electric motor
EMS	Energy management system
EU	European union
FC	Fuel consumption
GB	Gearbox
GHG	Greenhouse gases
HEVs	Hybrid electric vehicles
ICE	Internal combustion engine
LB	Learning based
NVH	Noise, vibration and harshness
OB	Optimization based
PHEVs	Plug in hybrid electric vehicles
PMP	Pontryagin's minimum principle
R	Correlation coefficient
RB	Rule based
RPA	Relative positive acceleration
SFC	Simulated fuel consumption
SOC	State of charge
WLTC	Worldwide harmonized Light vehicles Test Cycle

References

1. Ehsani, M.; Gao, Y.; Gay, S.E.; Emadi, A. *Modern Electric, Hybrid Electric, and Fuel Cell Vehicles, Fundamentals, Theory, and Design*; CRC Press: Boca Raton, FL, USA, 2004; ISBN 0849331544.
2. Küng, L.; Bütler, T.; Georges, G.; Boulouchos, K. How much energy does a car need on the road? *Appl. Energy* **2019**, *256*, 113948. [CrossRef]
3. Fontaras, G.; Valverde, V.; Arcidiacono, V.; Tsiakmakis, S.; Anagnostopoulos, K.; Komnos, D.; Pavlovic, J.; Ciuffo, B. The development and validation of a vehicle simulator for the introduction of Worldwide Harmonized test protocol in the European light duty vehicle CO_2 certification process. *Appl. Energy* **2018**, *226*, 784–796. [CrossRef]
4. Dipierro, G.; Millo, F.; Tansini, A.; Fontaras, G.; Scassa, M. An Integrated Experimental and Numerical Methodology for Plug-In Hybrid Electric Vehicle 0D Modelling. *SAE Tech. Pap.* **2019**, *24*, 72. [CrossRef]
5. CO_2 Emission Performance Standards for Cars and Vans. Available online: https://ec.europa.eu/clima/policies/transport/vehicles/regulation_en (accessed on 5 May 2022).
6. Mi, C.; Masrur, M.A.; Gao, D.W. *Hybrid Electric Vehicles: Principles and Applications with Practical Perspectives*; John Wiley & Sons, Ltd.: Hoboken, NJ, USA, 2011; ISBN 9780470747735. [CrossRef]
7. Mercedes-Benz S580 e L Saloon. Available online: https://www.mercedes-benz.co.uk/passengercars/mercedes-benz-cars/models/s-class/saloon-wv223/plugin-hybrid/key-stats.module.html (accessed on 5 May 2022).
8. Onori, S.; Serrao, L.; Rizzoni, G. *Hybrid Electric Vehicles_Energy Management Strategies*; Springer: London, UK, 2016; ISBN 9781447167792.
9. Guzzella, L.; Sciarretta, A. *Vehicle Propulson Systems*; Springer: Berlin/Heidelberg, Germany, 2013; ISBN 978-3-642-35912-5.
10. Wu, G.; Xuewei, Q.; Barth, M.; Boriboonsomsin, K. *Advanced Energy Management Strategy Development for Plug—In Hybrid Electric Vehicles*; A Research Report from the National Center for Sus; National Center for Sustainable Transportation: Davis, CA, USA, 2016.
11. Sciarretta, A.; Guzzella, L. Control of hybrid electric vehicles. *IEEE Control Syst.* **2007**, *27*, 60–70. [CrossRef]
12. Musardo, C.; Rizzoni, G.; Staccia, B. A-ECMS: An Adaptive Algorithm for Hybrid Electric Vehicle Energy Management. In Proceedings of the 44th IEEE Conference on Decision and Control, Seville, Spain, 12–15 December 2005; pp. 1816–1823.
13. Sun, C.; Sun, F.; He, H. Investigating adaptive-ECMS with velocity forecast ability for hybrid electric vehicles. *Appl. Energy* **2017**, *185*, 1644–1653. [CrossRef]
14. Lodaya, D.; Zeman, J.; Okarmus, M.; Mohon, S.; Keller, P.; Shutty, J.; Kondipati, N. Optimization of Fuel Economy Using Optimal Controls on Regulatory and Real-World Driving Cycles. *SAE Int. J. Adv. Curr. Pract. Mobil.* **2020**, *2*, 1705–1716.

15. Serrao, L.; Onori, S.; Rizzoni, G. A Comparative Analysis of Energy Management Strategies for Hybrid Electric Vehicles. *J. Dyn. Syst. Meas. Control* **2011**, *133*, 031012. [CrossRef]
16. Kim, N.; Cha, S.; Peng, H. Optimal control of hybrid electric vehicles based on Pontryagin's minimum principle. *IEEE Trans. Control Syst. Technol.* **2011**, *19*, 1279–1287. [CrossRef]
17. Xu, B.; Malmir, F.; Rathod, D.; Filipi, Z. Real-Time reinforcement learning optimized energy management for a 48V mild hybrid electric vehicle. *SAE Tech. Pap.* **2019**, *2019*, 1208. [CrossRef]
18. Hu, X.; Liu, T.; Qi, X.; Barth, M. Reinforcement Learning for Hybrid and Plug-In Hybrid Electric Vehicle Energy Management: Recent Advances and Prospects. *IEEE Ind. Electron. Mag.* **2019**, *13*, 16–25. [CrossRef]
19. Torreglosa, J.P.; Garcia-Triviño, P.; Vera, D.; López-García, D.A. Analyzing the improvements of energy management systems for hybrid electric vehicles using a systematic literature review: How far are these controls from rule-based controls used in commercial vehicles? *Appl. Sci.* **2020**, *10*, 8744. [CrossRef]
20. Wu, J.; Ruan, J.; Zhang, N.; Walker, P.D. An Optimized Real-Time Energy Management Strategy for the Power-Split Hybrid Electric Vehicles. *IEEE Trans. Control Syst. Technol.* **2019**, *27*, 1194–1202. [CrossRef]
21. Kim, N.; Rousseau, A.; Rask, E. Autonomie model validation with test data for 2010 Toyota Prius. *SAE Tech. Pap.* **2012**, *48*, 46. [CrossRef]
22. Hwang, H.Y.; Chen, J.S. Optimized fuel economy control of power-split hybrid electric vehicle with particle swarm optimization. *Energies* **2020**, *13*, 2278. [CrossRef]
23. Liu, W. *Hybrid Electric Vehicle System Modeling*; Wiley: Hoboken, NJ, USA, 2017; ISBN 9781119279327. [CrossRef]
24. Tsiakmakis, S.; Fontaras, G.; Cubito, C.; Pavlovic, J.; Anagnostopoulos, K.; Ciuffo, B. *From NEDC to WLTP: Effect on the Type-Approval CO$_2$ Emissions of Light-Duty Vehicles*; EUR 28724 EN; Publications Office of the European Union: Luxembourg, 2017; ISBN 978-92-79-71642-3. [CrossRef]
25. Franco, V. Evaluation and Improvement of Road Vehicle Pollutant Emission Factors Based on Instantaneous Emissions Data Processing. Ph.D. Thesis, Universitat Jaume, Castello, Spain, 2014; p. 234.
26. Doulgeris, S.; Tansini, A.; Dimaratos, A.; Fontaras, G.; Samaras, Z. Simulation-based assessment of the CO$_2$ emissions reduction potential from the implementation of mild-hybrid architectures on passenger cars to support the development of CO2MPAS. In Proceedings of the 23rd Transport and Air Pollution Conference, Thessaloniki, Greece, 15–17 May 2019; pp. 1–12.
27. Doulgeris, S.; Toumasatos, Z.; Prati, M.V.; Beatrice, C.; Samaras, Z. Assessment and design of real world driving cycles targeted to the calibration of vehicles with electrified powertrain. *Int. J. Engine Res.* **2021**, *22*, 3503–3518. [CrossRef]
28. Lee, W.; Kim, T.; Jeong, J.; Chung, J.; Kim, D.; Lee, B.; Kim, N. Control analysis of a real-world P2 hybrid electric vehicle based on test data. *Energies* **2020**, *13*, 4092. [CrossRef]
29. Sciarretta, A.; Serrao, L.; Dewangan, P.C.; Tona, P.; Bergshoeff, E.N.D.; Bordons, C.; Charmpa, L.; Elbert, P.; Eriksson, L.; Hofman, T.; et al. A control benchmark on the energy management of a plug-in hybrid electric vehicle. *Control Eng. Pract.* **2014**, *29*, 287–298. [CrossRef]
30. Li, L.; Yang, C.; Zhang, Y.; Zhang, L.; Song, J. Correctional DP-Based Energy Management Strategy of Plug-In Hybrid Electric Bus for City-Bus Route. *IEEE Trans. Veh. Technol.* **2015**, *64*, 2792–2803. [CrossRef]
31. Gong, Q.; Li, Y.; Peng, Z.R. Trip-based optimal power management of plug-in hybrid electric vehicles. *IEEE Trans. Veh. Technol.* **2008**, *57*, 3393–3401. [CrossRef]
32. Dynamometer Drive Schedules. Available online: https://www.epa.gov/vehicle-and-fuel-emissions-testing/dynamometer-drive-schedules (accessed on 5 May 2022).
33. Tanishita, M.; Kobayashi, T. Analysis of the Deviation Factors between the Actual and Test Fuel Economy. *Vehicles* **2021**, *3*, 162–170. [CrossRef]
34. Sundström, O.; Guzzella, L. A generic dynamic programming Matlab function. In Proceedings of the 2009 IEEE Control Applications, (CCA) & Intelligent Control, (ISIC), St. Petersburg, Russia, 8–10 July 2009; pp. 1625–1630. [CrossRef]
35. Hou, C.; Ouyang, M.; Xu, L.; Wang, H. Approximate Pontryagin's minimum principle applied to the energy management of plug-in hybrid electric vehicles. *Appl. Energy* **2014**, *115*, 174–189. [CrossRef]
36. Geng, W.; Lou, D.; Wang, C.; Zhang, T. A cascaded energy management optimization method of multimode power-split hybrid electric vehicles. *Energy* **2020**, *199*, 117224. [CrossRef]
37. Hao, L.; Wang, Y.; Bai, Y.; Zhou, Q. Energy management strategy on a parallel mild hybrid electric vehicle based on breadth first search algorithm. *Energy Convers. Manag.* **2021**, *243*, 114408. [CrossRef]

Article

Impact of Reactive Power from Public Electric Vehicle Stations on Transformer Aging and Active Energy Losses

Ana Pavlićević * and Saša Mujović

Faculty of Electrical Engineering, University of Montenegro, Džordža Vašingtona bb, 81000 Podgorica, Montenegro
* Correspondence: anaz@ucg.ac.me; Tel.: +382-67559535

Abstract: Climate change at the global level has accelerated the energy transition around the world. With the aim of reducing CO_2 emissions, the paradigm of using electric vehicles (EVs) has been globally accepted. The impact of EVs and their integration into the energy system is vital for accepting the increasing number of EVs. Considering the way the modern energy system functions, the role of EVs in the system may vary. A methodology for analyzing the impact of reactive power from public electric vehicle charging stations (EVCSs) on two main indicators of the distribution system is proposed as follows: globally, referring to active power losses, and locally, referring to transformer aging. This paper indicates that there is an optimal value of reactive power coming from EV chargers at EVCSs by which active energy losses and transformer aging are reduced. The proposed methodology is based on relevant models for calculating power flows and transformer aging and appropriately takes into consideration the stochastic nature of EV charging demand.

Keywords: electric vehicles; electric vehicle charging stations; active power losses; power distribution transformers; thermal aging; reactive power

Citation: Pavlićević, A.; Mujović, S. Impact of Reactive Power from Public Electric Vehicle Stations on Transformer Aging and Active Energy Losses. *Energies* **2022**, *15*, 7085. https://doi.org/10.3390/en15197085

Academic Editor: Wojciech Cieslik

Received: 16 August 2022
Accepted: 21 September 2022
Published: 27 September 2022

Publisher's Note: MDPI stays neutral with regard to jurisdictional claims in published maps and institutional affiliations.

1. Introduction

In July 2009, the leaders of the European Union and the G8 announced the aim to reduce greenhouse gas emissions to at least 80% below 1990s' levels by 2050. The automotive sector has a very important role to play in achieving this ambitious goal [1].

One of the best ways to promote the use of electric vehicles (EVs) is to increase the number of available public electric vehicle charging stations (EVCSs). Connecting a large EV fleet in a small area faces a number of challenges such as: overheating distribution transformers, increasing power losses, voltage instability and harmonic distortion [2,3]. Therefore, there is a need for modeling and mitigating the impacts of EVCSs on the distribution grid. Connecting EVCSs to the distribution system has a significant impact on the exploitation of the existing energy infrastructure. This especially refers to increasing the total load, peak load and changing the load profile of the distribution transformer where EVCSs are connected. Taking into account a large number of distribution transformers as well as the fact that in most cases they do not operate in parallel in low voltage networks, the prevention of failures or possible losses of life (LOL) is of great importance. Since charging EVs at EVCSs increases load on distribution transformers, it could cause overloading and therefore overheating of distribution transformers. Apart from overheating, there are other sources of transformer aging such as: fault currents caused by short-circuits or inrush currents, overvoltage caused by lightning and switching impulses, contamination of the oil that could shorten their lifetime and cause premature failure [4,5]. Bearing in mind that the subject of this paper is the influence of EVCS on the power system and on transformer aging under normal operating conditions, an overview of the literature related to the aging of transformers under such conditions is given.

Researches have studied the impact of distributed generation (DG) and energy storage (ES) on transformer aging [6,7]. It has been concluded that with a high penetration of

EVs, and no support from photovoltaic (PV) and ES, the transformers will dramatically age. Furthermore, researchers have dealt with the influence of the charging method on transformers' aging [8–13]. The main idea is to find an optimal charging scheme to minimize the impact of EVs' charging demand on distribution transformers. The [8] proposed model considers time-of-use rates in order to minimize energy consumption costs and avoid transformer overloading and LOL based on load and meteorological data. The proposed smart charging scheme together with PV with ES can prevent transformer overloading and LOL. PV generation can reduce the energy purchased from the grid, while ES can assist during peak hours. Three smart ways of charging (central, decentralized and hierarchical) as well as dump charging have been analysed [9]. Based on the obtained results, it has been concluded that hierarchical charging strategies are the most desirable in terms of LOL, while the centralized charging strategy (valley filling) has shown the greatest LOL of transformers. In one paper [10], a simple method is used to coordinate the charging process of PEVs to avoid the overloading of transformers by using the fact that idle time represents a significant percentage of EV parking time. In the paper, [11] reducing the impact on transformer aging by implementing the capability of delivering a variable charging rate has been proposed. Two charging algorithms have been used as strategic ways for reducing the overload of distribution transformers. The results suggest that for both uncontrolled and controlled charging, the load on the transformer will be reduced. Since the current commercial, residential charging technology is categorized, infrastructure changes are necessary in order to support this idea. Another paper [12] proposes a temperature-based smart charging strategy that reduces transformer aging without substantially reducing the frequency by which EVs obtain a full charge. These reductions are substantially larger at hot climate locations compared to cool climate ones. The results indicate that simple smart charging schemes, such as delaying charging until after midnight can actually increase, rather than decrease, transformer aging. The authors of [13] analyze the impacts of a price incentive-based demand response on neighburhood distribution transformer aging. The results indicate that the integration of EVs in residential premises may indeed cause accelerated aging of distribution transformers, while the need to investigate the effectiveness of dynamic pricing mechanisms is evident. In [14], the number of EVs has been optimized by calculating additional steady-state hottest-spot temperature rises by ensuring that distribution reliability requirements are met.

In order to reduce transformer aging, the reactive power potential from on-board EV chargers (OBCs) at EVCSs is proposed in this paper. OBCs in modern EVs are bidirectional, which allows them to work in all four quadrants [15]. Using reactive power support, as a part of the vehicle-to-grid concept, is already familiar and has been investigated as, e.g., part of on-line control for improving the voltage stability of a microgrid with PV, wind, battery storage, residential and commercial building feeders [3], the load control of active and reactive power [16], the compensation of reactive load of wind turbines near the station [17] and the revenue potential of providing reactive power service [18]. In this paper, the respective value of reactive power is obtained by small oversizing charging converters, which does not require an upgrade of the charging infrastructure. In relation to [15], where the influence of the oversizing of converters on voltage deviations, peak load and grid losses in the network has been analyzed, this paper's analysis is extended to include transformer aging. Besides the impact on transformer aging, also the influence of reactive power potential of EVCSs is analyzed with the aim of reducing active energy losses without additional devices and facilities. Locations of EVCSs are chosen from the aspect of increasing the efficiency of the network, which refers to reducing total active energy losses.

The reduction of losses in the power system in the presence of EVCSs is achieved in different ways. A large number of papers can be found in the literature that treat the problem of reducing energy losses in the presence of electric vehicles, e.g., optimal operation of vehicles with distributed generators [19–23], optimal reconfiguration of the distribution network [24–28], optimal operation of the battery storage, [29,30], optimal charging strategies [31–34] and optimal location and size of reactive power compensators [35,36].

The contribution of this paper is related to the impact analysis of the influence of reactive power on active power losses and transformer aging. In other words, the paper proposes a methodology that allows the determination of the optimal injection of reactive power of the EVCSs with the aim of reducing transformer aging and reducing active power losses. The proposed method is based on the well-known model for calculating transformers' aging, which is described in detail in [37]. It is applicable to an arbitrary network with an arbitrary number of EVCSs. Considering the stochastic nature of EV charging demand, the presented paper clearly indicates the benefits and limitations of reactive power from OBCs at EVCSs to energy losses reduction and transformers aging.

Numerical results, obtained in MATLAB and on the IEEE 33 bus radial distribution network, indicate that it is possible to achieve the optimal value of injecting reactive power from EV vehicles in order to enable reductions of transformers' aging and a reduction in energy losses in relation to the case where there is no reactive power injection or when reactive power injection is the maximum possible. This analysis may serve focus on local distributions in order to prevent shortening transformers' working life. Furthermore, the presented results may be useful for the analysis of global distribution system parameters such as energy losses and node voltages.

The rest of the paper is organized as follows. Section 2 is dedicated to the system components modeling. In Section 3, a description of methodology is given, while in the final Section 4, the most important conclusions and directions for further research have been stated.

2. Modeling System Components

2.1. Load Modeling

There is certain number of load models in the literature that can be used for various analyses. In addition, in recent times, new nodal classifications have been made and their characteristics have been reviewed [38]. Considering that a steady state analysis is performed in this paper, basic information about node loads have been used in this paper. Namely, loads in the system are modeled as PQ nodes. On the basis of the type of loads connected to the nodes busses, three types have been adopted: residential, commercial and industrial busses, as seen in Table 1 [19]. There are 18 residential busses, 5 commercial buses and 9 industrial buses. Load changing over time has been taken into account by introducing a load scaling factor. Load scaling factor changes over time as a percentage of nominal load, which varies depending on a load type, Figure 1, [39].

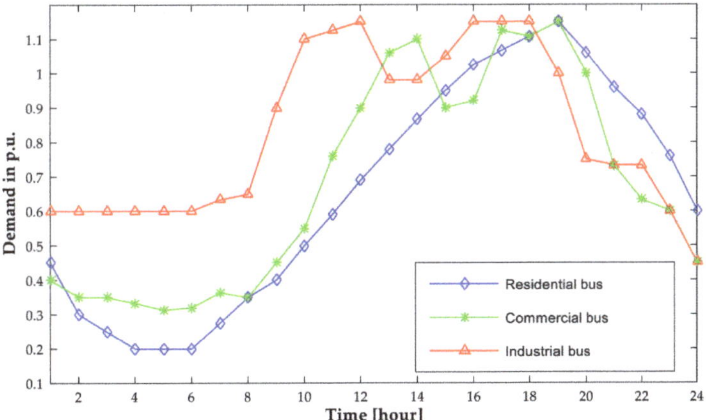

Figure 1. Typical buses load curve, [39].

Table 1. Grouping of buses data, [19].

Bus Type	Bus Numbers
Residential	2, 3, 5, 6, 7, 8, 9, 10, 13, 14, 15, 16, 17, 20, 21, 23, 24, 25
Commercial	4, 11, 12, 18, 19
Industrial buses	22, 26, 27, 28, 29, 30, 31, 32, 33

2.2. Public EVCS Modeling

According to the way EVs are connected to the electric grid, there are two types of charging strategy: conductive EV charging and inductive EV charging. In conductive charging, there is a physical connection between EV and the electrical grid. In inductive charging, also known as wireless power transfer [40], there is no galvanic connection between the vehicle and the electrical grid. Since conductive charging is the most common method of charging, this type of charging has been analyzed. There are two types of conductive charging in EVCSs: AC charging and DC charging with different power levels: Level 1, Level 2 and Level 3 charging. In this paper, the AC Level 2 type of charging station has been analyzed because of its simple infrastructure, low price and availability for the most commercial types of charging in public places.

There is a certain number of models for grid support by EVCSs [41]. Bearing in mind that vehicles are mostly stationary during the day, their role may be different. The literature recognizes that EVCSs could most effectively serve in the following ways: to ensure a fast frequency response [42,43] for shifting electricity [44,45], improve power quality [46] and as energy storage [44].

Significant work has been done towards minimizing grid losses due to PHEV charging through efficient charging algorithms with different objective functions [47–49]. All mentioned papers, in one way or another, with different stimulation techniques, have an influence on drivers' behavior. Nevertheless, this paper adopts an approach that does not impair the comfort of EV users, which means that two cases have been observed, which are as follows: active power is consumed during the charging of EVs; while in the second case, while charging with active power, vehicles inject reactive power.

2.2.1. Operation Region of EV Chargers

In order to obtain reactive power and not to jeopardize EVs drivers' habits, the concept of small oversizing of 5.3% of on-board chargers ratings has been used, [15]. With this, a significant amount of reactive power which is 32.9% of active power is obtained. What is important is that this does not require an upgrade of the charging infrastructure, and charging active power is the same in both cases. The operating region of on-board charger charging is region IV, Figure 2. It can be seen that with small oversizing of an EV charger, in Figure 2 shown as a larger circle, with the same active charging power, a representative value of reactive power is obtained.

The charging operation in the second and third quadrants of the P-Q diagram was not considered, since the management of the active power of the chargers affects the convenience of the consumers themselves as well as battery life. Furthermore, the operation of chargers in the first quadrant, wherein EV can also consume reactive power, is not suitable for the analyzed network and loads.

During the charging of EVs, we have analyzed the influence of reactive power from equal to zero to the maximum available. The national network operator in our country does not yet support the injection of active power or the injection and absorption of reactive power by the EVCSs. The operation of chargers in all four or two quadrants is discussed in the literature at the level of conceptual solutions for which there are numerical or real experiments. The results obtained both in this work and in the literature represent guidelines for changing the network requirements of new electricity consumers/producers. This paper demonstrates the benefits of reactive power from EV charger compared to the current regulation, which stipulates that they only consume active power. Finally, it is important to note that the network codes for the LV network, after looking at the possible positive impact of the operation of network inverters of photovoltanic power plants, have been changed in Germany. For example, in Germany, LV grid-connected PV installations rated above 3.68 kVA have to follow a specified PF droop curve as a function of their instantaneous power output [15].

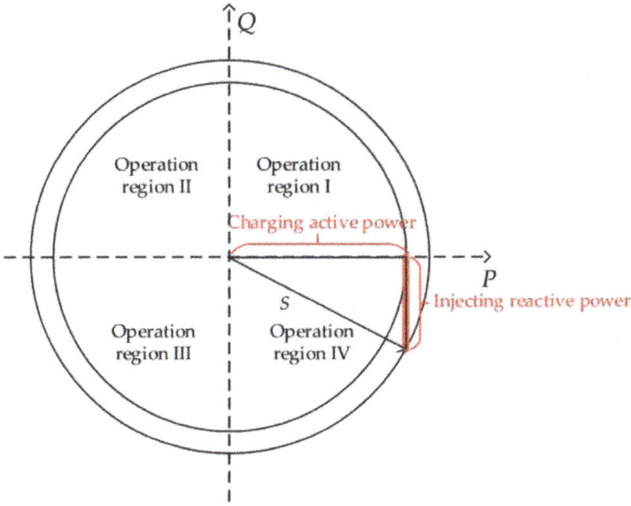

Figure 2. Operation region of EVs chargers.

Furthermore, it is very important to regulate the interaction between EVs the and distribution system. The design of the battery charger will be crucial in this effort to effectively control the power flows. One of the important requirements of an EV charger is the amount of current distortion that it draws from the grid [50]. Parameters such as total harmonic distortion (THD) and total demand distortion (TDD) can be used to evaluate the harmonic content of the charger. Additionally, the chargers should meet the individual harmonic limits as well [50–52]. The tasks of OBC system controller is to follow the charging power and reactive power commands controlled by the grid operator. In [50], the proposed design of OBC has shown analytically and experimentally that chargers are able to symmetrically operate at all four quadrants of the power plane. Furthermore, grid current requirements such as that the ac-dc converter input current THD should be limited to 5% have been satisfied.

Figure 3 shows the proposed application of on-board EV chargers used in this paper. The chargers form EVCS injecting reactive power at the point of common coupling (PCC) and decreases transformer overloading. In the case of the AC charging station, electric vehicle supply equipment (EVSE) serves for monitoring, management and communication with a vehicle during charging, while energy conversion from AC to DC power is suitable for charging battery performed via OBC.

In this paper, it is assumed that the power factor management of EVCSs is performed based on the previous day's consumption forecast of the analyzed network and EVCS demand. After processing the necessary data, the distribution network control centre sends a signal for the value of the power factor of the station. The power factor is constant during the 24 h period.

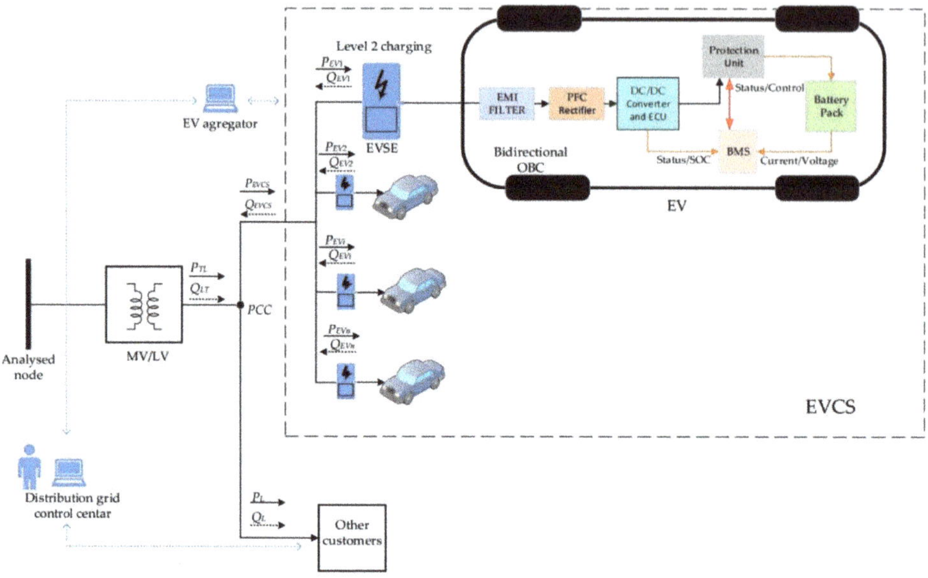

Figure 3. Proposed reactive power support diagram using OBC from EVCS.

2.2.2. Mobility Stochastic Behavior Model

There is a large number of papers dealing with the optimal structure of EVCSs, which refers to number of charging spots, types of charging and technology of chargers themselves [53–55]. In this paper, the charging demand of EVCSs is obtained using the number of EVs that arrive at the EVCS during 24 h as an input. Furthermore, the random nature of EV arrival time, electrical efficiency, battery capacity and daily miles driven are respected in order to obtain EVCS demand. In this paper, the stochastic nature of EV load is modeled using probability density functions (PDF) for two parameters: plug-in time and miles traveled before last charge.

- Behavior of start charging time

A PDF of vehicles' arrival time is obtained from [56], where the results of a large number of measurements from public charging stations have been analyzed. Based on actual data, the multimodality of distribution has been described using the Beta Mixture Model (BMM), which has proved to be appropriate for analysing EV data, [56].

A corresponding density function from the BMM model is obtained by summing up the beta distributions of different parameters and weight factors. The beta distribution of probability density is defined by the following equation:

$$\text{Beta}(x \mid a, b) = \frac{\Gamma(a+b)}{\Gamma(a)\Gamma(b)} x^{a-1} x^{b-1},\tag{1}$$

where $\Gamma(.)$ is gamma function, a and b are parameters which define the shape of distribution and x is a parameter from the interval between [0, 1].

The BMM is represented by the following sum:

$$\text{BMM}(x) = \sum_{m=1}^{M} (w_n \text{Beta}(x \mid a_m, b_m))\tag{2}$$

wherein the sum of weights are equal to one:

$$\sum_{m=1}^{M} w_n = 1\tag{3}$$

Obtained PDFs for working days and weekends plug in times are presented in Figure 4, while parameters of BMM are given in Table 2. The values in Table 2 are rounded to two decimal places [56].

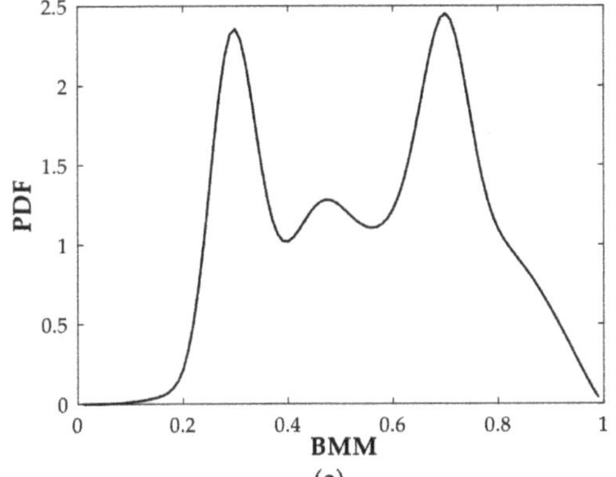

(a)

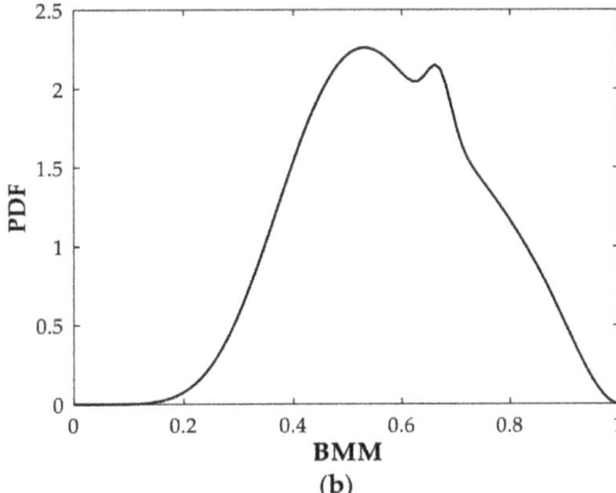

(b)

Figure 4. PDF for plug in time for: (**a**) Working day; (**b**) Weekend.

Table 2. Parameters for BMM model.

Bus Type	Working Day	Weekend
Number of modes	4	3
Parameter a	32.80 76.42 4.13 36.12	300.71 8.69 6.22
Parameter b	76.76 33.40 2.34 42.65	150.08 9.12 3.05
Weights	0.23 0.14 0.56 0.08	0.02 0.48 0.50

- Behavior of distance travelled and initial state of charge

In order to determine the energy required to charge vehicles, it is necessary to predict the level of the charge of a battery (SOC) of a single EV. An important factor for this is to determine daily miles traveled. The lognormal distribution of the distance travelled function has been shown to be suitable for describing the mileage distribution functions, and is represented by way of [6]:

$$g(d, \mu, \sigma) = \frac{1}{d \cdot \sqrt{2\pi\sigma^2}} e^{-\frac{(\ln d - \mu)^2}{2\sigma^2}}, \ d > 0 \tag{4}$$

The distribution parameters depend on a driver's habits in the area being analyzed. Statistics on the conventional vehicles are often taken into account due to insufficient data of EVs. This paper uses statistical data that 50% of drivers drive less than 25 miles during the day, and 80% of drivers travel less than 40 miles [57]. The corresponding lognormal distribution of daily traveled miles and coresponding CDF are presented at Figures 5 and 6, with mean value $\mu = 3.37$ and standard deviation $\sigma = 0.5$ [58].

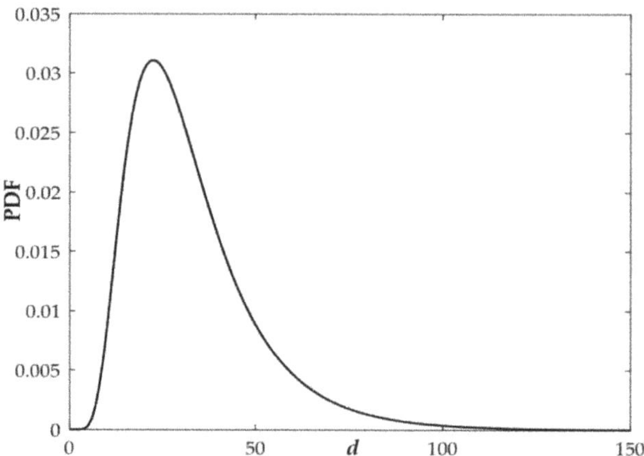

Figure 5. PDF of daily distance traveled of EV.

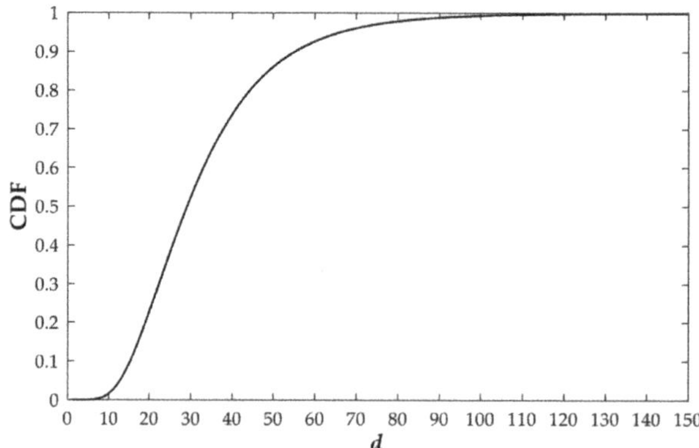

Figure 6. CDF of daily distance traveled of an EV.

Now, the initial state of the charge can be determined as shown below [6]:

$$SOC_i(x) = \left(1 - \frac{E_{cons} \cdot d}{C_b}\right) \cdot 100$$

(5)

where C_b battery capacity in kWh, E_{cons} is energy consumtion in kWh per miles and d estimated daily distance traveled in miles.

2.2.3. Charging Power Model

Considering the latest statistics of dominating shares of EV types, [59,60] all EVs charged in this study have 71% probability of being a battery electric vehicle (BEV) and 29% probability of being a plug-in hybrid electric vehicle (PHEVs).

Electric car economy changes all the time, depending on things such as road gradient and car speed. In this paper, electrical efficiency and battery capacity are taken as random numbers between values given in Table 3 depending on the type of EV.

Table 3. Parameters of BMM model.

EV Types	BEV	PHEV
Number of vehicles [%]	71	29
Battery capacity [kWh]	40–80	10–16
Electrical efficiency [kWh/mile]	0.25–0.40	0.25–0.40

In order to sustain a longevity, it is known that whole battery's energy capacity is usually not fully utilized. It has been found that between 78–95% of the full batteries capacity is used for different EV models [60]. The required value of SOC_{req} for the user is taken to be 95% [6,61,62]:

$$E_{req} = \frac{\left(SOC_{req} - SOC_i\right) \cdot C_b}{\eta \cdot 100} \tag{6}$$

where η is charging efficiency 0.95. The duration of EV charging depends on the required energy and charging power of EV and is equal to:

$$T_{ch} = \frac{E_{req}}{P_{EV}}, P_{EV} = \begin{cases} 3.4\text{kW, PHEV} \\ 7.2\text{kW, BEV} \end{cases} \tag{7}$$

2.3. Power Flow Analysis

Power flow or load flow calculations are very fundamental for all power system analyses. It is a very important tool for power systems planning, design, operations, maintenance, optimization and control. At the planing stage, load flow analysis is used to determine: the location of distributed generators, the location of capacitors, economic scheduling, power quality improvements, network reconfiguration, power systems optimization and other applications. Furthermore, in order to plan future growing load demands, power flow analysis is necessary. At the operation stage, it is run to explore system stability and to improve efficiency. Usage of power flow analysis enables that maintenance plans can proceed, without violating power system security. Load flow studies are performed for the determination of the steady state operating condition of a power system. Input parameters of power flow analysis of distribution network are network topology and network parameters, as well as load and generator models. Power flow calculations are performed by iterative methods. The most common power flow calculation methods are Gauss–Seidel Method, Newton–Raphson Method and Fast Decoupled Method. These methods are not appropriate for distribution systems due to their special characteristics such as low line X/R ratios, unbalanced load, radial or weakly meshed network structure and large number of nodes, etc. These features make distribution systems power flow computation different to analyze, as compared to the transmission systems. There is a number of methods proposed in the literature for the solution of power flow problem in radial distribution networks [63]. Back–forward sweeping (BFS) iterative method is suitable for calculations of radial distribution system load flow, which is analyzed in this paper [64]. The algorithm is very robust and numerically efficient for convergence. It is applicable for a wide variation of distribution networks. The algorithm begins with an initial solution for all node voltages and performs basic steps until a convergence criterion is satisfied. In the backward step, currents of each branch are calculated. Bus voltages are updated in a forward sweep starting from the first branch and moving towards end branches. After the convergence criterion has been satisfied, node voltages, currents in branches and active and reactive power losses are determined. This iterative method is implemented in Matlab for calculating power losses in radial IEEE 33 bus network. Different load types together with EVCS demands are implemented in this model. In this paper, power flow analysis has been used from the planning aspect, as a part of the optimization process for determinating optimal positions for EVCSs in an analyzed network. Additionally, it is used for calculating power losses for different EVCS demands level.

3. Methodology

3.1. Description

A flowchart of the proposed methodology is shown in Figure 7. The methodology includes both global and local solution approaches. The global one refers to the modeling and analysis of the power network and the determination of active power losses in the network, and the local one to the determination of the aging of the selected distribution transformer. The proposed algorithm consists of four steps. In the first step, network data and nodal loads are entered. Additionally, active and reactive nodal power are defined depending on a type of load (industrial, commercial or residential) according to Figure 1. Then, in this step, the number of stations are entered, as well as the data needed to determine the EVCS demand. The stochastic nature of EV charging is taken into account and explained in detail in Section 2. The second step of the proposed methodology includes the position of EVCSs in the power system. More precisely, in this step, a matrix that includes the possible positions of EVCSs is formed. In this step, the criterion that only one station can be placed in each part of the network has been met. The proposed methodology takes into account the fact that it is possible to divide the network into an arbitrary number of parts, and in each part of the network an arbitrary number of stations can be found. The third step represents the calculation of the electrical parameters of the network. In this step, the method for calculating power flows according to the characteristics of the network is implemented. In the fourth step, the locations of EVCSs are selected based on the criterion of minimum network losses, defined by the following equation:

$$J = \min P_l = \sum_{i=1}^{B} P_{li} \tag{8}$$

where i, B, P_l and P_{li}, are the index, total number of branches, total active power and active power losses of branch i, respectively.

In this paper, optimization is achieved by searching the obtained results, so that the focus of paper is not the development and improvement of optimization methods. Nevertheless, it is important to note that in the last few years, different methods have been developed and applied to account for optimal power flows in the energy system or for different types of examples that comply with different limitations. Some of them are detailed and tested in papers [65–68]. In the fifth step, the final value of reactive power injection of the EVCSs is selected. Namely, in this step, by choosing the final value of the reactive power injected by EVCSs, the influence of the reactive power on the losses of active energy and on transformer aging through which the station is connected to the grid is affected. In this paper, the GRA method is used for determining the final value of reactive power.

After obtaining the best positions for EVCSs, off-grid managing of the EVCSs has been proposed. Based on the predicted daily diagram of EVCSs and other consumption, the station operator evaluates the optimal daily reactive power injection from EVs, i.e., power factor of on-board chargers of EVs in EVCSs. The proposed methodology assume that power factor of OBC is constant during the 24 h period. Detailed post-optimization analysis of the reactive power injection from EVs at EVCSs to energy losses and transformer aging has been performed.

It is important to emphasize that the performed post-optimization analysis represents an incremental contribution in relation to the paper [15], since it is extended to include transformer aging analyzis. Based on the performed analyses, the correlation of the impact of reactive power on energy losses and the impact of reactive power on transformer aging can be clearly seen.

The aim of this paper is to explore the possibility of the impact of reactive power from oversized OBCs on active power losses and on transformer aging. With regard to experimental confirmation of the obtained results, at this stage it was primarily necessary to computerize a model and verify the simultaneous influence of active energy losses and transformer aging. On the other hand, the impact of chargers on losses in the distribution system in real-time circumstances is very difficult. Furthermore, one of the bigger obstacles so far is the number of EVs, because in whole of Montenegro there are only 126 EVs [69] and these are vehicles that have chargers that are not pre-dimensional. Relatively small oversizing of the OBCs is the basis of the assumption of this paper.

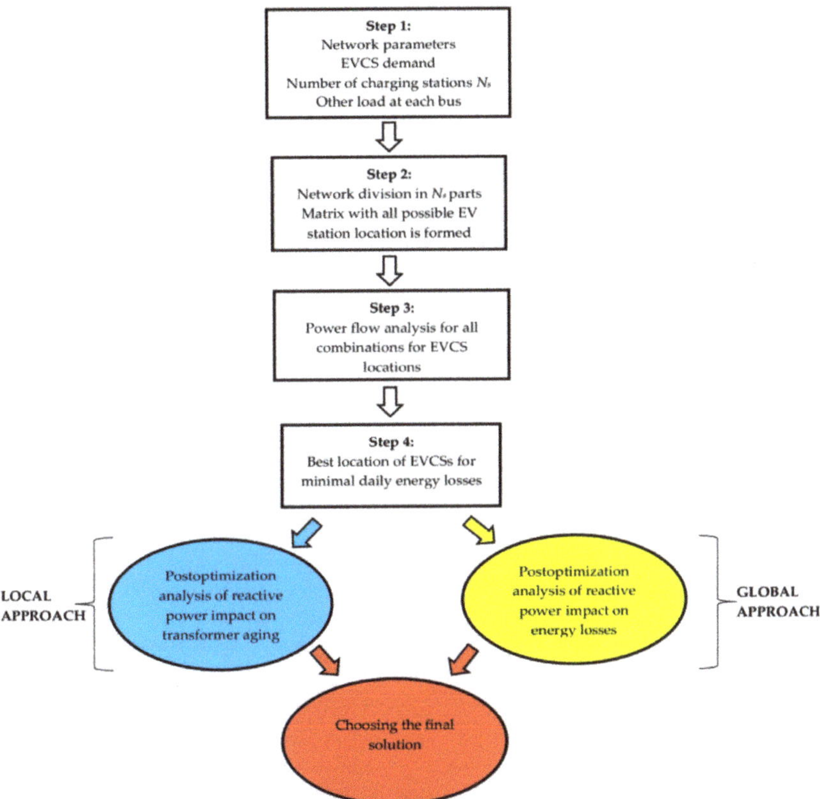

Figure 7. Flowchart of proposed methodology.

3.2. Sample Generation with Monte Carlo

The Monte Carlo (MC) statistical method was used to obtain the daily EVCS demand. The method requires that parameteres of the physical system, in this case EVCS demand, are described as PDFs.

When these functions are known, the MC simulation continues with a random selection of values from functions. For this purpose, the inverse transformation method was used, which states that any probability distribution can be obtained from a uniform probability distribution, if an inverse cumulative probability distribution can be determined [70]. A flowchart of the generation of a EVCS demand profile is given in Figure 8.

To terminate the MC process criteria number of MC iterations is used. The process was set to repeat until reaching 5000 iterations.

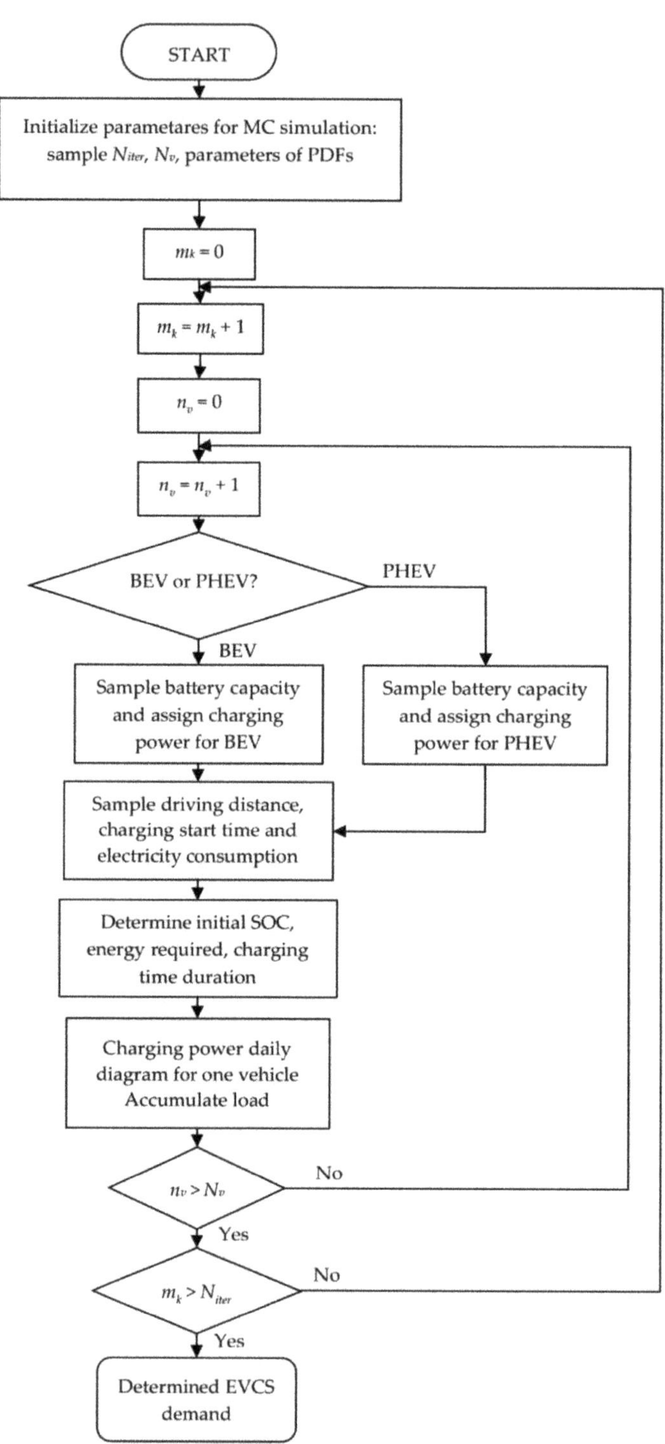

Figure 8. Flowchart for EVCS demand.

3.3. Transformer Aging Model

The insulation system of mineral oil transformers is composed of thermally upgraded oil-impregnated paper. Cellulose deterioration is influenced by hydrolysis, oxidation and pyrolysis which are the consequences of water, oxygen and heat, respectively. Taking into consideration that exposure to moisture and oxygen in transformers are generally reduced, the most significant determining factor to insulation deterioration is the heat. Transformer heating is caused primarily by energy losses. The majority of losses are located in magnetic core (no-load losses) and the windings (load losses). No-load are hysteresis and eddy current losses, while winding losses are primarily due to DC losses and stray load loss due to the eddy currents induced in other structural parts of the transformer. All these losses cause heating in the corresponding parts of the transformer. The most critical part for the transformer isolation system is placed where temperature has maximum value, which is known as hot spot temperature [4].

According to the loading guides, IEEE guide for loading mineral-oil-immersed transformers [37], the hot-spot temperature in a transformer winding consists of three components:

$$\theta_H = \theta_A + \Delta\theta_{TO} + \Delta\theta_H \tag{9}$$

where θ_H, is the winding hottest-spot temperature, °C, θ_A, is the average ambient temperature during the load cycle to be studied, °C, $\Delta\theta_{TO}$ is the top-oil rise over ambient temperature, °C, and $\Delta\theta_H$ is the winding hottest-spot rise over top-oil temperature, °C.

As is proposed in the IEEE guide, θ_A is constant and equals to 30 °C. The top-oil rise over ambient temperature is given in the following equation:

$$\Delta\theta_{TO} = (\Delta\theta_{TO,u} - \Delta\theta_{TO,i})\cdot\left(1 - e^{-t/\tau_{TO}}\right) + \Delta\theta_{TO,i} \tag{10}$$

where τ_{TO} top-oil time is constant, $\Delta\theta_{TO,i}$ and $\Delta\theta_{TO,u}$ are the initial and ultimate top-oil rises over ambient temperature, respectively, which equals to:

$$\Delta\theta_{TO,i} = \Delta\theta_{TO,R}\cdot\left[\frac{(K_i^2\cdot R + 1)}{R + 1}\right]^n \tag{11}$$

$$\Delta\theta_{TO,u} = \Delta\theta_{TO,R}\cdot\left[\frac{(K_u^2\cdot R + 1)}{R + 1}\right]^n \tag{12}$$

In the previous equations, $\Delta\theta_{TO,R}$ is the top-oil temperature rise over ambient temperature at rated load, k_i and k_u represent the ratio of initial and ultimate load to rated load, per unit, n is an empirically derived exponent which depends on a cooling mode and describes effects of change in oil resistance to change in load, while R is the ratio of a load loss at a rated load to a no-load loss. The top-oil time constant at a rated kVA is given in the following equation. The time constant equals to:

$$\tau_{TO} = \tau_{TO,R}\frac{\left(\frac{\Delta\theta_{TO,u}}{\Delta\theta_{TO,r}}\right) - \left(\frac{\Delta\theta_{TO,i}}{\Delta\theta_{TO,r}}\right)}{\left(\frac{\Delta\theta_{TO,u}}{\Delta\theta_{TO,r}}\right)^{\frac{1}{n}} - \left(\frac{\Delta\theta_{TO,i}}{\Delta\theta_{TO,r}}\right)^{\frac{1}{n}}} \tag{13}$$

Transient winding hottest-spot temperature rise over top-oil temperature is equal to:

$$\Delta\theta_H = (\Delta\theta_{H,u} - \Delta\theta_{H,i})\cdot\left(1 - e^{-t/\tau_w}\right) + \Delta\theta_{H,i} \tag{14}$$

In Equation (14), $\Delta\theta_{H,i}$ and $\Delta\theta_{H,u}$ represent the initial and ultimate winding hottest-spot rises over top-oil temperature in °C, which equals to:

$$\Delta\theta_{H,i} = \Delta\theta_{H,R}\cdot K_i^{2m} \tag{15}$$

$$\Delta\theta_{H,u} = \Delta\theta_{H,R} \cdot K_u^{2m} \tag{16}$$

where $\Delta\theta_{H,R}$ is the winding hottest-spot rise over top-oil temperature at rated load. Parameter m describes changes in resistance and oil viscosity with changes in load.

Calculating the temperature of the hottest spot allows the determination of important parameters that describe transformers' aging, which are the accelerated transformer aging factor F_{AA} and LOL.

The aging acceleration factor for a given load and hottest spot temperature can be obtained as shown in Equation (17):

$$F_{AA} = e^{\left[\frac{15000}{383} - \frac{15000}{\theta_H + 273}\right]} \tag{17}$$

If $F_{AA} > 1$, the transformer is experiencing accelerated aging. According to [37], normal aging occurs at the reference hottest-spot temperature of 110 °C, where $F_{AA} = 1$. The equivalent aging factor F_{EQA} is further obtained as indicated below:

$$F_{EQA} = \frac{\sum_{j=1}^{N} F_{AA,j} \Delta t_j}{\sum_{n=j}^{N} \Delta t_j} \tag{18}$$

where j is the time intervals index, N is the total number of time intervals, Δt is the time interval, $F_{AA,j}$ is the aging acceleration factor for j-th time interval. The percentage of shortening the isolation life of a transformer is calculated as shown in the relation below:

$$\%LOL = \frac{F_{EQA} \cdot t \cdot 100}{\text{Normal insulation life}} \tag{19}$$

The value of the normal duration of insulation, [37], is taken to be 18,000 h or 20.55 years. Therefore, when the temperature of the hottest point is 110 °C for 24 h, the percentage of the daily shortening of the lifespan is equal to:

$$\%LOL = \frac{24 \cdot 100}{180,000} = 0.01333\% \tag{20}$$

3.4. Transformer Loading with EVCS

Adding EVCS demand at node n to the distribution transformer affects its loading, and therefore can cause further transformer aging. The available reactive power of inverters of EVs can be utilized for reactive power compensation in order to improve power loss reduction and prevent transformer aging. The total active load of the transformer at node k, P_{TL_k}, is equal to:

$$P_{TL_k} = P_{L_k} + P_{EVCS_k} \tag{21}$$

where P_{L_k} and P_{EVCS_k} re the sum of load without EVs and active power demand of EVCSs at node k. In case, during the charging of EVs, reactive power is injected, the total reactive load of transformer Q_{TL_n} is equal to:

$$Q_{TL_k} = Q_{L_k} - Q_{EVCS_k} \tag{22}$$

where Q_{L_k} represents the sum of the load without EVs and Q_{EVCS_i} is reactive power injected from EVCSs obtained from EV charging inverters at node k. Injecting reactive power to the transformer at node k has an influence on reducing the apparent power of the total load and therefore the loading and temperature of the transformer as well.

3.5. Grey Relational Analysis (GRA)

The selection of the final solution is done using Grey Relational Analysis (GRA). The GRA is a multiple-attribute decision-making method and it is widely applied in electric power systems [71–74]. GRA is performed in several steps. In the first step, n alternatives sequences with m criteria are formed. Y_i represents i-th alternative sequence, and value y_{ij} represents the value of attributes j of alternative i:

$$Y_i = \left(y_{i1}, y_{i2}, \ldots y_{ij}, \ldots, y_{in}\right) \tag{23}$$

The second step of the GRA procedure is normalization, which converts the data to values between [0,1]. The smaller the normalization used in this case, the better, since smaller values are prefered in the problem desribed in this paper. The comparability sequence is as follows:

$$X_i = \left(x_{i1}, x_{i2}, \ldots x_{ij}, \ldots, x_{in}\right) \tag{24}$$

is obtain using the equation below:

$$x_{ij} = \frac{max\left\{y_{ij}, i = 1, 2, \ldots, m\right\} - y_{ij,}}{max\left\{y_{ij}, i = 1, 2, \ldots, m\right\} - min\left\{y_{ij}, i = 1, 2, \ldots, m\right\}} \tag{25}$$

where $i = 1, 2, \ldots, m$, and $j = 1, 2, \ldots, n$.

Reference sequence X_0 is defined as $(x_{01}, x_{01}, \ldots, x_{01}, \ldots x_{01}) = (1, 1, \ldots, 1, \ldots 1)$. The aim is to find an alternative whose comparability sequence is closest to the reference sequence. For this purpose, the Grey Relational Coefficient (GRC) between x_{ij} and x_{0j}, is calculated using the equation:

$$\gamma\left(x_{0j}, x_{ij}\right) = \frac{\Delta_{min} + \zeta \cdot \Delta_{max}}{\Delta_{ij} + \zeta \cdot \Delta_{max}} \tag{26}$$

where $i = 1, 2, \ldots, m$, and $j = 1, 2, \ldots, n$.

The values of Δ_{ij}, Δ_{min}, Δ_{max}, ζ are defined as:

$$\Delta_{ij} = \left| x_{0j} - x_{ij} \right| \tag{27}$$

$$\Delta_{min} = Min\left\{\Delta_{ij}, i = 1, 2, \ldots, m; j = 1, 2, \ldots n\right\}, \tag{28}$$

$$\Delta_{max} = Max\left\{\Delta_{ij}, i = 1, 2, \ldots, m; j = 1, 2, \ldots n\right\}, \tag{29}$$

$$\zeta \epsilon [0, 1] \tag{30}$$

For this purpose, 0.5 is usually used as the distinguishing coefficient value ζ.

For the problem described in this paper, the total number of alternatives which are compared equals the number of optimal solutions obtained from Pareto, while the elements of the sequence are the values of two criteria: transformer daily aging and active power losses. Now, in order to determine the best solution, it is necessary to determine the Grey Relation Grade (GRG). GRG is calculated by using the following equation:

$$\Gamma\left(X_0, X_i\right) = \sum_{j=1}^{n} w_j \cdot \gamma\left(x_{oj}, x_{ij}\right) \tag{31}$$

where w_j represents the weight factor for the corresponding index. The alternative with the highest GRG would be the best choice.

4. Analysis and Results

For the purpose of the analysis, the IEEE medium voltage radial distribution network with 33 bus has been modeled, Figure 9. The base voltage of the network is 12.66 kV, and the base power is 100 MVA. Line parameters are taken from [75]. Daily active power demand, without public stations, is 62,556 kWh. Total daily active energy losses, obtained from the power flow analysis amount to 2924 kWh. The lowest value of voltage in the system occurs during the 18th hour at bus 18 with the value of 0.9016 p.u.

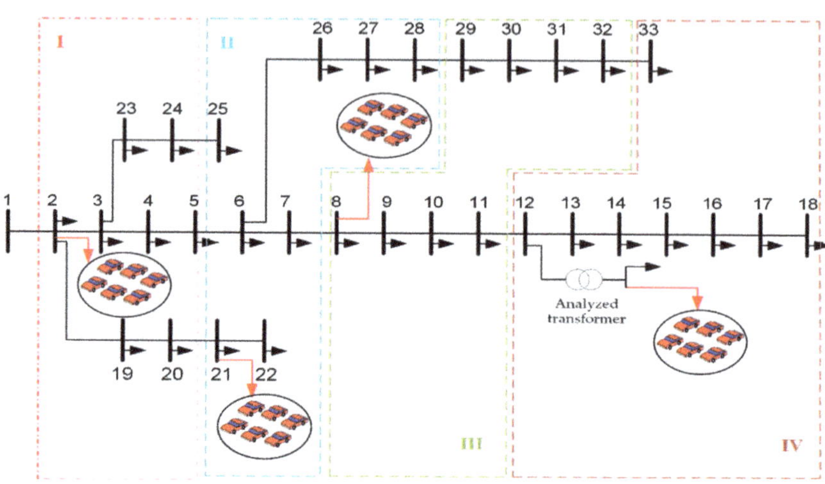

Figure 9. Single-line diagram of 33-bus test system.

4.1. EVCSs Location Impact of Active Energy Losses

The proposed methodology, Figure 7, enables us to perform an analysis of the impact of EVs on losses for an arbitrary number of EVCSs. In this paper, it is adopted that it is necessary to build four EVCSs. Based on the calculation of power flow, the best and worst cases in terms of energy losses are determined by the "brute force" algorithm. The results of the power flow calculations and algorithm are shown in Table 4 below. OBCs of EVs are assumed to operate with a unit power factor.

Table 4. Best and worst cases for EVCSs locations.

Parameters	Without PEV Load	With Public Charging Stations	
		Best Case	**Worst Case**
Location of charging station	-	2, 19, 20, 21	15, 16, 17, 18
Daily active power demand [kWh]	62,557	72,152	72,152
Daily active energy loss [kWh]	2924	3008	4486
Lowest voltage value in p.u.	0.9016 (18 h, bus 18)	0.9012 (18 h, bus 18)	0.8422 (18 h, bus 18)

Table 4 shows a comparison between the parameters with or without charging stations and a comparison between the best and worst cases for charging station positions when energy losses are concerned.

There is a clear difference in losses depending on the location of EVCSs. In the best case, the increase in daily losses is 3% and in the worst case, it is 53% compared to the case without charging stations. These results show a significant influence of location of charging stations on network losses.

4.2. Impact of Reactive Power from EVCSs on Energy Losses

From the previously obtained results shown in Table 4, it can be seen that the obtained locations of EVCSs are in a row, so it can be assumed that they will not satisfy the needs of EVs that are far from such station. For this purpose, the spatial division of the network into four parts is proposed, so that each separate part includes eight nodes, see Figure 9 [76].

The proposed methodology, Figure 7, also allows an arbitrary division of the network with an arbitrary number of nodes.

After the division of the network, it is necessary to determine one location in each zone of the network where a station for electric vehicles would be placed. In this part of the paper, the impact of the power factor of on-board chargers on active energy losses is analyzed.

Accordingly, two scenarios are considered: the first scenario is that EVs are charged with a unit power factor; the second scenario is that vehicles are charged with the same active power as in the first case, and the maximum reactive power capacity from on-board chargers is used. As is mentioned in Section 2, it is assumed that on-board charges are oversized by 5.3%. The results are summarized in Table 5 below.

Table 5. Locations for EVCSs and network parameters for the two scenarios.

Parameter	Scenario 1 No Reactive Power Support	Scenario 2 Reactive Power Support
Location of charging station	2, 21, 8, 12	2, 21, 8, 12
Daily active energy loss [kWh]	3419.34	3320.00
Lowest voltage value in p.u	0.8889	0.8920
(time and number of bus)	(18 h, bus 18)	(18 h, bus 18)

As can be seen in Table 5, for analyses of the nework and proposed network division, reactive power injection from on-board chargers does not have influence on EVCSs location. On the other hand, reactive power injection has an influence on daily acitive energy losess. Namely, it is shown that with a relatively small oversizing of the inverter, it is possible to reduce the loss of active power by about 3% without any interruption of an EV driver's habits or charging behavior.

Bearing in mind that voltage level of 12.33 kV, at which the R/X ratio is not large, produces the expected results obtained by these voltages. More precisely, the lowest voltage in the network in case the stations provide reactive power, is slightly higher compared to the scenario where the stations' power factor is 1.

4.3. Impact of Reactive Power from EVCS on Transformer Aging

The effect of reactive power injection on transformer aging is shown in Tables 6 and 7 below. The percentages in the tables refers to the gradual increase in power of EVCSs in relation to higher EV penetration. The results were obtained based on the formulas from Section 3.2. Two standardized distribution transformers 11/0.433 kV of different rated power, T_1 (200 kVA) and T_2 (250 kVA) have been compared in order to determine the appropriate power transformer for the analysed bus 12. Thermal parameters of transformers are taken from [33].

Table 6. LOL for different levels of reactive power for T_1 for different additional penetration level.

Q/Q_{max}	Working Day EV Additional Penetration [%]					Weekend EV Additional Penetration [%]				
	0	4	7	10	20	0	4	7	10	20
	LOL %					LOL %				
1	0.0082	0.0129	0.0183	0.0260	0.0831	0.0065	0.0102	0.0142	0.0198	0.0613
2/3	0.0076	0.0119	0.0168	0.0235	0.0728	0.0061	0.0094	0.0130	0.0180	0.0537
1/3	0.0080	0.0124	0.0174	0.0242	0.0740	0.0063	0.0097	0.0133	0.0184	0.0540
1/5	0.0084	0.0131	0.0182	0.0253	0.0771	0.0066	0.0102	0.0139	0.0192	0.0563
0	0.0094	0.0145	0.0202	0.0282	0.0856	0.0074	0.0112	0.0155	0.0213	0.0623

Table 7. LOL for different levels of reactive power for the T_2 for different additional penetration levels.

Q/Q_{max}	Working Day EV Additional Penetration [%]					Weekend EV Additional Penetration [%]				
	0	20	40	50	60	0	20	40	50	60
	LOL %					LOL %				
1	0	0.0024	0.0148	0.0379	0.0977	0	0.0020	0.0118	0.0294	0.0729
2/3	0	0.0021	0.0127	0.0316	0.0787	0	0.0018	0.0101	0.0244	0.0591
1/3	0	0.0022	0.0124	0.0305	0.0749	0	0.0018	0.0099	0.0234	0.0559
1/5	0	0.0022	0.0128	0.0313	0.0768	0	0.0019	0.0102	0.0241	0.0572
0	0	0.0024	0.0139	0.0341	0.0835	0	0.0020	0.0110	0.0261	0.0620

There is a large number of papers the optimal transformer sizing [77] and especially transformer sizing with the presence of electric vehicle charging [78]. In this paper, we only take these to values for rated power for purpose of analysis based on the overall load which is sum of EVCS load and the other load for the analysed bus 12.

In Table 6, LOL percentage for a workday and weekend are represented for different values of reactive power injection during the EV charging for T_1.

Since the daily energy consumption from EVCS during the weekend, and mainly concentrated in the central part of the day, is about 18% lower than during the weekday [56], the values for LOL are expected to be lower than during the weekday.

From Table 6, it can be concluded that different values of reactive power injection by the chargers also have different values of LOL. Namely, on the basis of Table 6, it is shown that, with an increased penetration of 4% per working day, LOL values change between 0.0129% and 0.0145 %. From Equation (20), it can be concluded that if the LOL value is greater than 0.01333%, additional aging of the transformer occurs. As we can see, with increased penetration of 4% per working day, with some injection level of reactive power, there is no additional aging. If reactive power injection is zero, than LOL = 0.0145% which means (0.0148 × 180,000/100 = 26.1 h) 26.1 h over the 24 h period. That means 2.1 h of additional ageing every day. It can also be noted that for values of reactive power $Q = 2/3Q_{max}$, LOL has a minimum value for each considered level of EV penetration. Within this case, with a higher degree level of reactive power, LOL is reduced compared to the case when there is no injection of reactive power from the vehicle charger. For this case, in Figure 10, the reactive load of the transformer per hour without reactive power from the vehicle (in blue color) is shown, and the reactive load of the transformer with the reactive power injection from EVs (in yellow) is shown as well. As it can be seen, there is a compensation of reactive power in the largest portion of the day. This has resulted in reduced apparent power for almost the entire duration of the day, Figure 10.

There is similar situation with the increase of 7% for the weekend, with adequate reactive power support there will be no aging during the weekend. If there is no reactive power support there will be 27.9 h daily aging over the 24 h period. That means 3.9 h of additional ageing every day.

As is presented in Table 6 for the example of the EV penetration increase of 7% at a EVCS, the daily aging value is 36.6 h (for $\cos\varphi = 1$) to 30.24 h (for $Q = 2/3Q_{max}$) representing a decrease of 17.37%. During the working day with 7% additional EV penetration, daily transformer LOL reaches 0.0202%. With reactive power injection of $2/3Q_{max}$, there is a decrease of daily aging from 36.36 h to 30.24 h, which is a 17% decrease. Even though it is significant reduction, hot spot temperature reaches 140 °C, which represents a critical value at which gas bubbles appear [35].

From Table 7, it can be seen that transformer T_2 with an increased penetration of 40% increases aging on the working day with 25 h aging over the 24 h period with no reactive power support. What is interesting is that LOL is higher at maximum reactive power support compared to where $\cos\varphi = 1$ and it is 26.64 h.

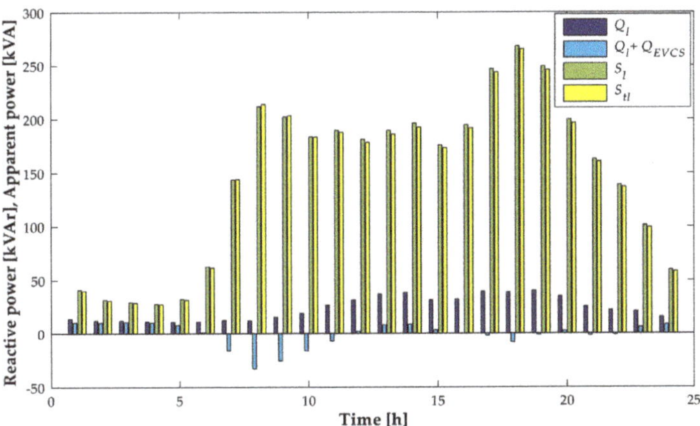

Figure 10. Reactive and apparent power loading of T_1 with (S_{tl}) and without reactive power (Q_{EVCS}) from EVCSs (S_l), with 4% increase in EVCS demand.

As is shown in Table 7, with an adequate value of reactive power, the value of LOL is less than 0.0133%. Furthermore, as in the previous case, there is an improvement in decreasing the aging, using reactive power from EVs. The change in daily aging of the transformer and daily energy losses in the function of reactive power injection are presented in Figures 11 and 12, for transformer T_2 with EV adiditional penetration of 40%. From the obtained characteristics, it can be seen that for some power injection there is aditional aging (above 24 h). So, it can be concluded that there is an optimal value of reactive power support that can have a positive influence on both transformer aging and energy losses.

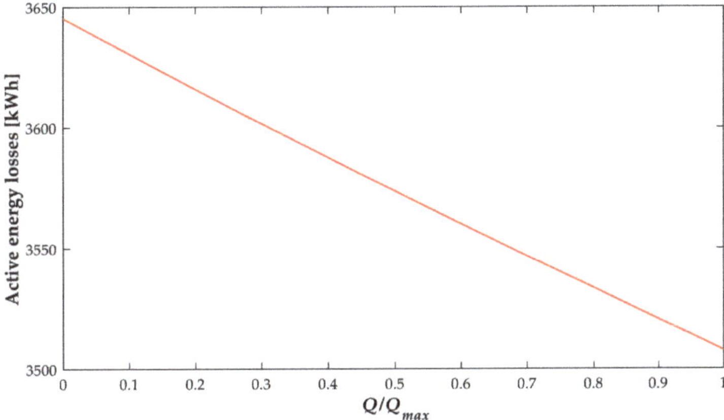

Figure 11. Change in active daily losses in function of reactive power from public EVCSs.

In order to better understand the mutual influence of reactive power on transformer aging and active energy losses, solutions that satisfy the Pareto principle of non-dominance are shown in Figure 13. The presented solutions have been obtained by changing the reactive power range from 0 to Q_{max} with a step of $0.05Q_{max}$.

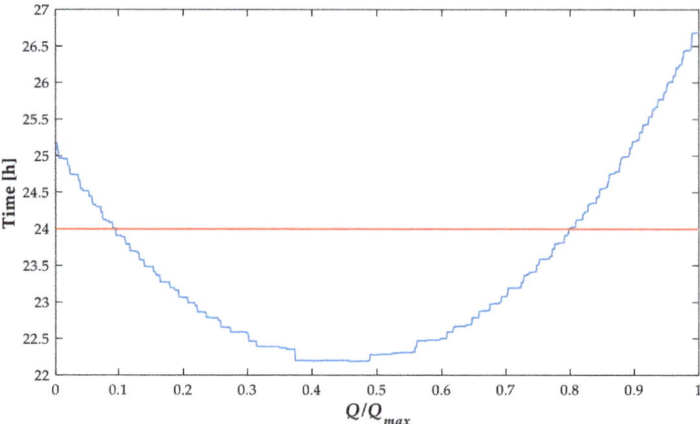

Figure 12. Change in daily transformer aging in function of reactive power factor of the EVCS.

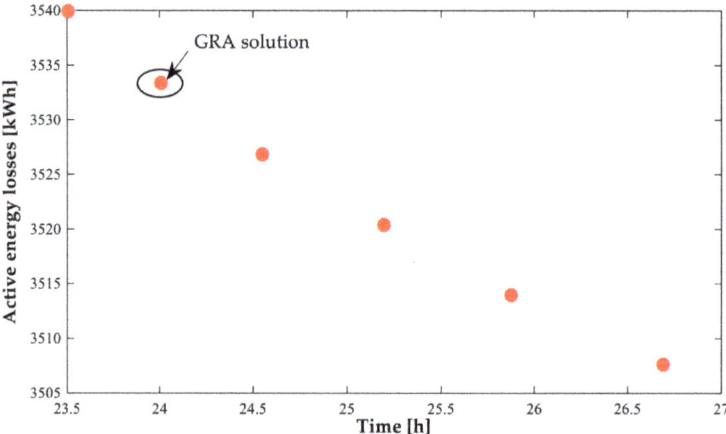

Figure 13. Pareto front optimal solutions.

In order to choose the final solution, a post-optimized analysis based on GRA has been performed. In this paper, the total number of sequences compared equals to the number of optimal solutions obtained by analysed Pareto front solutions. The elements of the sequence are the values of two criteria functions. The results from the GRA analysis are shown in Table 8, the while the optimal solution is pointed out in Figure 13.

Table 8. Obtained Γ_{0i} for the set of optimal solutions.

Daily Transformer Aging [h]	Active Daily Power Losses [kWh]	Q/Q_{max}	Γ_{0i}	Ranking Solution
24.000	3540.0	0.75	0.66667	2
24.012	3533.4	0.80	0.68824	1
24.553	3526.9	0.85	0.58257	4
25.200	3520.4	0.90	0.54352	6
25.880	3514.0	0.95	0.56736	5
26.689	3507.7	1.00	0.66667	2

A set of Pareto front optimal solutions, with points, is shown in Figure 12. For the selected solutions, it is shown that for the analyzed range of reactive power injection, the reduction of active energy losses leads to a faster aging of the transformer.

The solution involves reactive power injection of $0.8Q_{max}$, which means that the losses of active energy amount to 3533.4 kWh and that the daily aging of the transformer is 24.012 h. In the end, the adopted solution enables the reduction of active daily energy loss in the amount of 3.1%. Furthermore, with these values of reactive power injection, the reduction of transformer LOL amounts to 4% compared to the case when an EV operates with the unit power factor and 10.3% when an EV injects maximum available reactive power Q_{max}.

5. Conclusions

In this paper, the potential of reactive power from OBCs in EVCSs to reduce losses of active energy and the aging of transformers is analyzed in detail. For the analyzed network and the number of EVCSs, it is shown explicitly that the injection of reactive power contributes to a reduc-tion in the active energy losses, but contributes significantly more to the reduction of transformer aging. Namely, for the analyzed IEEE 33 bus distribution network and obtained EVCS demand, it is shown that if on-board EV chargers in EVCSs operate in the capacitive mode and with maximum possible reactive power, the losses of active energy are reduced by 3% compared to the scenario when stations operate with the unit power factor. On the other hand, the impact of reactive injection power on the aging of transformers is significantly more pronounced.

The main contribution of this paper is that it has been shown that a relatively small oversizing of the OBCs enables a significant reduction in transformer aging in addition to the reduction of active energy losses in the system. The proposed methodology is based on generally known models for calculating power flows and the widely used model for transformer aging. It is important to emphasize that the proposed methodology can be applied to any network and any number of EVCSs. Furthermore, the proposed methodology takes into account the stochastic nature of the EVs, using appropriate PDFs and MC simulations.

It is important to point out that the contribution of reactive power from EVs at EVCSs depends on the transformer rating and the EVCS demand. In particular, it was shown that if the EVCS demand is 4% higher than planned, it is possible to prevent additional aging of a 200 kVA transformer by injecting reactive power in the range from $0.154Q_{max}$ to Q_{max}. Furthermore, for a transformer of 250 kVA, for a 40% higher EVCS demand, additional aging of a transformer could be prevented by injecting reactive power in the range from $0.096Q_{max}$ to $0.799Q_{max}$.

For different levels of the EVCS demand, there are different values of improvement in LOL. For example, for the increase in the EVCS demand for 7% 200 kVA transformers, there is an 18% improvement in LOL % in comparison to the case where there is no reactive power injecting from EVs. Moreover, with a 60% increase in the EVCS demand for 250 kVA transformers, for optimal reactive power injection, there is a 23.33% improvement of LOL compared to a case where there is a maximum reactive power injection and a 10.3% improvement when an EV operates with the unit power factor.

Author Contributions: All the authors have contributed equally to this work. All authors have read and agreed to the published version of the manuscript.

Funding: This Ministry of Science and Technological Development, Montenegro.

Acknowledgments: The authors are thankful to the anonymous reviewers for comments that helped us improve the paper content.

Conflicts of Interest: The authors declare no conflict of interest.

References

1. Communication from the Commision to the European Parliment, the Council, the European Economic Social Committee a 7nd the Committee of the Regions—A Roadmap for Moving to a Competitive Low Carbon Economy in 2050. 2011. Available online: https://eravisions.archiv.zsi.at/stocktaking/7.html (accessed on 20 September 2022).
2. Alame, D.; Azzouz, M.; Kar, N. Assessing and Mitigating Impacts of Electric Vehicle Harmonic Currents on Distribution Systems. *Energies* **2020**, *13*, 3257. [CrossRef]

3.   Khalid, M.S.; Lin, X.; Zhuo, Y.; Kumar, R.; Rafique, M.K. Impact of Energy Management of Electric Vehicles on Transient Voltage Stability of Microgrid. *World Electr. Veh. J.* **2015**, *7*, 577–588. [CrossRef]

4.   Gao, Y. *Assessment of Future Adaptability of Distribution Transformer Population under EV Scenarios*; Univeristy of Manchester: Manchester, UK, 2016.

5.   Thermal Performance of Transformers. Available online: https://e-cigre.org/publication/ELT_246_4-thermal-performance-of-transformers (accessed on 7 September 2022).

6.   Soleimani, M.; Affonso, C.M.; Kezunovic, M. Transformer Loss of Life Mitigation in the Presence of Energy Storage and PV Generation. In Proceedings of the 2019 IEEE PES Innovative Smart Grid Technologies Europe (ISGT-Europe)), Bucharest, Romania, 29 September–2 October 2019; pp. 1–5.

7.   Hong, S.-K.; Lee, S.G.; Kim, M. Assessment and Mitigation of Electric Vehicle Charging Demand Impact to Transformer Aging for an Apartment Complex. *Energies* **2020**, *13*, 2571. [CrossRef]

8.   Affonso, C.d.M.; Kezunovic, M. Technical and Economic Impact of PV-BESS Charging Station on Transformer Life: A Case Study. *IEEE Trans. Smart Grid* **2019**, *10*, 4683–4692. [CrossRef]

9.   Sanchez, A.; Romero, A.; Rattá, G.; Rivera, S. Smart Charging of PEVs to Reduce the Power Transformer Loss of Life. In Proceedings of the 2017 IEEE PES Innovative Smart Grid Technologies Conference—Latin America (ISGT Latin America), Quito, Ecuador, 20–22 September 2017; pp. 1–6.

10. Konstantinidis, G.; Karapidakis, E.; Paspatis, A. Mitigating the Impact of an Official PEV Charger Deployment Plan on an Urban Grid. *Energies* **2022**, *15*, 1321. [CrossRef]

11. Smith, T.; Garcia, J.; Washington, G. Novel PEV Charging Approaches for Extending Transformer Life. *Energies* **2022**, *15*, 4454. [CrossRef]

12. Hilshey, A.D.; Hines, P.D.H.; Rezaei, P.; Dowds, J.R. Estimating the Impact of Electric Vehicle Smart Charging on Distribution Transformer Aging. *IEEE Trans. Smart Grid* **2013**, *4*, 905–913. [CrossRef]

13. Paterakis, N.; Pappi, I.; Erdinc, O.; Godina, R.; Rodrigues, E.; Catalão, J. Consideration of the Impacts of a Smart Neighborhood Load on Transformer Aging. *IEEE Trans. Smart Grid* **2015**, *76*, 2793–2802. [CrossRef]

14. Aravinthan, V.; Jewell, W. Controlled Electric Vehicle Charging for Mitigating Impacts on Distribution Assets. *IEEE Trans. Smart Grid* **2015**, *6*, 999–1009. [CrossRef]

15. Leemput, N.; Geth, F.; Van Roy, J.; Büscher, J.; Driesen, J. Reactive Power Support in Residential LV Distribution Grids through Electric Vehicle Charging. *Sustain. Energy Grids Netw.* **2015**, *3*, 24–35. [CrossRef]

16. Gao, X.; De Carne, G.; Langwasser, M.; Liserre, M. Online Load Control in Medium Voltage Grid by Means of Reactive Power Modification of Fast Charging Station. In Proceedings of the 2019 IEEE Milan PowerTech, Milan, Italy, 23–27 June 2019; pp. 1–6.

17. Wu, C.; Mohsenian-Rad, H.; Huang, J. PEV-Based Reactive Power Compensation for Wind DG Units: A Stackelberg Game Approach. In Proceedings of the 2012 IEEE Third International Conference on Smart Grid Communications (SmartGridComm), 5–8 November 2012; pp. 504–509.

18. Mojdehi, M.N.; Ghosh, P. Modeling and Revenue Estimation of EV as Reactive Power Service Provider. In Proceedings of the 2014 IEEE PES General Meeting | Conference & Exposition, National Harbor, MD, USA, 27–31 July 2014; pp. 1–5.

19. Injeti, S.K.; Thunuguntla, V.K. Optimal Integration of DGs into Radial Distribution Network in the Presence of Plug-in Electric Vehicles to Minimize Daily Active Power Losses and to Improve the Voltage Profile of the System Using Bio-Inspired Optimization Algorithms. *Prot. Control Mod. Power Syst.* **2020**, *5*, 3. [CrossRef]

20. Jamian, J.J.; Mustafa, M.W.; Mokhlis, H.; Baharudin, M.A. Minimization of Power Losses in Distribution System via Sequential Placement of Distributed Generation and Charging Station. *Arab. J. Sci. Eng.* **2014**, *39*, 3023–3031. [CrossRef]

21. Chippada, D.; Reddy, M.D. Optimal Planning of Electric Vehicle Charging Station along with Multiple Distributed Generator Units. *IJISA* **2022**, *14*, 40–53. [CrossRef]

22. Dharavat, N.; Sudabattula, S.K.; Velamuri, S. Integration of Distributed Generation and Electric Vehicles in a Distribution Network Using Political Optimizer. In Proceedings of the 2021 4th International Conference on Recent Developments in Control, Automation & Power Engineering (RDCAPE), Noida, India, 7–8 October 2021; pp. 521–525.

23. Velamuri, S.; Cherukuri, S.H.C.; Sudabattula, S.K.; Prabaharan, N.; Hossain, E. Combined Approach for Power Loss Minimization in Distribution Networks in the Presence of Gridable Electric Vehicles and Dispersed Generation. *IEEE Syst. J.* **2022**, *16*, 3284–3295. [CrossRef]

24. Chang, R.-F.; Chang, Y.-C.; Lu, C.-N. Loss Minimization of Distribution Systems with Electric Vehicles by Network Reconfiguration. In Proceedings of the 2012 International Conference on Control Engineering and Communication Technology, Shenyang, China, 7–9 December 2012; pp. 551–555.

25. Wang, J.; Wang, W.; Wang, H.; Zuo, H. Dynamic Reconfiguration of Multiobjective Distribution Networks Considering DG and EVs Based on a Novel LDBAS Algorithm. *IEEE Access* **2020**, *8*, 216873–216893. [CrossRef]

26. Reddy, M.S.K.; Selvajyothi, K. Optimal Placement of Electric Vehicle Charging Stations in Radial Distribution System along with Reconfiguration. In Proceedings of the 2019 IEEE 1st International Conference on Energy, Systems and Information Processing (ICESIP), Chennai, India, 4–6 July 2019; pp. 1–6.

27. Melo, D.F.R.; Leguizamon, W.; Massier, T.; Beng Gooi, H. Optimal Distribution Feeder Reconfiguration for Integration of Electric Vehicles. In Proceedings of the 2017 IEEE PES Innovative Smart Grid Technologies Conference—Latin America (ISGT Latin America), Quito, Ecuador, 20–22 September 2017; pp. 1–6.

28. Reddy, M.S.K.; Panigrahy, A.K.; Selvajyothi, K. Minimization of Electric Vehicle Charging Stations Influence on Unbalanced Radial Distribution System with Optimal Reconfiguration Using Particle Swarm Optimization. In Proceedings of the 2021 International Conference on Sustainable Energy and Future Electric Transportation (SEFET), Hyderabad, India, 21–23 January 2021; pp. 1–6.
29. Wang, C.; Dunn, R.; Lian, B. Power Loss Reduction for Electric Vehicle Penetration with Embedded Energy Storage in Distribution Networks. In Proceedings of the 2014 IEEE International Energy Conference (ENERGYCON), Cavtat, Croatia, 13–16 May 2014; pp. 1417–1424.
30. Yang, Y.; Zhang, W.; Jiang, J.; Huang, M.; Niu, L. Optimal Scheduling of a Battery Energy Storage System with Electric Vehicles' Auxiliary for a Distribution Network with Renewable Energy Integration. *Energies* **2015**, *8*, 10718–10735. [CrossRef]
31. Acha, S.; Green, T.C.; Shah, N. Effects of Optimised Plug-in Hybrid Vehicle Charging Strategies on Electric Distribution Network Losses. In Proceedings of the IEEE PES T&D 2010, New Orleans, LA, USA, 19–22 April 2010; pp. 1–6.
32. Sortomme, E.; Hindi, M.M.; MacPherson, S.D.J.; Venkata, S.S. Coordinated Charging of Plug-In Hybrid Electric Vehicles to Minimize Distribution System Losses. *IEEE Trans. Smart Grid* **2011**, *2*, 198–205. [CrossRef]
33. Sinha, R.; Moldes, E.; Zaidi, A.; Mahat, P.; Pillai, J.; Hansen, P. An Electric Vehicle Charging Management and Its Impact on Losses. In Proceedings of the IEEE PES ISGT Europe 2013, Lyngby, Denmark, 6–9 October 2013; pp. 1–5.
34. Nafisi, H.; Agah, S.M.M.; Askarian Abyaneh, H.; Abedi, M. Two-Stage Optimization Method for Energy Loss Minimization in Microgrid Based on Smart Power Management Scheme of PHEVs. *IEEE Trans. Smart Grid* **2016**, *7*, 1268–1276. [CrossRef]
35. Rajesh, P.; Shajin, F.H. Optimal Allocation of EV Charging Spots and Capacitors in Distribution Network Improving Voltage and Power Loss by Quantum-Behaved and Gaussian Mutational Dragonfly Algorithm (QGDA). *Electr. Power Syst. Res.* **2021**, *194*, 107049. [CrossRef]
36. Suresh, V.; Sudabattula, S.; Cherukuri, S.H.C. Coordinated Power Loss Minimization Technique for Distribution Systems in the Presence of Electric Vehicles. In Proceedings of the 2019 National Power Electronics Conference (NPEC), Tiruchirappalli, India, 13–15 December 2019; pp. 1–5.
37. IEEE SA—IEEE C57.91-2011. Available online: https://standards.ieee.org/ieee/C57.91/5297/ (accessed on 10 August 2022).
38. Waseem, M.; Sajjad, I.A.; Haroon, S.S.; Amin, S.; Farooq, H.; Martirano, L.; Napoli, R. Electrical Demand and Its Flexibility in Different Energy Sectors. *Electr. Power Compon. Syst.* **2020**, *48*, 1339–1361. [CrossRef]
39. Abdelaziz, A.Y.; Hegazy, Y.G.; El-Khattam, W.; Othman, M.M. A Multi-Objective Optimization for Sizing and Placement of Voltage-Controlled Distributed Generation Using Supervised Big Bang–Big Crunch Method. *Electr. Power Compon. Syst.* **2015**, *43*, 105–117. [CrossRef]
40. Mo, T.; Li, Y.; Lau, K.-t.; Poon, C.; Wu, Y.; Luo, Y. Trends and Emerging Technologies for the Development of Electric Vehicles. *Energies* **2022**, *15*, 6271. [CrossRef]
41. Ravi, S.; Aziz, M. Utilization of Electric Vehicles for Vehicle-to-Grid Services: Progress and Perspectives. *Energies* **2022**, *15*, 589. [CrossRef]
42. Germanà, R.; Liberati, F.; De Santis, E.; Giuseppi, A.; Delli Priscoli, F.; Di Giorgio, A. Optimal Control of Plug-In Electric Vehicles Charging for Composition of Frequency Regulation Services. *Energies* **2021**, *14*, 7879. [CrossRef]
43. Neofytou, N.; Blazakis, K.; Katsigiannis, Y.; Stavrakakis, G. Modeling Vehicles to Grid as a Source of Distributed Frequency Regulation in Isolated Grids with Significant RES Penetration. *Energies* **2019**, *12*, 720. [CrossRef]
44. Lan, T.; Jermsittiparsert, K.; Alrashood, S.T.; Rezaei, M.; Al-Ghussain, L.; Mohamed, M.A. An Advanced Machine Learning Based Energy Management of Renewable Microgrids Considering Hybrid Electric Vehicles' Charging Demand. *Energies* **2021**, *14*, 569. [CrossRef]
45. Aziz, M.; Oda, T.; Mitani, T.; Watanabe, Y.; Kashiwagi, T. Utilization of Electric Vehicles and Their Used Batteries for Peak-Load Shifting. *Energies* **2015**, *8*, 3720–3738. [CrossRef]
46. Tovilović, D.M.; Rajaković, N.L.J. The Simultaneous Impact of Photovoltaic Systems and Plug-in Electric Vehicles on the Daily Load and Voltage Profiles and the Harmonic Voltage Distortions in Urban Distribution Systems. *Renew. Energy* **2015**, *76*, 454–464. [CrossRef]
47. Deilami, S.; Masoum, A.; Moses, P.; Masoum, M. Real-Time Coordination of Plug-In Electric Vehicle Charging in Smart Grids to Minimize Power Losses and Improve Voltage Profile. Smart Grid. *IEEE Trans. Smart Grid* **2011**, *2*, 456–467. [CrossRef]
48. Khatiri-Doost, S.; Amirahmadi, M. Peak Shaving and Power Losses Minimization by Coordination of Plug-in Electric Vehicles Charging and Discharging in Smart Grids. In Proceedings of the 2017 IEEE International Conference on Environment and Electrical Engineering and 2017 IEEE Industrial and Commercial Power Systems Europe (EEEIC/I&CPS Europe), Milan, Italy, 6–9 June 2017; pp. 1–5.
49. Wang, J.; Bharati, G.R.; Paudyal, S.; Ceylan, O.; Bhattarai, B.P.; Myers, K.S. Coordinated Electric Vehicle Charging With Reactive Power Support to Distribution Grids. *IEEE Trans. Ind. Inf.* **2019**, *15*, 54–63. [CrossRef]
50. Kisacikoglu, M.C. *Vehicle-to-Grid (V2G) Reactive Power Operation Analysis of the EV/PHEV Bidirectional Battery Charger*; Univeristy of Tennessee: Knoxville, TN, USA, 2013.
51. *IEEE Standards 519-1992*; IEEE Recommended Practices and Requirements for Harmonic Control in Electrical Power Systems. InSMute of Elmeal and Ehxtronics Engineers, Inc.: New York, NY, USA, 1993; pp. 1–112. [CrossRef]
52. *IEEE Standards 1547-2003*; IEEE Standard for Interconnecting Distributed Resources with Electric Power Systems. InSMute of Elmeal and Ehxtronics Engineers, Inc.: New York, NY, USA, 2003; pp. 1–28. [CrossRef]

53. Xu, D.; Pei, W.; Zhang, Q. Optimal Planning of Electric Vehicle Charging Stations Considering User Satisfaction and Charging Convenience. *Energies* **2022**, *15*, 5027. [CrossRef]

54. Campaña, M.; Inga, E.; Cárdenas, J. Optimal Sizing of Electric Vehicle Charging Stations Considering Urban Traffic Flow for Smart Cities. *Energies* **2021**, *14*, 4933. [CrossRef]

55. Baik, S.H.; Jin, Y.G.; Yoon, Y.T. Determining Equipment Capacity of Electric Vehicle Charging Station Operator for Profit Maximization. *Energies* **2018**, *11*, 2301. [CrossRef]

56. Flammini, M.G.; Prettico, G.; Julea, A.; Fulli, G.; Mazza, A.; Chicco, G. Statistical Characterisation of the Real Transaction Data Gathered from Electric Vehicle Charging Stations. *Electr. Power Syst. Res.* **2019**, *166*, 136–150. [CrossRef]

57. Sanna, L. Driving the Solution the Plug-in Hybrid Electric Vehicle. *EPRI J.* **2005**, *1*, 8–17.

58. Mohamed, A.; Salehi, V.; Ma, T.; Mohammed, O. Real-Time Energy Management Algorithm for Plug-In Hybrid Electric Vehicle Charging Parks Involving Sustainable Energy. *IEEE Trans. Sustain. Energy* **2014**, *5*, 577–586. [CrossRef]

59. EV-Volumes—The Electric Vehicle World Sales Database. Available online: https://www.ev-volumes.com/ (accessed on 10 August 2022).

60. Worldwide Electric Vehicle Sales by Model. 2021. Available online: https://www.statista.com/statistics/960121/sales-of-all-electric-vehicles-worldwide-by-model/ (accessed on 10 August 2022).

61. Grunditz, E.A.; Thiringer, T. Performance Analysis of Current BEVs Based on a Comprehensive Review of Specifications. *IEEE Trans. Transp. Electrif.* **2016**, *2*, 270–289. [CrossRef]

62. De Freige, M.; Ross, M.; Joos, G.; Dubois, M. Power & Energy Ratings Optimization in a Fast-Charging Station for PHEV Batteries. In Proceedings of the 2011 IEEE International Electric Machines & Drives Conference (IEMDC), Niagara Falls, ON, Canada, 15–18 May 2011; pp. 486–489.

63. Kawambwa, S.; Mwifunyi, R.; Mnyanghwalo, D.; Hamisi, N.; Kalinga, E.; Mvungi, N. An Improved Backward/Forward Sweep Power Flow Method Based on Network Tree Depth for Radial Distribution Systems. *J. Electr. Syst. Inf. Technol.* **2021**, *8*, 7. [CrossRef]

64. Ghosh, S.; Das, D. Method of Load Flow Solution of Radial Distribution Network. *Gener. Transm. Distrib. IEE Proc.* **1999**, *146*, 641–648. [CrossRef]

65. Rizwan, M.; Waseem, M.; Liaqat, R.; Sajjad, I.A.; Dampage, U.; Salmen, S.H.; Obaid, S.A.; Mohamed, M.A.; Annuk, A. SPSO Based Optimal Integration of DGs in Local Distribution Systems under Extreme Load Growth for Smart Cities. *Electronics* **2021**, *10*, 2542. [CrossRef]

66. Waseem, M.; Lin, Z.; Liu, S.; Zhang, Z.; Aziz, T.; Khan, D. Fuzzy Compromised Solution-Based Novel Home Appliances Scheduling and Demand Response with Optimal Dispatch of Distributed Energy Resources. *Appl. Energy* **2021**, *290*, 116761. [CrossRef]

67. Waseem, M.; Lin, Z.; Liu, S.; Sajjad, I.A.; Aziz, T. Optimal GWCSO-Based Home Appliances Scheduling for Demand Response Considering End-Users Comfort. *Electr. Power Syst. Res.* **2020**, *187*, 106477. [CrossRef]

68. Abdmouleh, Z.; Gastli, A.; Ben-Brahim, L.; Haouari, M.; Al-Emadi, N.A. Review of Optimization Techniques Applied for the Integration of Distributed Generation from Renewable Energy Sources. *Renew. Energy* **2017**, *113*, 266–280. [CrossRef]

69. Smart City—INVEST IN PODGORICA. Available online: http://investinpodgorica.me/?page_id=1040 (accessed on 18 September 2022).

70. Rubinstein, R.Y.; Kroese, D.P. *Simulation and the Monte Carlo Method*, 3rd ed.; Wiley Series in Probability and Statistics; John Wiley & Sons, Inc.: Hoboken, NJ, USA, 2017; ISBN 978-1-118-63220-8.

71. Malekpoor, H.; Chalvatzis, K.; Mishra, N.; Mehlawat, M.K.; Zafirakis, D.; Song, M. Integrated Grey Relational Analysis and Multi Objective Grey Linear Programming for Sustainable Electricity Generation Planning. *Ann. Oper. Res.* **2018**, *269*, 475–503. [CrossRef]

72. Durkovic, V.; Savić, A.S. ATC Enhancement Using TCSC Device Regarding Uncertainty of Realization One of Two Simultaneous Transactions. *Int. J. Electr. Power Energy Syst.* **2020**, *115*, 105497. [CrossRef]

73. Xu, G.; Yang, Y.; Lu, S.; Li, L.; Song, X. Comprehensive Evaluation of Coal-Fired Power Plants Based on Grey Relational Analysis and Analytic Hierarchy Process. *Energy Policy* **2011**, *39*, 2343–2351. [CrossRef]

74. Liu, G.; Baniyounes, A.; Rasul, M.; Than Oo, A.; Khan, M.M. General Sustainability Indicator of Renewable Energy System Based on Grey Relational Analysis. *Int. J. Energy Res.* **2013**, *37*, 1928–1936. [CrossRef]

75. Dharageshwari, K.; Nayanatara, C. Multiobjective Optimal Placement of Multiple Distributed Generations in IEEE 33 Bus Radial System Using Simulated Annealing. In Proceedings of the 2015 International Conference on Circuits, Power and Computing Technologies [ICCPCT-2015], Nagercoil, India, 19–20 March 2015; pp. 1–7.

76. Karadimos, D.I.; Karafoulidis, A.D.; Doukas, D.I.; Gkaidatzis, P.A.; Labridis, D.P.; Marinopoulos, A.G. Techno-Economic Analysis for Optimal Energy Storage Systems Placement Considering Stacked Grid Services. In Proceedings of the 2017 14th International Conference on the European Energy Market (EEM), Dresden, Germany, 6–9 June 2017; pp. 1–6.

77. Chen, C.-S.; Wu, T.-H. Optimal Distribution Transformer Sizing by Dynamic Programming. *Int. J. Electr. Power Energy Syst.* **1998**, *20*, 161–167. [CrossRef]

78. Aravinthan, V.; Argade, S. Optimal Transformer Sizing with the Presence of Electric Vehicle Charging. In Proceedings of the 2014 IEEE PES T&D Conference and Exposition, Chicago, IL, USA, 14–17 April 2014; pp. 1–5.

Article

Multi-Criteria Decision Analysis during Selection of Vehicles for Car-Sharing Services—Regular Users' Expectations

Katarzyna Turoń

Department of Road Transport, Faculty of Transport and Aviation Engineering, Silesian University of Technology, 8 Krasińskiego Street, 40-019 Katowice, Poland; katarzyna.turon@polsl.pl

Abstract: Car-sharing systems, i.e., automatic, short-time car rentals, are among the solutions of the new mobility concept, which in recent years has gained popularity around the world. With the growing interest in services in society, their demands for the services offered to them have also increased. Since cars play a key role in car-sharing services, the fleet of vehicles should be properly adapted to the needs of customers using the systems. Due to the literature gap related to the procedure of proper selection of vehicles for car sharing and the market need for car-sharing service operators, this work has been devoted to the selection of car models for car sharing from the perspective of users constantly using the systems (regular users). This paper considered the case of the Polish who are constantly using car-sharing service systems. Vehicle selection was classified as a multi-faceted, complex problem, which is why one of the ELECTRE III multi-criteria decision support methods was used for this study. This study focused on the classification of vehicles from the user's perspective. Twelve modern and most popular car models in 2021 with internal combustion, electric and hybrid engines were considered. The results indicate that the best choice from the point of view of regular customers is large cars (representing vehicle classes C and D), with a large luggage compartment capacity, the highest possible ratio of engine power to vehicle weight, and the ratio of engine power to energy consumption. Importantly, small urban vehicles, which ideologically should be associated with car-sharing services due to occupying as little urban space as possible, were classified as the worst in the ranking. The results support car-sharing operators during the process of completing or upgrading their vehicle fleets.

Keywords: car sharing; car-sharing services; e-car-sharing systems; electric car sharing; hybrid car sharing; short-term car rentals; shared mobility; modern mobility; sustainable transport systems; multi-criteria decision analysis; fleet management

Citation: Turoń, K. Multi-Criteria Decision Analysis during Selection of Vehicles for Car-Sharing Services —Regular Users' Expectations. *Energies* **2022**, *15*, 7277. https://doi.org/10.3390/en15197277

Academic Editor: Wojciech Cieslik

Received: 8 September 2022
Accepted: 28 September 2022
Published: 4 October 2022

Publisher's Note: MDPI stays neutral with regard to jurisdictional claims in published maps and institutional affiliations.

1. Introduction

Modern cities are developing at a very fast pace. Currently, 55% of the world's population is urban residents, and statistics indicate that this percentage is expected to increase to 68% by 2050 [1]. It is projected that the phenomena of globalization and urbanization, as well as the gradual shift of public habitation from rural areas, could add another 2.5 billion people to urban areas [2]. Such an increase is to be particularly noticeable in the case of cities with a population of less than 1 million [3], which makes the issue important for both small urban centers and large agglomerations.

The dynamic development of urban centers, in addition to several advantages, is also associated with many problems, including difficulties with one of the key factors of their economic development—transport and the elementary need of society, which is mobility [4]. To ensure efficient, cost-effective, and, above all, sustainable urban mobility, so-called new transport mobility services are offered [5]. As part of new mobility, many different forms of transport are offered. These include, inter alia, all services offering shared mobility services [6]. These services derive from the trend of the sharing economy, according to which, using publicly available cooperation platforms (websites, mobile applications), it is possible to temporarily use given goods or services by several different people [7]. In addition to the use of web-based systems, companies offering services within the sharing economy also rely on three main assumptions [8]:

- Ensuring the greatest possible time flexibility in terms of the availability of a full range of services for the user;
- Having a rating system for users, aimed at increasing trust in the user's offer;
- Basing this mainly on rented, shared, or borrowed resources.

Of all the forms of shared mobility offered, car-sharing services are among the most affordable in terms of convenience and autonomy [9]. Car-sharing services are systems that give the possibility of renting a motor vehicle for a short time via a website or mobile application. There are four main types of car sharing [10,11]:

- Roundtrip car sharing (roundtrip station-based, back-to-base car sharing)—when the vehicle is rented and returned always in the same location—a dedicated parking space;
- Roundtrip home-zone-based car sharing—when the vehicle is rented and returned in specific zones of operation of the operator of a given system in the city;
- One-way (station-based) car sharing—when the vehicle is rented, e.g., at point A, and is returned at another point, e.g., at point B, but limited only to rental points established by the system operator;
- Free-floating car sharing—when the vehicle is rented and returned anywhere in the city, within the entire area of operation of the car sharing.

From year to year, car-sharing services are gaining more and more popularity. The latest data indicate that vehicle-sharing systems are currently in operation in 59 countries around the world [12]. They are offered by 236 operators and are available in 3128 cities [12]. Statistics estimate that the fleet of vehicles will grow from the current 380,000 available cars to almost 7.5 million units in 2025 [13], and the global car-sharing market will be worth over USD 11 billion [12].

Since car-sharing services are developing very dynamically both in terms of the growing number of operators, vehicles, and users, there are also more and more problems related to their proper and, above all, effective functioning in cities. The literature review indicates the occurrence of numerous problems covering a very wide range of issues. These include, inter alia [14–21]:

- Economic and technical problems (e.g., the problem of proper adjustment of systems to a given area of operation in terms of business model; the problem of defining operating rules and the need for system location restrictions for a given area of the city; the problem of inadequate pricing policy);
- Transport problems (e.g., the problem of appropriate adjustment of the number of vehicles to the given system of services offered; the problem of determining the location of system operation areas, the location of parking spaces or charging stations for electric vehicles; problems with the technical maintenance of vehicles);
- Environmental problems (e.g., problems related to exhaust emissions of conventionally powered vehicles used in car sharing);
- Social problems (e.g., problems with meeting society's expectations of the services offered);
- Legal problems (e.g., the problem of identifying privileges that can promote this way of traveling, and at the same time not adversely affect other pro-ecological solutions—sharing bus lanes or entering zones available only for public transport).

Making a detailed analysis of the indicated problems, it can be seen that most of them are directly related to one aspect—the fleet of vehicles offered for rental in car-sharing services. This is a factor directly related to the quality of services provided and attractiveness for potential users. The fleet of vehicles available in car sharing has been the research topic of many scientific studies. These studies involved various research areas. One of them was the issue of vehicle rotation within the scope of the zone provided and the appropriate number of vehicles offered. This aspect was called the 'Fleet Position

Problem' and was considered for appropriate vehicle optimization, accessibility of cars for customers in the city area, or proper placement of cars using specially dedicated parking spaces [22–26].

Another current trend was the analysis of all kinds of improvements and systemic changes implemented in car sharing during the COVID-19 pandemic. At that time, the researchers focused on defining the right framework for disinfecting vehicles and actions that users believe increase their comfort and safety level [27–29].

A separate group of studies was the issue of the impact of vehicles used in car sharing on society, the economy, and the environment. The change, i.a., social decisions in the field of giving up one's vehicle in favor of cars from car sharing was analyzed. For example, Jain et al. have shown that Melbourne residents are able to give up having a second car in their family in favor of a properly functioning car-sharing system [30]. In turn, Liao et al. emphasized that in their analyses of Dutch society, 20% of respondents are able to give up buying the first car in the family in favor of car sharing, which would be close to their place of residence [31]. Similar results with a result of 50% abandonment of car purchase were achieved by Hui et al. for Hangzhou in China [32]. Restrictions on purchasing decisions are topics that were also directly related to the impact of car sharing on the environment. Many scientists also took up topics related to vehicles equipped with alternative drives and the possibilities of their use in car sharing. For example, Shaheen et al. showed that users in the U.S. increased their interest in car sharing after the deployment of electric and hybrid vehicles [33]. Migliore et al. pointed to the numerous benefits of making changes to car-sharing fleets and reducing the harmful impact of cars on the environment by significantly reducing exhaust emissions [34]. Many studies were also closely related to the spectra of the direct impact of vehicles on the environment through, for example, exhaust emissions analysis, the possibility of replacing internal combustion engine cars with hybrid or electric vehicles, as well as many detailed studies on electric vehicle batteries and their tests in various road conditions [35–46].

In a broad literature review, one can notice a research gap related to the selection of the right type of vehicle models for the car-sharing fleet. In the maze of car-sharing research, the factors influencing car-sharing systems are widely considered, but the factors determining the selection of given models are ignored. It should be borne in mind that car models, and hence their detailed equipment, are the main element needed to provide the car-sharing service. Meeting the appropriate requirements of the society by the vehicle may become among the factors that will change their transport behavior and, as a result, allow them to use car sharing instead of their own car. Moreover, the car can become among the main factors that will determine whether the car-sharing service is successful and whether the customer will use the services of a given operator more often than once. Analyzing the literature on the specific social needs of vehicles, one can find my previous research focused on the requirements of operators [42] or research on users who use car sharing up to ten times a month [47]. Noticing this niche, I dedicated a research cycle to the subject of fleet selection for car-sharing vehicles, considering the needs in its scope from the point of view of various groups of users of car-sharing systems. This article was devoted to the analysis of vehicle selection from the perspective of users constantly using car-sharing systems, i.e., people who regularly rent cars from car-sharing systems more often than ten times a month.

This study was conducted in the case of a car-sharing operator providing services in the territory of Poland. The Polish car-sharing market is considered to be among the most dynamically developing in terms of shared mobility [12,44]. Concerning the European market, car-sharing systems in Poland appeared relatively late—in 2016—despite this, since their appearance on the market, they have gained great interest, to the extent that at the peak of market development in Poland there were 17 car-sharing operators, and the services could be used in over 250 cities [44]. This type of expansion also translated into significant financial results. Annual revenues in 2019 amounted to over PLN 50 million, while in 2021 it was already over PLN 100 million [44]. The Polish car-sharing market, despite many good practices, is also associated with numerous imperfections, which in many cases led to the closure of numerous service systems or a significant reduction in the zones of operation of the systems [42,45,46]. The literature states that the causes of market failures were often inadequately adapted to the needs of customers' rental service, which was based on the use of vehicles that did not meet the expectations of customers [42,45,46]. This work was therefore a response to a real market needs and an attempt to improve the functioning of car-sharing systems operating in the Polish area.

This article consists of four chapters. The first chapter refers to the literature review and the definition of the purpose of the work. The second part was devoted to presenting the research process and the detailed methodology used to achieve the results of this study. The third part presents the obtained research results. The fourth chapter contains a discussion of the results and a summary, as well as limitations on the conducted research and future research plans about vehicle selection for car sharing.

2. Methodology

2.1. Multi-Criteria Decision Making

Choosing the right type of vehicle for the needs of users of car-sharing systems is a problem that requires making the right decisions. Decision making is a difficult task for the person responsible. Usually, along with the question of choice, there are thoughts about other possible alternatives or ways in which you can check whether a given decision will have a positive impact on the analyzed issue or not. The problem becomes even more complicated if it turns out that many different factors can affect the accuracy of a given decision. Then, all kinds of methods for performing decision analysis come to the rescue. These include i.a., multi-criteria decision support methods.

Multi-criteria decision making (MCDM) or multi-criteria decision analysis (MCDA) methods are a subdiscipline of operations research [48]. Their task is to provide tools that, in the presence of many, often contradictory, criteria, will be used to evaluate and rank decision options, to facilitate the decision-making process [48–50]. What is more, methods help to structure and formalize decision-making processes transparently and consistently [51].

The methods are based on elements of knowledge from such fields as decision theory, mathematics, economics, computer science or information systems [51]. Many methods can be used for solving problems and they can be arranged according to different parameters and different stakeholders [51]. Due to the high level of utilitarianism, the methods are successfully used in the case of individuals, enterprises, and government institutions [51]. In the case of transport issues, the methods were used for, inter alia, during selection the Paris Metro project to choose the right types of scooters for sharing companies, to decide what type of car-sharing services should be provided in Shanghai, to assess the overall state of transport in Istanbul, to improve the bike rental station for the city located in Beijing, China, when making decisions on the availability of air connections with Pittsburgh, analyzing the functioning of shared mobility services in the post-COVID-19 era, or improving the quality of bike-sharing services in the Chinese city of Xi'an [52–58].

There are many different methods of multi-criteria decision support that are widely used. There are three main groups of methods—methods based on overshooting rations, aggregate measures, and utility functions. Each MCDA method has its calculation method by which alternatives are queued [51]. Since different types of vehicles are considered when selecting vehicles, the ELECTRE III method is often contradictory to comparing and evaluating them in pairs and used to obtain a ranking of variants in the work.

The ELECTRE III is among the ELECTRE collection methods, named after *Elimination Et Choix Traduisant la Realitè*. The ELECTRE III method owes its widespread popularity to the fact that among other methods of the ELECTRE family, it is possible to perform analyses with the indication of a ranked final ranking [59]. It is the possibility of obtaining a hierarchy among the objects under consideration that makes the method widely popular [59–61]. The ELECTRE III method is a particularly frequently used tool when solving various types of transport issues. It has been used, inter alia, during the evaluation of urban transportation projects [62], during the selection of means of urban passenger transport [63], safety analysis in a suburban road network [64,65], evaluation of environmental indicators for transport [64], evaluation service quality of international airports in Sicily [66] or during choosing a route for Dublin port motorway [67]. Due to the possibility of obtaining an ordered final ranking, detailed pairwise comparisons of individual criteria, and the universality of application to the problem of selection for analysis, the use of the ELECTRE III method was proposed.

2.2. Research Process

To obtain results on the selection of appropriate vehicle models for the needs of regular users of car-sharing systems, a four-stage research process was proposed, which was shown in Figure 1.

RESEARCH PROCESS

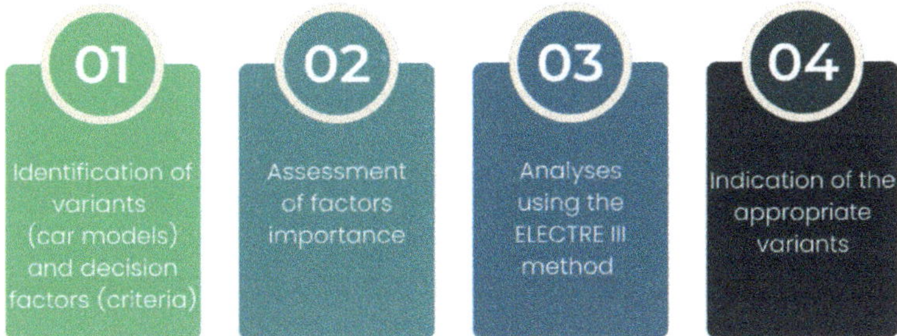

Figure 1. Research process.

The proposed research process was directly related to the algorithm of conduct in the ELECTRE III method. In the first step, the decision variants and a set of factors (criteria characterizing vehicle models parameters), which were used for detailed analyses were identified [60,61]. In the analyzed case study, the variants were car models considered for implementation in car-sharing systems. A detailed list of vehicles and the criteria considered is presented in the Results section. In the second step, research was carried out with the participation of users who constantly use car sharing. Their task was to assess the importance of individual factors considered when selecting vehicles. The evaluation was carried out by performing a pairwise comparison of each of the analyzed factors.

The criteria were assessed by comparing them in pairs and giving ratings from 1 to 9, according to Saaty's scale. The values of the scale ratings are presented in Table 1. Then, by comparing the two analyzed criteria, the exceedance index was calculated. A detailed pairwise comparison matrix is presented in the Results section.

Table 1. Saaty's Scale.

Weight	Detailed Description
"1"	Equal importance of the criteria
"2"	Very weak advantage of one criterion over the other
"3"	Weak advantage of one criterion over the other
"4"	More than a weak advantage of one criterion over the other, but less than a strong advantage
"5"	Strong advantage of one criterion over the other
"6"	More than a strong advantage of one criterion over the other, but less than very strong
"7"	Very strong advantage of one criterion over the other
"8"	More than a very strong advantage of one criterion over the other, less than an extreme
"9"	Extreme, total advantage of one criterion over the other

In the third stage, detailed analyses were carried out using the ELECTRE III method. Based on the calculated exceedance index, it was determined whether the first of analyzed variant is not worse than the second analyzed variant due to the indicated factor. Consequently, calculations of the compliance rate should be performed to obtain the level of advantage of one variant over the other in terms of all analyzed factors [53,54]. The compliance rate is the sum of the weights of the criteria for which the evaluation value of one variant is greater than or equal to the evaluation value of the other variant [52,53]. ELECTRE III introduces three main parameters that allow determining the relationships between the analyzed variants [61]:

- The maximum difference of factors values Δ—the difference between the highest and lowest value in the assessment of two variants;
- Indifference threshold Q—is the biggest difference between the performance of the variants and profiles on the factors;
- Preference threshold p—the greatest difference between the performance of the variants and profiles such that one is preferable to the other on the considered factor;
- Veto threshold V—the difference in the assessment of two variants concerning a given factor.

Sequentially, an altitude difference matrix is created. Variants should be arranged sequentially, starting from their initial ordering using classification procedures of ascend distillation and descend distillation [51–53]. Both distillations rate the variants from best to worst [51–53]. Ascend distillation is a planning process that begins with selecting the best variant and placing it at the top of the ranking [51–53]. The best variant is selected one by one from the remaining variants and placed in the next position in the classification. This procedure is repeated until all possible variants have been analyzed [51–53]. Descend distillation is a planning process that starts with selecting the worst variant and placing it at the end of the ranking. Subsequently, similarly, to ascending distillation, further analyses should be performed, bearing in mind that in the subsequent iterations of the variants to be considered, the worst variant is always selected and placed in the next positions from the end of the ranking [55,56]. After the distillation has been performed, a final ranking is made. The results are presented in the next chapter.

3. Results

This study was conducted in June 2022 for a case study of a car-sharing company operating in Poland. Currently, the company has a fleet of 2000 cars. Cars owned by the operator are vehicles of one type constituting urban, small-sized cars, which are equipped with three or five doors. The operator expected to receive indications as to the possibility of modernizing the fleet, and to check what factors were the most crucial for users during the process of choosing vehicles.

To determine the fleet of implementable cars, twelve modern vehicles were selected to represent different types of drives including cars with internal combustion engine, cars with hybrid engines and cars with electric engines. To consider the vehicles attractive to users, the focus was on the selection of vehicles that were the most popular cars in Europe in 2021 according to the list published by Automotive News Europe [68]. The vehicles published in the report were representatives of different classes of cars. Car classes are a standard used in Europe that specifies the regulations, description, and detailed categorization of vehicles according to ISO 3833-1977. The standard distinguishes nine classes of vehicles from A to M which distinguish cars from small and city to medium, large, family, vans, off-road and luxury vehicles. From a wide range of vehicles presented in the report, vehicles representing car classes from A to D were selected, following the fact that vehicles of these classes are the cars most often chosen by Poles [69]. Moreover, vehicles from classes A to D are also the most frequently used cars in European car sharing involving passenger cars [45]. Van and combivan cars are offered in Poland in cargo car-sharing systems. Since the operator does not provide this type of service, this study was limited to models of classes from A to D. Detailed characteristics of the selected car classes are presented in Figure 2.

A class

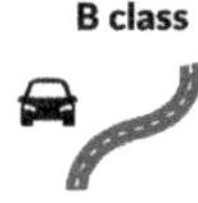

Cars designed for urban driving; are characterized by small dimensions and low operating costs. Impractical to travel on extra-urban routes. They can be two- or four-seater, and five-seaters usually allocate three rear seats for children.

B class

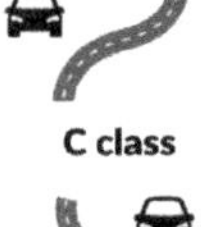

Small cars that offer more than the A segment space for passengers and a practical boot. These features allow them to be driven on routes outside the city, but they are more intended for use in the city as "another car" in the family.

C class

Medium-sized cars; designed for city and highway driving. They offer space for five adults and a luggage compartment, as well as relatively comfortable travel conditions. Selected as both the first and the next vehicle in the family.

D class

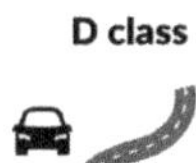

Cars that provide comfortable travel conditions for five adults (with luggage) over longer distances. Most often in body versions of sedans (or similar in size to hatchback sedans) and station wagons. Many of them are available in coupé versions, most often as sporty, exclusive versions of a given model.

Figure 2. Characteristics of the analyzed vehicle classes.

Since the preferences of Poles who most often choose vehicles from classes A to D also correspond to the fleets of cars used in the Polish car-sharing market, twelve vehicle models were selected, which were considered in further analyses. The selection of vehicles did not favor any of the specific brands of vehicles. A detailed list of vehicle models considered in the analyses is presented in Table 2.

Table 2. Analyzed car models.

Vehicle Model Number	Car Class	Type of Engine
VM1	C class	Internal Combustion Engine
VM2	B class	Internal Combustion Engine
VM3	B class	Hybrid Engine
VM4	D class	Hybrid Engine
VM5	B class	Internal Combustion Engine
VM6	C class	Hybrid Engine
VM7	C class	Internal Combustion Engine
VM8	A class	Electric Engine
VM9	D class	Hybrid Engine
VM10	A class	Electric Engine
VM11	D class	Electric Engine
VM12	D class	Electric Engine

Subsequently, a detailed list of factors that were used to evaluate the various variants was indicated. A detailed list is presented in Table 3.

Table 3. Set of factors considered during car-sharing fleet selection analysis.

Factor Number	Factor Characteristics
F1	Rental fee [€]
F2	The ratio of engine power to vehicle weight [kW/kg]
F3	The ratio of engine power to consumption [kW/kWh]
F4	Time of battery charging/time of refueling [min]
F5	Boot capacity [l]
F6	Number of doors in the vehicle [-]
F7	Vehicle length [m]
F8	Euro NCAP rating [-]
F9	Safety equipment [-]
F10	Warranty period in years [-]

The list of factors for the assessment of individual vehicles was prepared based on the author's previous research [47], in which slight modifications were made, inter alia, in terms of costs, considering the cost of renting cars instead of the cost of purchasing vehicles. In the case of vehicle rental costs, due to the lack of use in the current Polish car-sharing systems indicated in Table 2 of vehicle models, Formula (1) has been developed, which considers the costs of renting vehicles depending on the time of their use and the distance traveled. Stopover costs are also included.

$$rental_{fee}(a, b) = (f_{min} + s_{min})a + f_{km}b \; [€] \tag{1}$$

where a—rental time [min], b—travel distance [km],

$$f_{min} = \begin{cases} 0.14 \; € \; for \; A - class \; cars \\ 0.17 \; € \; for \; B - class \; cars \\ 0.21 \; € \; for \; C - class \; cars \\ 0.27 \; € \; for \; D - class \; cars \end{cases} \text{—rental cost for 1 min,}$$

$$f_{km} = \begin{cases} 0.24 \; € \; for \; A - class \; cars \\ 0.24 \; € \; for \; B - class \; cars \\ 0.24 \; € \; for \; C - class \; cars \\ 0.28 \; € \; for \; D - class \; cars \end{cases} \text{—rental cost for 1 km,}$$

$$s_{min} = 0.03 \; € \; for \; A, B, C, D - class \; car\text{—stopover fee for 1 min}$$

The next step was to establish the importance of individual indicators when selecting vehicles. For this purpose, pairwise comparisons of all factors were developed. The factors were assessed by the car-sharing service users. The questionnaire was available online by using the CAWI (Computer-Assisted Web Interview) method. The aim of this study was to obtain a pairwise comparison of each of the factors and assign an appropriate weighting according to the Saaty scale presented in Table 1. The respondents indicated appropriate weights by filling in the matrix presented in Figure 3. This study included 250 people who use car-sharing systems very often (more often than ten times a month) and were considered regular customers of the systems. The survey was conducted anonymously in June 2022. The users who participated in this study represented a population of 200,000 users of the system of the analyzed enterprise. For the research sample, the confidence level was 95% ($\alpha = 0.95$). The fraction size was 0.5 and the maximum error was estimated at 7%.

The next step was to prepare a summary of the values of individual criteria for each of the analyzed variants. A detailed list is presented in Table 4.

Table 4. Summary of the values of individual factors for each of the considered car models.

Variant	Rental Cost	The Ratio of Engine Power to Vehicle Weight	The Ratio of Engine Power to Energy Consumption	Charging Time/ Refueling Time	Boot Capacity	Number of Doors	Vehicle Length	Euro NCAP Rating	Safety Equipment	The Warranty Period in Years
	F1 [€]	F2 [kW/kg]	F3 [kW/kWh]	F4 [min]	F5 [l]	F6 [-]	F7 [m]	F8 [-]	F9 [-]	F10 [-]
VM1	0.48	0.051	0.475	2	380	5	4.28	5	10	2
VM2	0.44	0.077	0.511	2	311	5	4.05	4	9	2
VM3	0.44	0.078	0.388	1.5	286	3	3.94	5	8	3
VM4	0.58	0.154	0.062	2	480	4	4.70	5	11	2
VM5	0.44	0.049	0.613	1.5	391	5	4.05	5	10	2
VM6	0.48	0.078	0.327	2.5	361	4	4.37	5	10	3
VM7	0.48	0.075	0.420	2.5	600	5	4.68	5	10	3
VM8	0.41	0.034	0.421	90	300	5	3.73	1	6	2
VM9	0.58	0.056	0.229	2	443	5	4.47	5	8	5
VM10	0.41	0.070	0.157	240	363	3	3.63	4	8	2
VM11	0.58	0.051	0.132	360	585	5	4.49	5	8	2
VM12	0.58	0.063	0.133	450	543	5	4.58	5	8	3

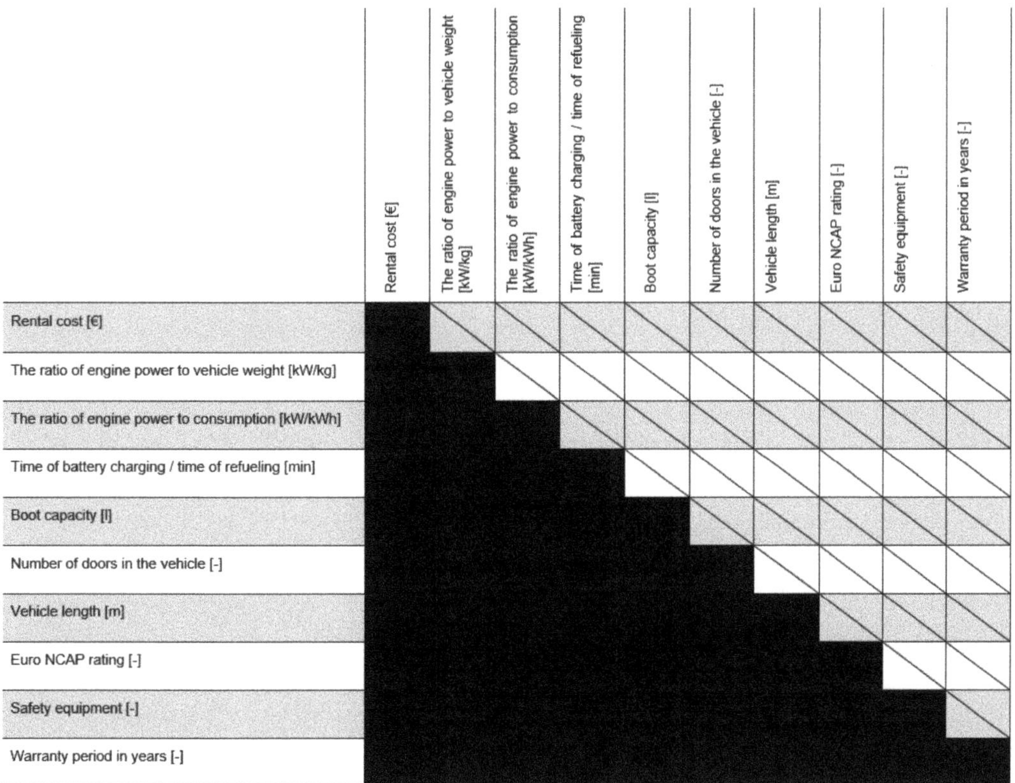

Figure 3. Pairwise comparison matrix provided to respondents.

Then, based on the ELECTRE III methodology, the maximum difference of criteria values, equivalence threshold, preference threshold, and veto threshold values were determined. Detailed data were presented in Table 5.

Table 5. The set of equivalence, preference, and veto thresholds.

Factor Number	Maximum Difference of Factors Values Δ	Indifference Threshold Q	Preference Threshold p	Veto Threshold V
F1	0.17	0.0425	0.085	0.17
F2	182	45.5	91	182
F3	27.5	6.875	13.75	27.5
F4	448.5	112.125	224.25	448.5
F5	314	78.5	157	314
F6	2	0.5	1	2
F7	1.07	0.2675	0.535	1.07
F8	4	1	2	4
F9	5	1.25	2.5	5
F10	3	0.75	1.5	3

The next step according to the ELECTRE III methodology was to create the concordance matrix. The matrix is presented in the form of Table 6.

Table 6. Concordance matrix values.

Variants	VM1	VM2	VM3	VM4	VM5	VM6	VM7	VM8	VM9	VM10	VM11	VM12
VM1	-	1.0	0.9977	0.8564	0.9994	0.9977	0.8578	1.0	0.911	0.916	0.7617	0.8694
VM2	1.0	-	0.9977	0.7317	1.0	0.9765	0.7707	1.0	0.8452	0.916	0.6454	0.7122
VM3	0.8491	0.918	-	0.5999	0.7322	0.8128	0.6514	0.918	0.6955	0.916	0.525	0.609
VM4	0.6875	0.6875	0.7672	-	0.6875	0.7844	0.6852	0.6875	0.8619	0.916	0.834	0.9157
VM5	1.0	1.0	0.9977	0.81	-	0.9765	0.8136	1.0	0.8494	0.916	0.7023	0.7868
VM6	0.9007	0.8404	1.0	0.8928	0.6875	-	0.7968	0.918	0.829	0.916	0.6619	0.7851
VM7	1.0	1.0	1.0	0.918	0.9073	1.0	-	1.0	0.911	0.916	0.834	0.918
VM8	0.5719	0.7435	0.8121	0.543	0.6208	0.5696	0.5696	-	0.5624	0.7028	0.4854	0.5671
VM9	0.779	0.7695	0.9643	0.934	0.7299	0.9604	0.8436	0.909	-	0.916	0.9025	1.0
VM10	0.4868	0.6259	0.8242	0.662	0.5863	0.6609	0.4029	0.7065	0.721	-	0.6113	0.6767
VM11	0.7299	0.7695	0.7995	0.934	0.7299	0.8621	0.7276	0.7695	0.993	1.0	-	0.9977
VM12	0.7299	0.7695	0.8035	0.934	0.7299	0.8661	0.7299	0.7695	0.993	0.9434	0.916	-

The next stage in the ELECTRE III method was to perform the ascend and descend distillation against each of the variants and create and in the final step create a dominance matrix. The dominance matrix was presented in Table 7.

Table 7. Dominance matrix values.

Variants	VM1	VM2	VM3	VM4	VM5	VM6	VM7	VM8	VM9	VM10	VM11	VM12
VM1	-	B+	B+	B+	W−	B+	W−	B+	W−	B+	W−	B+
VM2	W−	-	B+	B+	W−	B+	W−	B+	W−	B+	W−	B+
VM3	W−	W−	-	W−	W−	W−	W−	W−	W−	B+	W−	W−
VM4	W−	W−	B+	-	W−	W−	W−	B+	W−	B+	W−	W−
VM5	B+	B+	B+	B+	-	B+	W−	B+	W−	B+	R	B+
VM6	W−	W−	B+	B+	W−	-	W−	B+	W−	B+	W−	R
VM7	B+	B+	B+	B+	B+	B+	-	B+	R	B+	R	B+
VM8	W−	W−	B+	W−	W−	W−	W−	-	W−	B+	W−	W−
VM9	B+	B+	B+	B+	B+	B+	R	B+	-	B+	B+	B+
VM10	W−	W−	W−	W−	W−	W−	W−	W−	W−	-	W−	W−
VM11	B+	B+	B+	B+	R	B+	R	B+	W−	B+	-	B+
VM12	W−	W−	B+	B+	W−	R	W−	B+	W−	B+	W−	-

where (B+)—the first variant is better than the second variant; (R) — a pair of variants are equivalent; (W−)—the first variant is worse than the second variant.

The last step was to prepare the final ranking presenting the ranking of variants in terms of the preferences of experts and the adopted factors. The final ranking was presented in Table 8.

Table 8. Variants final ranking.

Dominance Matrix	Ascend Distillation	Descend Distillation
VM1	3.0	5.0
VM2	4.0	5.0
VM3	6.0	9.0
VM4	5.0	7.0
VM5	3.0	3.0
VM6	5.0	5.0
VM7	2.0	1.0
VM8	6.0	8.0
VM9	1.0	2.0
VM10	7.0	9.0
VM11	1.0	4.0
VM12	4.0	6.0

The graphical arrangement of the variants is shown in Figure 4.

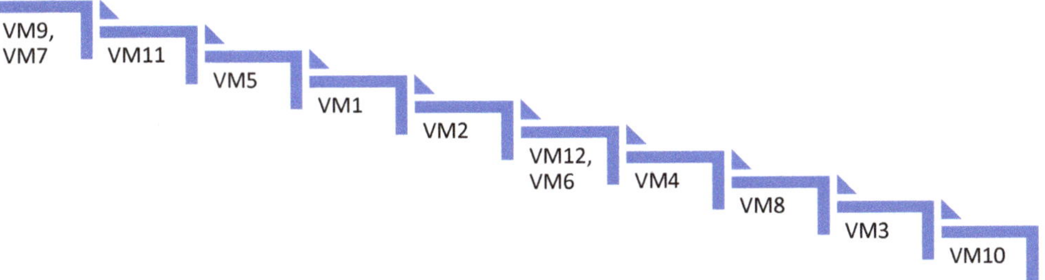

Figure 4. Ranking of car models best suited to car sharing from the perspective of regular users.

4. Discussion and Conclusions

Research carried out using the ELECTRE III multi-criteria decision support method allowed obtaining the final ranking of ranked vehicle models for the car-sharing system, which best meet the expectations of people regularly using car-sharing systems. In the first place, ex aequo placed two models of vehicles—VM9 and VM11. The second place was taken by VM5.

When making detailed analyses in terms of the size of the winning vehicles, it should be stated that the models represented C and D class cars. They are therefore vehicles with medium and large dimensions providing comfortable travel conditions simultaneously for five adults on urban routes, but also over long distances. The vehicles are also equipped with large cargo space. In the case of class D, these were family cars. Interestingly, the worst place in the ranking was achieved by a vehicle representing class A, the smallest of the car models under consideration.

Considering the obtained results in terms of the car propulsion, it should be stated that hybrid and conventionally powered vehicles ranked highest in the ranking. In turn, the second place was taken by an electric vehicle. Interestingly, the last places in the ranking were also taken by vehicles with conventional and hybrid drives. This means that for the respondents, the type of drive was not a key factor, and the ecological thread is debatable. Research indicates that it was not the type of power supply but only more detailed technical parameters, inter alia, the ratio of engine power to fuel consumption or engine power to vehicle weight characterizing specific car models played a key role. This indicates that over time if electric vehicles are equipped with more and more capacious batteries and achieve greater ranges, these vehicles will reach higher places in the rankings.

Analyzing the obtained results from the point of view of the importance of individual criteria for users, it should be mentioned that the most important issues were the ratio of engine power to energy consumption, the ratio of engine power to vehicle weight, boot capacity, and vehicle length. In the case of the ratio of engine power to energy consumption, this means that for users, the issue of eco-friendliness and economy of cars is important. In turn, the ratio of engine power to vehicle weight is directly related to the dynamics of the vehicle. The higher the ratio, the greater the dynamics and driving comfort for the users. Analysis of the most important factors shows that that regular customers of car-sharing systems prefer vehicles with high engine power, which at the same time are economy cars, providing the opportunity to overcome the longest possible reach. What is more, it is worth emphasizing that it was particularly important for users that the vehicles were large, comfortable spacious, and roomy cars. Therefore, the relatively smallest cars were placed in the worst positions. Such conclusions show that regular users of car-sharing systems treat rental vehicles as classic, large family cars owned, which means that in their case car-sharing cars can replace ownership of their car. On the one hand, this is a very interesting conclusion, because it indicates that car sharing fulfills the basic task of exchanging a single vehicle for a rented one. On the other hand, it is worth emphasizing that the idea of carsharing was to ensure the high availability of small, urban cars that do not take up much public space [70,71]. However, the conducted research indicates the opposite. This kind of conclusion is in line with the realities of the Polish car-sharing market because many systems that were based on fleets comprised of small, city cars have been closed.

Comparing the results on the choice of vehicles by regular users using the systems more than 10 times a month, to less using cars up to ten times a month, it should be emphasized that they have similar preferences in terms of size and type of vehicles. Such results can be an important indication for operators when composing the composition of their fleet because frequent customers and regular customers can be included in one group of service recipients. It is also an important tip for researchers when further considering the segmentation of car-sharing customers.

To sum up, based on the research carried out, operators of Polish car-sharing services, when composing their fleet tailored to the needs of users constantly using the systems, should focus on large and long C or D class vehicles, equipped with engines with high parameters and at the same time low energy consumption and equipped with the largest luggage space. This type of fleet should also find interest among customers who often use car-sharing systems. Since the systems are also used by occasional customers who rarely use the systems, it is recommended to use fleet differentiation. This article has some limitations. This article focuses on analyses concerning only one group of users—regular users. The analyses were performed exclusively for the Polish market and focused on vehicles representing classes from A to D. The respondents assessed the criteria indicated arbitrarily by the author, without the possibility of indicating their own proposals of factors that could affect the choice of vehicles.

In the following articles, the author plans to analyze the composition of the car-sharing fleet, considering the opinions of people who rarely use the systems, to obtain the full social perspective. Moreover, the author would also like to consider in the analyses the vehicle classes that were not included in this article, i.e., E, F, J, and M. This will allow for a possible consideration of the operators' approach to the implementation of cargo services. The author also plans to introduce the possibility for users to indicate their own factors which, in their opinion, affect the choice of car-sharing vehicles. The author would also like to perform similar research for other countries to obtain a comparison of the approach to the car-sharing fleet on the world market.

Funding: The publication received no external funding.

Institutional Review Board Statement: According to our University Ethical Statement, following, the following shall be regarded as research requiring a favorable opinion from the Ethic Commission in the case of human research (based on document in polish: https://prawo.polsl.pl/Lists/Monitor/Attachments/7291/M.2021.501.Z.107.pdf (accessed on 21 March 2022): research in which persons with limited capacity to give informed or research on persons whose capacity to give informed or free consent to participate in research and who have a limited ability to refuse research before or during their implementation, in particular: children and adolescents under 12 years of age, persons with intellectual disabilities persons whose consent to participate in the research may not be fully voluntary prisoners, soldiers, police officers, employees of companies (when the survey is conducted at their workplace), persons who agree to participate in the research on the basis of false information about the purpose and course of the research (masking instruction, i.e., deception) or do not know at all that they are subjects (in so-called natural experiments); research in which persons particularly susceptible to psychological trauma and mental health disorders are to participate mental health, in particular: mentally ill persons, victims of disasters, war trauma, etc., patients receiving treatment for psychotic disorders, family members of terminally or chronically ill patients; research involving active interference with human behavior aimed at changing it research involving active intervention in human behavior aimed at changing that behavior without direct intervention in the functioning of the brain, e.g., cognitive training, psychotherapy psychocorrection, etc. (this also applies if the intended intervention is intended to benefit (this also applies when the intended intervention is o benefit the subject (e.g., to improve his/her memory); research concerning controversial issues (e.g., abortion, in vitro fertilization, death penalty) or requiring particular delicacy and caution (e.g., concerning religious beliefs or attitudes towards minority groups) minority groups); research that is prolonged, tiring, physically or mentally exhausting. Our research is not done on people meeting the mentioned condition. Any of the researched people: any of them had limited capacity to be informed, any of them had been susceptible to psychological trauma and mental health disorders, the research did not concern the mentioned-above controversial issues, the research was not prolonged, tiring, physically or mentally exhausting.

Informed Consent Statement: Informed consent was obtained from all subjects involved in this study.

Data Availability Statement: The data presented in this study are available on request from the author.

Conflicts of Interest: The author declares no conflict of interest.

References

1. United Nations. Revision of World Urbanization Prospects. Available online: https://population.un.org/wup/Publications/Files/WUP2018-Report.pdf (accessed on 18 August 2022).
2. United Nations. Analysis and Policy Recommendations from the United Nations Secretary-General's High-Level Advisory Group on Sustainable Transport, Mobilizing Sustainable Transport for Development High-level Advisory Group on Sustainable Transport. 2016. Available online: https://sustainabledevelopment.un.org/index.php?page=view&type=400&nr=2375&menu=1515 (accessed on 18 August 2022).
3. United Nations. Population Facts. The Speed of Urbanization around the World. Available online: https://population.un.org/wup/Publications/Files/WUP2018-PopFacts_2018-1.pdf (accessed on 18 August 2022).
4. Hoerler, R.; Stünzi, A.; Patt, A.; Del Duce, A. What Are the Factors and Needs Promoting Mobility-as-a-Service? Findings from the Swiss Household Energy Demand Survey (SHEDS). *Eur. Transp. Res. Rev.* **2020**, *12*, 27. [CrossRef]
5. Long, Z.; Axsen, J. Who Will Use New Mobility Technologies? Exploring Demand for Shared, Electric, and Automated Vehicles in Three Canadian Metropolitan Regions. *Energy Res. Soc. Sci.* **2022**, *88*, 102506. [CrossRef]
6. Kamargianni, M.; Li, W.; Matyas, M.; Schäfer, A. A Critical Review of New Mobility Services for Urban Transport. *Transp. Res. Procedia* **2016**, *14*, 3294–3303. [CrossRef]
7. Quirós, C.; Portela, J.; Marín, R. Differentiated Models in the Collaborative Transport Economy: A Mixture Analysis for Blablacar and Uber. *Technol. Soc.* **2021**, *67*, 101727. [CrossRef]
8. Luri Minami, A.; Ramos, C.; Bruscato Bortoluzzo, A. Sharing Economy versus Collaborative Consumption: What Drives Consumers in the New Forms of Exchange? *J. Bus. Res.* **2021**, *128*, 124–137. [CrossRef]
9. Jung, J.; Koo, Y. Analyzing the Effects of Car Sharing Services on the Reduction of Greenhouse Gas (GHG) Emissions. *Sustainability* **2018**, *10*, 539. [CrossRef]
10. Ciari, F.; Bock, B.; Balmer, M. Modeling station-based and free-floating carsharing demand: A test case study for Berlin, Germany. In *Emerging and Innovative Public Transport and Technologies*; Transportation Research Board of the National Academies: Washington, DC, USA, 2014.
11. Ferrero, F.; Perboli, G.; Rosano, M.; Vesco, A. Car-sharing services: An annotated review. *Sustain. Cities Soc.* **2018**, *37*, 501–518. [CrossRef]
12. Global Market Insights. Car Sharing Market Size by Model (P2P, Station-Based, Free-Floating), by Business Model (Round Trip, One Way), by Application (Business, Private), COVID-19 Impact Analysis, Regional Outlook, Application Potential, Price Trend, Competitive Market Share & Forecast, 2021–2027. Available online: https://www.gminsights.com/industry-analysis/carsharing-market (accessed on 12 July 2022).
13. ING Forecast. Car-Sharing Unlocked. Available online: https://www.ing.nl/zakelijk/kennis-over-de-economie/uw-sector/automotive/car-sharing-unlocked-english.html (accessed on 15 August 2022).
14. Dowling, R.; Kent, J. Practice and public–private partnerships in sustainable transport governance: The case of car sharing in Sydney, Australia. *Transp. Policy* **2015**, *40*, 58–64. [CrossRef]
15. Becker, H.; Ciari, F.; Axhausen, K. Comparing car-sharing schemes in Switzerland: User groups and usage patterns. *Transp. Res. Part A* **2017**, *97*, 17–29. [CrossRef]
16. Hui, Y.; Ding, M.; Qian, C.; Wang, W.; Xu, Q. Research on the operational characteristics of car sharing service stations: A case study of a car sharing program in Hangzhou. In *World Conference on Transport Research-WCTR 2016 Shanghai*; Elsevier: Amsterdam, The Netherlands, 2016; pp. 140–4152.
17. Lansley, G. Cars and socioeconomics: Understanding neighbourhood variations in car characteristics from administrative data. *J. Reg. Stud. Reg. Sci.* **2016**, *3*, 264–285.
18. Nijland, H.; Meerkerk, J. Mobility and environmental impacts of car sharing in the Netherlands. *Environ. Innov. Soc. Transit.* **2017**, *23*, 84–91. [CrossRef]
19. Schwieterman, J.; Bieszczat, A. The cost to carshare: A review of the changing prices and taxation levels for carsharing in the United States 2011–2016. *Transp. Policy* **2017**, *57*, 1–9. [CrossRef]
20. Wappelhorst, S.; Sauer, M.; Hinkeldein, D.; Bocherding, A.; Glaß, T. Potential of Electric Carsharing in Urban and Rural Areas. *Transp. Res. Procedia* **2014**, *4*, 374–386. [CrossRef]

21. Wieliński, G.; Trepanier, M.; Morency, C. Electric and hybrid car use in a free-floating carsharing system. *Int. J. Sustain. Transp.* **2017**, *11*, 161–169. [CrossRef]

22. Changaival, B.; Lavangnananda, K.; Danoy, G.; Kliazovich, D.; Guinand, F.; Brust, M.; Musial, J.; Bouvry, P. Optimization of Carsharing Fleet Placement in Round-Trip Carsharing Service. *Appl. Sci.* **2021**, *11*, 11393. [CrossRef]

23. Ströhle, P.; Flath, C.M.; Gärttner, J. Leveraging Customer Flexibility for Car-Sharing Fleet Optimization. *Transp. Sci.* **2019**, *53*, 42–61. [CrossRef]

24. Monteiro, C.M.; Machado, C.A.S.; Lage, M. de O.; Berssaneti, F.T.; Davis, C.A.; Quintanilha, J.A. Optimization of Carsharing Fleet Size to Maximize the Number of Clients Served. *Comput. Environ. Urban Syst.* **2021**, *87*, 101623. [CrossRef]

25. Lemme, R.F.F.; Arruda, E.F.; Bahiense, L. Optimization Model to Assess Electric Vehicles as an Alternative for Fleet Composition in Station-Based Car Sharing Systems. *Transp. Res. Part D Transp. Environ.* **2019**, *67*, 173–196. [CrossRef]

26. Carlier, A.; Munier-Kordon, A.; Klaudel, W. Optimization of a one-way carsharing system with relocation operations. In Proceedings of the 10th International Conference on Modeling, Optimization and SIMulation MOSIM 2014, Nancy, France, 5–7 November 2014.

27. Alonso-Almeida, M.D.M. To Use or Not Use Car Sharing Mobility in the Ongoing COVID-19 Pandemic? Identifying Sharing Mobility Behaviour in Times of Crisis. *IJERPH* **2022**, *19*, 3127. [CrossRef]

28. Faiyetole, A.A. Impact of Covid-19 on Willingness to Share Trips. *Transp. Res. Interdiscip. Perspect.* **2022**, *13*, 100544. [CrossRef]

29. Kim, S.; Lee, S.; Ko, E.; Jang, K.; Yeo, J. Changes in Car and Bus Usage amid the COVID-19 Pandemic: Relationship with Land Use and Land Price. *J. Transp. Geogr.* **2021**, *96*, 103168. [CrossRef] [PubMed]

30. Jain, T.; Rose, G.; Johnson, M. Changes in Private Car Ownership Associated with Car Sharing: Gauging Differences by Residential Location and Car Share Typology. *Transportation* **2022**, *49*, 503–527. [CrossRef] [PubMed]

31. Liao, F.; Molin, E.; Timmermans, H.; van Wee, B. Carsharing: The Impact of System Characteristics on Its Potential to Replace Private Car Trips and Reduce Car Ownership. *Transportation* **2020**, *47*, 935–970. [CrossRef]

32. Hui, Y.; Wang, Y.; Sun, Q.; Tang, L. The Impact of Car-Sharing on the Willingness to Postpone a Car Purchase: A Case Study in Hangzhou, China. *J. Adv. Transp.* **2019**, *2019*, 1–11. [CrossRef]

33. Shaheen, S.; Martin, E.; Totte, H. Zero-Emission Vehicle Exposure within U.S. Carsharing Fleets and Impacts on Sentiment toward Electric-Drive Vehicles. *Transp. Policy* **2020**, *85*, A23–A32. [CrossRef]

34. Migliore, M.; D'Orso, G.; Caminiti, D. The environmental benefits of carsharing: The case study of Palermo. *Transp. Re-Search Procedia* **2020**, *48*, 2127–2139. [CrossRef]

35. Nowak, M.; Kamińska, M.; Szymlet, N. Determining If Exhaust Emission from Light Duty Vehicle During Acceleration on the Basis of On-Road Measurements and Simulations. *J. Ecol. Eng.* **2021**, *22*, 63–72. [CrossRef]

36. Szałek, A.; Pielecha, I.; Cieslik, W. Fuel Cell Electric Vehicle (FCEV) Energy Flow Analysis in Real Driving Conditions (RDC). *Energies* **2021**, *14*, 5018. [CrossRef]

37. Pielecha, I.; Cieślik, W.; Szałek, A. Energy Recovery Potential through Regenerative Braking for a Hybrid Electric Vehicle in a Urban Conditions. *IOP Conf. Ser.: Earth Environ. Sci.* **2019**, *214*, 012013. [CrossRef]

38. Szymlet, N.; Lijewski, P.; Kurc, B. Road Tests of a Two-Wheeled Vehicle with the Use of Various Urban Road Infrastructure Solutions. *J. Ecol. Eng.* **2020**, *21*, 152–159. [CrossRef]

39. Pigłowska, M.; Kurc, B.; Galiński, M.; Fuć, P.; Kamińska, M.; Szymlet, N.; Daszkiewicz, P. Challenges for Safe Electrolytes Applied in Lithium-Ion Cells—A Review. *Materials* **2021**, *14*, 6783. [CrossRef] [PubMed]

40. Cieslik, W.; Szwajca, F.; Rosolski, S.; Rutkowski, M.; Pietrzak, K.; Wójtowicz, J. Historical Buildings Potential to Power Urban Electromobility: State-of-the-Art and Future Challenges for Nearly Zero Energy Buildings (nZEB) Microgrids. *Energies* **2022**, *15*, 6296. [CrossRef]

41. Szymlet, N.; Lijewski, P.; Sokolnicka, B.; Siedlecki, M.; Domowicz, A. Analysis of Research Method, Results and Regulations Regarding the Exhaust Emissions from Two-Wheeled Vehicles under Actual Operating Conditions. *J. Ecol. Eng.* **2020**, *21*, 128–139. [CrossRef]

42. Turoń, K.; Kubik, A.; Chen, F. What Car for Car-Sharing? Conventional, Electric, Hybrid or Hydrogen Fleet? Analysis of the Vehicle Selection Criteria for Car-Sharing Systems. *Energies* **2022**, *15*, 4344. [CrossRef]

43. Turoń, K. Selection of Car Models with a Classic and Alternative Drive to the Car-Sharing Services from the System's Rare Users Perspective. *Energies* **2022**, *15*, 6876. [CrossRef]

44. Statista. Forecast Revenues from Carsharing Services in Poland from 2019 to 2025. Available online: https://www.sta-tista.com/statistics/1059362/poland-carsharing-revenues/ (accessed on 5 June 2022).

45. Turoń, K.; Kubik, A.; Łazarz, B.; Czech, P.; Stanik, Z. Car-sharing in the context of car operation. *IOP Conf. Ser. Mater. Sci. Eng.* **2018**, *421*, 032027. [CrossRef]

46. Turoń, K.; Kubik, A.; Chen, F. Operational Aspects of Electric Vehicles from Car-Sharing Systems. *Energies* **2019**, *12*, 4614. [CrossRef]

47. Turoń, K. Carsharing Vehicle Fleet Selection from the Frequent User's Point of View. *Energies* **2022**, *15*, 6166. [CrossRef]

48. Cinelli, M.; Kadziński, M.; Miebs, G.; Gonzalez, M.; Słowiński, R. Recommending Multiple Criteria Decision Analysis Methods with a New Taxonomy-Based Decision Support System. *Eur. J. Oper. Res.* **2022**, *302*, 633–651. [CrossRef]

49. Martyn, K.; Kadziński, M. Deep Preference Learning for Multiple Criteria Decision Analysis. *Eur. J. Oper. Res.* **2022**, S0377221722005422. [CrossRef]

50. Athanasakis, K.; Igoumenidis, M.; Boubouchairopoulou, N.; Vitsou, E.; Kyriopoulos, J. Two Sides of the Same Coin? A Dual Multiple Criteria Decision Analysis of Novel Treatments Against Rheumatoid Arthritis in Physicians and Patients. *Clin. Ther.* **2021**, *43*, 1547–1557. [CrossRef] [PubMed]

51. Zlaugotne, B.; Zihare, L.; Balode, L.; Kalnbalkite, A.; Khabdullin, A.; Blumberga, D. Multi-Criteria Decision Analysis Methods Comparison. *Environ. Clim. Technol.* **2020**, *24*, 454–471. [CrossRef]

52. Ziemba, P.; Gago, I. Compromise Multi-Criteria Selection of E-Scooters for the Vehicle Sharing System in Poland. *Energies* **2022**, *15*, 5048. [CrossRef]

53. Liu, A.; Wang, R.; Fowler, J.; Ji, X. Improving Bicycle Sharing Operations: A Multi-Criteria Decision-Making Approach. *J. Clean. Prod.* **2021**, *297*, 126581. [CrossRef]

54. Ma, F.; Shi, W.; Yuen, K.F.; Sun, Q.; Guo, Y. Multi-Stakeholders' Assessment of Bike Sharing Service Quality Based on DEMATEL–VIKOR Method. *Int. J. Logist. Res. Appl.* **2019**, *22*, 449–472. [CrossRef]

55. Torbacki, W. Achieving Sustainable Mobility in the Szczecin Metropolitan Area in the Post-COVID-19 Era: The DEMATEL and PROMETHEE II Approach. *Sustainability* **2021**, *13*, 12672. [CrossRef]

56. Saaty, T. How to make decision: The analytic hierarchy process. *Eur. J. Oper. Res.* **1990**, *48*, 9–26. [CrossRef]

57. Li, W.; Li, Y.; Fan, J.; Deng, H. Siting of Carsharing Stations Based on Spatial Multi-Criteria Evaluation: A Case Study of Shang-hai EVCARD. *Sustainability* **2017**, *9*, 152. [CrossRef]

58. Awasthi, A.; Breuil, D.; Singh Chauhan, S.; Parent, M.; Reveillere, T. A Multicriteria Decision Making Approach for Carsharing Stations Selection. *J. Decis. Syst.* **2007**, *16*, 57–78. [CrossRef]

59. Kobryń, A. *Wielokrotne Wspomaganie Decyzji w Gospodarowaniu Przestrzenią*; Difin: Warszaw, Poland, 2014.

60. Figueira, J.R.; Greco, S.; Roy, B.; Słowiński, R. ELECTRE methods: Main features and recent developments. In *Handbook of Multicriteria Analysis*; Springer: Berlin/Heidelberg, Germany, 2010; pp. 51–89.

61. Norese, M.F. ELECTRE III as a support for participatory decision-making on the localisation of waste-treatment plants. *Land Use Policy* **2006**, *23*, 76–85. [CrossRef]

62. Żak, J.; Kruszyński, M. Application of AHP and ELECTRE III/IV Methods to Multiple Level, Multiple Criteria Evaluation of Urban Transportation Projects. *Transp. Res. Procedia* **2015**, *10*, 820–830. [CrossRef]

63. Dudek, M.; Solecka, K.; Richter, M. A Multi-Criteria Appraisal of the Selection of Means of Urban Passenger Transport Using the Electre and AHP Methods. *Czas. Tech.* **2018**, *6*, 79–93. [CrossRef]

64. Fancello, G.; Carta, M.; Fadda, P. A Decision Support System Based on Electre III for Safety Analysis in a Suburban Road Network. *Transp. Res. Procedia* **2014**, *3*, 175–184. [CrossRef]

65. Borken, J. Evaluation of environmental indicators for transport with ELECTRE III. In *Actes du Seminaire PIE*; INRETS. INRETS Seminar PIE: Lyon, France, 2005.

66. Lupo, T. Fuzzy ServPerf Model Combined with ELECTRE III to Comparatively Evaluate Service Quality of International Airports in Sicily. *J. Air Transp. Manag.* **2015**, *42*, 249–259. [CrossRef]

67. Rogers, M.; Bruen, M. Using ELECTRE III to Choose Route for Dublin Port Motorway. *J. Transp. Eng.* **2000**, *126*, 313–323. [CrossRef]

68. Auto Magazine. Top 20: The Most Popular Models of the 20 Brands in Europe. Available online: https://magazynauto.pl/wiadomosci/top-20-najpopularniejsze-modele-20-marek-w-europie,aid,1335 (accessed on 10 July 2022).

69. Auto World Portal. The Most Popular Car Classes in Poland. Available online: https://www.auto-swiat.pl/wiadomosci/aktualnosci/najpopularniejsze-klasy-samochodow-w-polsce/gmkcf8w (accessed on 23 August 2022).

70. Roblek, V.; Meško, M.; Podbregar, I. Impact of Car Sharing on Urban Sustainability. *Sustainability* **2021**, *13*, 905. [CrossRef]

71. Glotz-Richter, M. Car-Sharing—"Car-on-call" for reclaiming street space. *Procedia-Soc. Behav. Sci.* **2012**, *48*, 1454–1463. [CrossRef]

Article

Selection of Car Models with a Classic and Alternative Drive to the Car-Sharing Services from the System's Rare Users Perspective

Katarzyna Turoń

Department of Road Transport, Faculty of Transport and Aviation Engineering, Silesian University of Technology, 8 Krasińskiego Street, 40-019 Katowice, Poland; katarzyna.turon@polsl.pl

Abstract: Short-term, automated car rental services, i.e., car sharing, are a solution that has been improving in urban transportation systems over the past few years. Due to the intensive expansion of the systems, service providers face increasing challenges in their competitiveness. One of them is to meet the customer expectations for the fleet of vehicles offered in the system. Although this aspect is noted primarily in the literature review on fleet optimization and management, there is a gap in research on the appropriate selection of vehicle models. In response, the article aimed to identify the vehicles best suited for car-sharing systems from the customer's point of view. The selection of suitable vehicles was treated as a multi-criteria decision-making issue; therefore, the study used ELECTRE III—one of the multi-criteria decision-making methods. The work focuses on researching the opinions of users who rarely use car-sharing services in Poland. The most popular car models in 2021, equipped with internal combustion, hybrid, and electric engines, were selected for the analysis. The results indicate that the best suited cars are relatively large, spacious, and equipped with electric drive and represent the D segment of vehicles in Europe. In addition, these vehicles are to be equipped with a powerful engine, a spacious boot, and a fast battery charging time. Interestingly, small city cars, so far associated with car sharing, ranked the worst in the classification method. In addition, factors such as the warranty period associated with the quality of the vehicles, or the number of car doors, are not very important to users. The results support car-sharing operators in the process of selecting or modernizing a fleet of vehicles.

Keywords: car sharing; shared mobility; sustainable transportation; fleet management; mobility management; vehicle selection; transportation engineering; multi-criteria decision analysis; ELECTRE III; MCDA; electromobility

Citation: Turoń, K. Selection of Car Models with a Classic and Alternative Drive to the Car-Sharing Services from the System's Rare Users Perspective. *Energies* **2022**, *15*, 6876. https://doi.org/10.3390/en15196876

Academic Editor: Wojciech Cieslik

Received: 24 August 2022
Accepted: 16 September 2022
Published: 20 September 2022

1. Introduction

Car-sharing systems, that is, short-term automated car rental services, are solutions that are becoming more and more popular around the world. The systems' popularity and intensive development are mainly related to our high convenience and self-commissioning [1]. Furthermore, the systems also benefit from the fact that the vehicles of the systems have free access to parking lots within the operating zones of operation, and in most systems, it is possible to return the vehicle anywhere within the zones located in the city [2]. The great interest in car-sharing services also translates into international statistics. In 2020, the global car-sharing market exceeded USD 2 billion [3]. By 2027, the market value is projected to exceed USD 3 billion [4].

The significant development of car-sharing services in the world has led to many changes in the rules of their operation. For example, operators have made improvements in the operation and optimization of their systems, service management, and the implementation of new transport or area solutions to all innovations related to the COVID-19 pandemic, with a need to adapt the vehicle fleet to a higher level of safety for users [5–8].

All of these aspects are of great interest to scientists around the world. However, from the point of view of scientists, one issue is considered relatively often—the fleet of cars used in car sharing. When the international literature is analyzed, four main thematic areas can be distinguished from the point of view from which a car-sharing fleet is considered.

The first area concerns the relocation of vehicles. The relocation of vehicles is particularly important due to the limited parking space in cities [9]. The analysis and recommendations for the relocation of cars in systems are particularly important due to the functioning of various types of car-sharing systems on the market. These include [10–14]:

- Round-trip car-sharing (round-trip station-based, back-to-base car-sharing)—when the vehicle is rented and always returned to the same location—a dedicated parking space;
- Round-trip home zone-based—when the vehicle is rented and returned to specific zones of operation by the operator of a given system in the city;
- One-way (station-based car-sharing)—when the vehicle is rented, e.g., at point A, and is returned to another point, e.g., at point B, but limited only to the rental points established by the system operator;
- Free-floating car-sharing—when the vehicle is rented and returned anywhere in the city, within the entire area of operation of the car-sharing system.

Various forms of rentals and returns generate the need for a proper rotation of vehicles within the available zones, which is emphasized by Changaival et al., defining the placement of the vehicles as a fleet placement problem (FPP) [15]. Ströhle et al. showed a relationship between leveraging the customer's flexibility for car sharing and fleet optimization, indicating that a customer's flexibility in the range of 1 km allows a fleet reduction of 12% [16]. In turn, Monteiro et al. analyzed the distribution of the zones in car sharing and proved that settling more parking spaces and vehicles near each other is more effective than having parking spaces located in the city but distant from each other [17]. In turn, Lemme et al. focused on the creation of an optimization model to evaluate electric vehicles as an alternative to a fleet composition in station-based car-sharing systems, demonstrating that it is possible to rotate vehicles properly in zones; although, in the case of electric vehicles being implemented for the first time in a fleet, this should be checked in pilot programs due to the main disadvantage of vehicles, which is the economic dimension [18]. In turn, Carlier et al. proposed a programming-oriented mathematical approach and introduced a simple linear model based on total flow variables [19]. Their solution was based on three optimization criteria: maximizing the met car-sharing requirements while minimizing the vehicle fleet and relocation operations [19].

The second of the thematic areas dedicated to the fleet of cars in car-sharing systems is devoted to issues related to the size of the fleets owned in the systems. For example, Nourinejad and Roorda devoted their research to fleet size decision support and showed that the number of cars is related to specific user demand patterns [20]. In turn, Barrios and Godier simulated the appropriate number of vehicles to achieve a flexible car-sharing system, stressing that the periodic redistribution of vehicles, which is not carried out continuously, is of particular importance [21]. In comparison, Lu et al., conducting research on optimizing the profitability and quality of service in car-sharing systems under demand uncertainty, showed that exogenously given one-way car-sharing demand can increase car-sharing profitability under a given one-way and round-trip price difference and vehicle relocation cost, while an endogenously generated one-way demand, due to pricing and strategic customer behavior, may decrease car-sharing profitabilities [22].

The third area of fleet research is devoted to considering vehicles from car-sharing systems and their impact on economic and social issues. For example, Hui et al. considered the impact of car sharing on the willingness to postpone a car purchase, indicating that 50% of respondents in the Chinese city of Hangzhou will postpone their car purchases by participating in car sharing [23]. For comparison, Jain et al., in their research on the Australian city of Melbourne, showed that residents of densely populated inner suburbs used a shared car to avoid or delay owning a car, while residents of the middle suburbs used car sharing to avoid buying a second car [24].

In turn, Liao et al., who performed research in the Netherlands, obtained results that around 40% of the respondent's car drivers indicated that they are willing to replace some of their private car trips with car sharing, and 20% indicated that they could abandon a planned purchase or lose a current car if car sharing becomes available near them [25].

Another identified research topic was devoted strictly to the use of alternatively powered vehicles and all pro-ecological solutions affecting the improvement in the level of sustainability of car-sharing systems. In this area, many research works were carried out. Many works have been devoted to the idea of alternative drives and eco-friendly issues, including the application of an alternative power supply for vehicles through the possibility of urban power electromobility from historical buildings, the use of vehicles-to-grid or research on the real energy consumption of vehicles that can be used in car-sharing services in urban conditions [26–30]. For example, Migliore et al. dealt with the definition of the environmental benefits related to car-sharing systems, indicating the possibility of achieving limits in pollutants emission by 25% for PM10 and 38% for CO_2 [31]. Shaheen et al., examining the approach of system users to the fleet of alternatively powered vehicles, indicated that pairing shared electric or plug-in hybrid vehicles increased user sympathy for the use of car sharing [32]. In turn, Liao and Correia showed that electric vehicles in car sharing are mainly used for short trips, and their current users are mostly middle-aged men with relatively high incomes and education [33].

The last of the identified thematic groups is the research on the operational and technical aspects of a fleet of vehicles made available in car-sharing systems. In this area, together with our co-authors, we carried out various types of research aimed at identifying the main technical aspects that are important for the proper functioning of the services [34]. We also carried out research on the determination of the types of fleets used in car-sharing systems in Europe [35], as well as analyzing the type of vehicle tailored to the requirements of car-sharing system operators [36]. However, this research focused on the needs of the service providers and how this translated into business profitability, not on checking the real expectations of society. Noticing this research gap, the author proposed a research cycle devoted to the selection of vehicles for the car-sharing fleet from the point of view of various types of users. This article aimed to analyze the types of vehicles best suited to the needs of customers who rarely use car-sharing systems.

The research was proposed in a case study of a company operating in the Polish car-sharing services market. The Polish car-sharing market has not been selected by chance, as Poland is considered one of the fastest-growing shared-mobility markets [3]. Although car-sharing systems in Poland were relatively late compared to other European countries (in 2016), the market is considered dynamic and valuable [37,38]. At the highest stage of the development of the systems, 17 service providers offered car-sharing services in 250 Polish cities [38]. From a financial point of view, car-sharing services generated revenues of more than PLN 50 million in 2019 and more than PLN 100 million in 2021 [38]. In Poland, car-sharing services, despite many superlatives, have also suffered many failures. These included, in addition to the financial problems of the operators, an unsuitable vehicle fleet or the type of car-sharing services offered in cities [34–36]. In many cases, changes to the vehicle fleet appear only as pilots, such as the introduction of several electric vehicles [34–36]. In response to the appropriate adaptation of the vehicles to the needs of society using car sharing, our own research was proposed. The results of the research are presented in this article.

The work was divided into five chapters. The first section is an introduction with a review of the literature. In the second chapter, the research methodology is presented. The third chapter indicates the obtained research results, which are discussed in the fourth part of the article. The fifth chapter presents a summary, research limitations, and further research plans of the author.

2. Methodology

Choosing appropriate vehicle models for a car-sharing fleet is a multifaceted problem. In situations related to complex decision-making problems, one of the methods of support in analytical processes is the method of multi-criteria decision support, called Multiple-Criteria Decision Making (MCDM) or Multiple-Criteria Decision Analysis (MCDA). These methods can provide a wide range of tools to help identify the best options for the criteria under consideration or a full ranking of the possible solutions [39]. The methods are based on elements of knowledge in such fields as decision theory, mathematics, economics, computer science, or information systems [31]. The widespread interest in these methods is related to their wide utilitarianism [31]. Transport processes, due to their multi-criteria nature and complexity [40–42], seem to be an excellent application for MCDA methods. From the point of view of transport issues, multi-criteria decision support methods were used in selecting the Paris Metro project, car-sharing services in Shanghai, assessment of the state of transport in Istanbul, or air connections with Pittsburgh [43–45].

There are many different methods of multi-criteria decision support. According to the classification, the methods based on superiority ratios, function, utility, and aggregate measures can be distinguished [46–49]. One method that allows for making a detailed comparison of the analyzed criteria and, on its basis, obtaining a ranking of the solutions (given variants) chosen for analysis is the ELECTRE III method.

ELECTRE III is a method that derives from the French name *Elimination Et Choix Traduisant la Realitè*. It owes its popularity to the fact that among all the methods of the ELECTRE group, performing analyses with an indication of a ranked final ranking is possible [50]. The ELECTRE III method introduces parameters that determine the relationship between individual variants—the preference threshold, the equivalence threshold, and the veto threshold [51].

The ELECTRE III method is based on the use of society's opinion to assess the importance of individual factors that influence the choice of a given variant [52]. Individual criteria in the ELECTRE III method may be strongly or slightly better than each other, respectively. Therefore, by using this method, it is possible to determine the insignificant or very significant differences between the analyzed variants [53]. The ELECTRE III method is based on a three-stage algorithm presented in Figure 1.

ELECTRE III
STEPS OF THE PROCEDURE

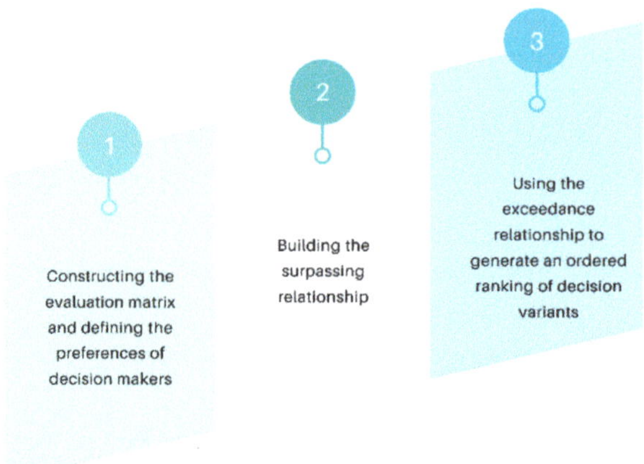

Figure 1. ELECTRE III steps of the procedure.

In the first stage, it is necessary to identify the variants of the decision and then define a set of criteria that will be used to evaluate each of the variants [50–55]. For each of the criteria, a weight is determined, which is indicated by experts. The respondents compared each pair of criteria according to Saaty's scale, giving grades from 1 to 9, where [46]:

- 1—same meaning;
- 2—very weak advantage;
- 3—weak advantage;
- 4—more than a weak advantage, less than strong;
- 5—strong advantage;
- 6—more than a strong advantage, less than very strong;
- 7—a very strong advantage;
- 8—more than a very strong advantage, less than an extreme;
- 9—extreme, total advantage.

Then, by comparing the two decision variants, the exceedance index was calculated [50–55].

In the second stage, using the calculated exceedance index, it was determined whether the first variant was better than the second due to the selected criterion. Consequently, the calculations of the compliance rate should be performed to obtain an answer with the level of advantage of one variant over the other in terms of all criteria [50–55]. The compliance rate is the sum of the criteria weights for which the evaluation value of one variant is greater than or equal to the evaluation value of the other variant [50–55].

In the third stage, an altitude difference matrix was created. The variants should be arranged sequentially, starting from their initial ordering using the classification procedures of ascending and descending distillation [50–55]. Both distillations rate the variants from best to worst [50–55]. Ascend distillation is a planning process that begins with selecting the best variant and placing it at the top of the ranking [50,51]. The best variant is selected one by one from the remaining variants and placed in the next position in the classification. This procedure is repeated until all possible variants have been analyzed [50,51]. Descend distillation is a planning process that begins with selecting the worst variant and placing it at the end of the ranking. Subsequently, similar to ascending distillation, further analyses should be performed, bearing in mind that in the subsequent iterations of the variants to be considered, the worst variant is always selected and placed in the next positions from the end ranking [53,54]. After the distillation has been completed, a final ranking is made.

The results are presented in the next chapter.

3. Calculation Procedure

The proposed study was carried out for the case study of one car-sharing company operating in the Polish area. The company currently has about 2000 vehicles, focusing on cars of one type: small-sized cars and urban hatchback cars equipped with three or five doors. The research aimed to analyze and indicate what type of fleet would be best suited to the needs of system users who rarely use rental cars, that is, from 5 to 10 times a year.

Twelve new vehicle models equipped with internal combustion, hybrid, and electric engines were selected for the study. The proposed models were chosen among the most popular cars in Europe in 2021, based on the *Automotive News Europe* report [56]. The car models selected for the analysis represented different vehicle classes (car segments). The car classes are car-scheme classifications used in Europe, standardized following ISO Standard 3833–1977. They categorize vehicles in terms of size and equipment. The standard distinguishes nine main classes marked with the letters A to M that characterize the type of vehicle. A detailed breakdown of the vehicle classes is presented in Table 1.

Among the car models included in the report, the focus was on the vehicles representing the four most popular car segments in Poland, which are the A, B, C, and D classes [57]. A list of the vehicle models included in the analysis is presented in Table 2.

Table 1. Characteristics of vehicle classes.

Segment	Description
A	Cars designed for urban driving; are characterized by small dimensions and low operating costs. Impractical to travel on extra-urban routes. They can be two- or four-seater, and five-seaters usually allocate three rear seats for children.
B	Small cars that offer more than the A segment space for passengers and a practical boot. These features allow them to be driven on routes outside the city, but they are more intended for use in the city as "another car" in the family.
C	Medium-sized cars; designed for city and highway driving. They offer space for five adults and a luggage compartment, as well as relatively comfortable travel conditions. Selected as both the first and the next vehicle in the family.
D	Cars that provide comfortable travel conditions for five adults (with luggage) over longer distances. Most often in body versions of sedans (or similar in size to hatchback sedans) and station wagons. Many of them are available in coupé versions, most often as sporty, exclusive versions of a given model.
E	Large, comfortable, and well-equipped cars, the purpose of which is not only to be used by families but also as representative limousines for companies. The technology and equipment contained in them allow for long journeys, and the technical data of the leading versions can often compete, even with typical sports cars.
F	Limousines with the highest level of equipment and the best (often the largest) engines. Their features allow for a very comfortable journey for both the driver and passengers. Often used as representative limos for heads of state, companies, etc., these cars are often better driven as rear seat passengers rather than as drivers.
J	Sport utility cars or cars have features that allow off-road driving.
M	Multipurpose cars. A class of spacious cars that can carry at least five people along with large luggage.
S	A class of cars that includes a very large group of vehicles considered being sports, sporting, and extravagant coupé style or very high-performance vehicles, designed either as models designed to achieve high speeds and high accelerations or as road versions of performance cars.

Table 2. Variants included in the analysis.

Variant Number	Car-Class	Type of Engine
V1	C	ICE
V2	B	ICE
V3	B	Hybrid
V4	D	Hybrid
V5	B	ICE
V6	C	Hybrid
V7	C	ICE
V8	A	Electric
V9	D	Hybrid
V10	A	Electric
V11	D	Electric
V12	D	Electric

ICE—Internal Combustion Engine (ICE).

Following the methodology to proceed using the ELECTRE III method, the next step was to develop a set of criteria from which individual variants were evaluated. Due to the lack of literature devoted to analyzing the impact of the individual criteria on fleet selection, the factors were arbitrarily indicated. When defining the set of criteria, the desire was made to indicate the measurable factors directly related to the specification of individual vehicles. A set of factors is presented in Table 3.

Table 3. Set of criteria considered during car-sharing fleet selection analysis.

Criteria Number	Name of the Criterion	Characteristics of the Criterion
C1	Rental cost [€]	The cost of renting a car from the car-sharing system, considering rental time, rental distance, and stop-over fee, expressed by the Formula (1) $$rental_{cost}(i,j) = (r_{min} + s_{min})i + r_{km}j \qquad (1)$$ where: i—rental time [min], j—rental distance [km], $$r_{min} = \begin{cases} 0.14 \ \text{€} \ for \ A - class \ cars \\ 0.17 \ \text{€} \ for \ B - class \ cars \\ 0.21 \ \text{€} \ for \ C - class \ cars \\ 0.27 \ \text{€} \ for \ D - class \ cars \end{cases}$$ —rental cost for 1 min, $$r_{km} = \begin{cases} 0.24 \ \text{€} \ for \ A - class \ cars \\ 0.24 \ \text{€} \ for \ B - class \ cars \\ 0.24 \ \text{€} \ for \ C - class \ cars \\ 0.28 \ \text{€} \ for \ D - class \ cars \end{cases}$$ —rental cost for 1 km, $s_{min} = 0.03 \ \text{€} \ for \ A, B, C, D - class \ car$—stop-over fee for 1 min
C2	Engine power [kW]	The power generated by the vehicle's engine.
C3	Energy consumption/fuel consumption [kWh/100 km]	The amount of fuel or electricity required for a car to travel 100 km.
C4	Time of battery charging/time of refueling [min]	Minutes needed to top up fuel/electricity to maximum fuel tank capacity or car battery capacity.
C5	Boot capacity [l]	The number of liters of luggage that can fit in the boot of a car.
C6	Number of doors in the vehicle [-]	The number of doors the vehicle is equipped with.
C7	Vehicle length [m]	Distance from the front to the rear of the vehicle in meters is one of the main dimensions describing the vehicle.
C8	Euro NCAP rating [-]	*Vehicle Safety Ranking*, published by the European New Car Assessment Program (Euro NCAP)—an independent and non-profit vehicle safety assessment organization. Euro NCAP has created the five-star safety rating system to help consumers, their families, and businesses compare vehicles more easily and to help them identify the safest choice for their needs. The safety rating is determined from a series of vehicle tests designed and carried out by Euro NCAP. These tests represent, in a simplified way, important real-life accident scenarios that could result in injured or killed car occupants or other road users. The number of stars reflects how well the car performs in the Euro NCAP tests, but it is also influenced by the safety equipment that the vehicle manufacturer is offering in each market.
C9	Safety equipment [-]	Vehicle equipment to increase the level of safety is one of the Euro NCAP system assessment categories considering factors, such as the frontal crash protection systems (front airbag, belt pre-tensioner, belt-load limiter, knee airbag), lateral crash protection (side head airbag, side chest airbag, side pelvis), airbag, center airbag), child protection (Isofix/i-size, integrated child seat, airbag cut-off switch), safety assist (seatbelt reminder), and other safety systems.
C10	Warranty period in years [-]	One of the institutions of contract law. In Polish law, this refers to certifying the quality of the item sold. It is expressed in years.

The developed criteria were used for the analysis of the vehicles. Each of the vehicles considered in the analysis (variants presented in Table 2) was represented by the technical parameters that characterized them as corresponding to the assumed criteria. Therefore, the next step was to assign each of the criteria values of the individual parameters based on the technical specifications of the vehicles and the *Euro NCAP* reports. A detailed list is presented in Table 4.

Table 4. Criteria values for individual car variants.

No.	C1	C2	C3	C4	C5	C6	C7	C8	C9	C10
	[€]	[kW]	[kWh/100 km]	[min]	[l]	[-]	[m]	[-]	[-]	[-]
V1	0.48	81	38.5	2	380	5	4.28	5	10	2
V2	0.44	74	37.8	2	311	5	4.05	4	9	2
V3	0.44	74	28.7	1.5	286	3	3.94	5	8	3
V4	0.58	215	13.3	2	480	4	4.70	5	11	2
V5	0.44	48	29.4	1.5	391	5	4.05	5	10	2
V6	0.48	90	29.4	2.5	361	4	4.37	5	10	3
V7	0.48	110	37.8	2.5	600	5	4.68	5	10	3
V8	0.41	33	13.9	90	300	5	3.73	1	6	2
V9	0.58	104	23.8	2	443	5	4.47	5	8	5
V10	0.41	70	11	240	363	3	3.63	4	8	2
V11	0.58	109	14.4	360	585	5	4.49	5	8	2
V12	0.58	128	17	450	543	5	4.58	5	8	3

The next step was to establish the importance of the individual criteria when determining the vehicles by respondents. For this purpose, a survey was conducted among users of the car-sharing system. Among the users of the analyzed operator, 200 car-sharing users were selected for the study, who use the systems rarely, that is, from 5 to 10 times a year. The survey was conducted anonymously in June 2022. The respondents who participated in the survey represented a population of 200,000 users of the system of the analyzed enterprise. For the research sample, the confidence level was 95% ($\alpha = 0.95$). The fraction size was 0.5, and the maximum error was estimated at 8%. The respondents filled in the questionnaire, which was made available via the internet using the Computer-Assisted Web Interview (CAWI) method. The questionnaire was fully anonymous and focused on obtaining only the answers needed to perform the ELECTRE III analyses, i.e., receiving pairwise comparisons of each of the criteria. The respondents assessed the importance of each criterion on Saaty's scale, assigning values from 1 to 9 and entering them into the appropriate field of the matrix. The matrix of pairwise comparisons is shown in Figure 2.

Based on the assessments given by the respondents, a list was created showing the average importance of each of the criteria. The score values were used for further analysis using the ELECTRE III method. The summary is presented in Table 5.

Table 5. Weight values.

Criteria Number	Weights
C1	0.133
C2	0.176
C3	0.066
C4	0.1225
C5	0.1395
C6	0.084
C7	0.082
C8	0.082
C9	0.108
C10	0.007

According to the ELECTRE III methodology, the next step was to determine the maximum difference in the criteria values, the equivalence threshold, the preference threshold, and the veto threshold. Detailed data are presented in Table 6.

The next step, according to the ELECTRE III methodology, was to create the concordance matrix. The matrix is presented in the form of Table 7.

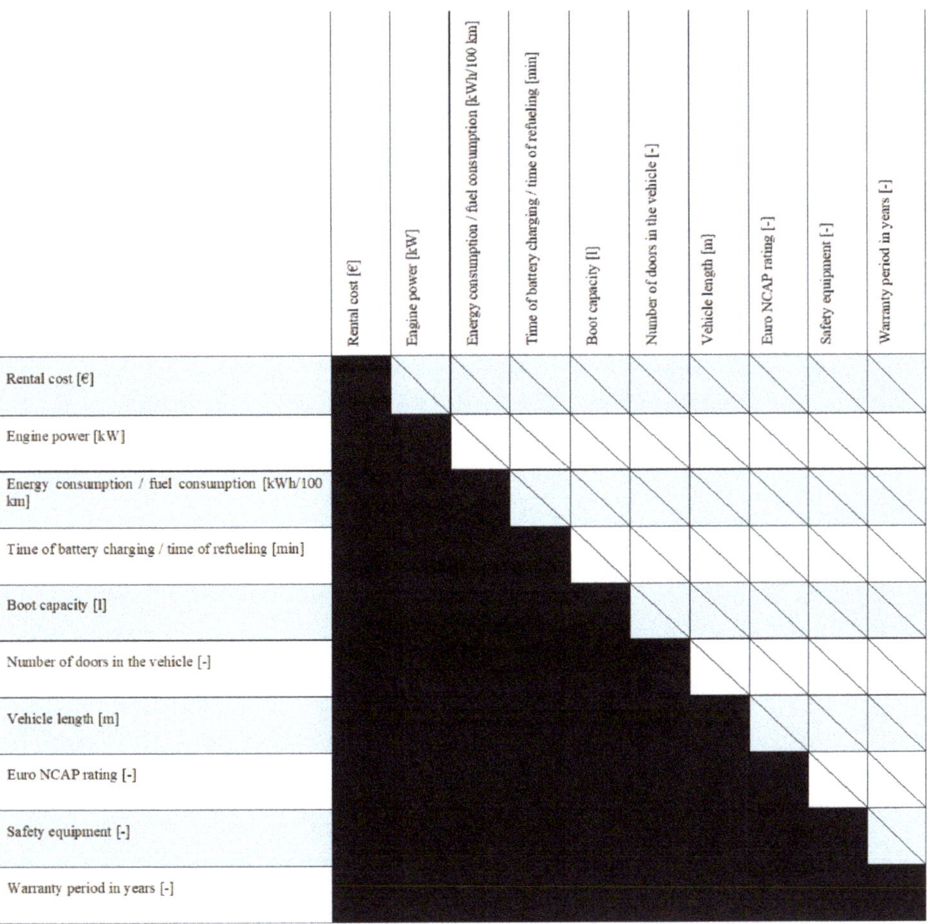

Figure 2. Matrix of pairwise comparisons.

Table 6. The set of thresholds for equivalence, preference, and veto.

Criteria Number	Maximum Difference of Criteria Values	Equivalence Threshold	Preference Threshold	Veto Threshold
	$\Delta = max - min$	$Q = 0.25 \times \Delta$	$p = 0.5 \times \Delta$	$V = \Delta$
C1	0.17	0.0425	0.085	0.17
C2	182	45.5	91	182
C3	27.5	6.875	13.75	27.5
C4	448.5	112.125	224.25	448.5
C5	314	78.5	157	314
C6	2	0.5	1	2
C7	1.07	0.2675	0.535	1.07
C8	4	1	2	4
C9	5	1.25	2.5	5
C10	3	0.75	1.5	3

Table 7. Concordance matrix values.

Variants	V1	V2	V3	V4	V5	V6	V7	V8	V9	V10	V11	V12
V1	-	1.0	0.9977	0.606	1.0	0.9977	0.8175	1.0	0.803	0.8775	0.548	0.5318
V2	1.0	-	0.9977	0.4047	0.9973	0.9816	0.7762	1.0	0.6612	0.8775	0.4951	0.443
V3	0.7734	0.8946	-	0.2775	0.8041	0.8014	0.6083	0.916	0.499	0.8775	0.382	0.3598
V4	0.85	0.85	0.9317	-	0.85	0.9317	0.7739	0.916	0.8742	0.8775	0.7464	0.7912
V5	0.9786	0.9854	0.9977	0.5903	-	0.9816	0.7184	1.0	0.7288	0.8775	0.4546	0.3839
V6	0.8946	0.9014	1.0	0.5999	0.916	-	0.7488	0.916	0.7128	0.8775	0.464	0.464
V7	1.0	1.0	1.0	0.691	1.0	1.0	-	1.0	0.803	0.8775	0.6875	0.6875
V8	0.5299	0.7279	0.7849	0.2795	0.7057	0.5066	0.3148	-	0.3638	0.8118	0.2394	0.1979
V9	0.8692	0.934	1.0	0.773	0.9352	0.9352	0.7297	1.0	-	0.8775	0.7647	0.8393
V10	0.5803	0.8033	0.9186	0.3486	0.7385	0.5779	0.4384	0.916	0.5775	-	0.4959	0.3625
V11	0.8692	0.934	0.9317	0.773	0.8692	0.8669	0.8669	1.0	0.9688	1.0	-	0.9977
V12	0.8692	0.934	0.9537	0.7835	0.8822	0.8822	0.8692	1.0	0.993	1.0	1.0	-

The next stage in the ELECTRE III method was to perform the ascending and descending distillations against each of the variants and create, in the final step, a dominance matrix. The dominance matrix is presented in Table 8.

Table 8. Dominance matrix values.

Variants	V1	V2	V3	V4	V5	V6	V7	V8	V9	V10	V11	V12
V1	-	E	R+	R−	E	E	R−	R+	R−	R+	R−	R−
V2	E	-	R+	R−	E	E	R−	R+	R−	R+	R−	R−
V3	R−	R−	-	R−	R−	R−	R−	R−	R−	R+	R−	R−
V4	R+	R+	R+	-	R+	R+	R+	R+	R	R+	R−	R−
V5	E	E	R+	R−	-	E	R−	R+	R−	R+	R−	R−
V6	E	E	R+	R−	E	-	R−	R+	R−	R+	R−	R−
V7	R+	R+	R+	R−	R+	R+	-	R+	R−	R+	R−	R−
V8	R−	R−	R+	R−	R-	R−	R-	-	R−	R+	R−	R−
V9	R+	R+	R+	R	R+	R+	R+	R+	-	R+	R−	R−
V10	R−	R−	R−	R−	R−	R−	R−	R−	R−	-	R−	R−
V11	R+	R+	R+	R+	R+	R+	R+	R+	R+	R+	-	R−
V12	R+	R+	R+	R+	R+	R+	R+	R+	R+	R+	R+	-

(E)—a pair of variants are equivalent; (R+)—the first variant is better than the second variant; (R−)—the first variant is worse than the second variant.

The last step was to prepare the final ranking that presents the variants in terms of the preferences of experts and the adopted factors. The final ranking is presented in Table 9.

Table 9. Final ranking.

Doinance Matrix	Ascend Distillation	Descend Distillation
V1	5	5
V2	5	5
V3	6	7
V4	2	4
V5	5	5
V6	5	5
V7	4	4
V8	6	6
V9	3	3
V10	7	7
V11	1	2
V12	1	1

The graphical arrangement of the variants is shown in Figure 3.

Figure 3. Final ranking—graphic visualization.

4. Discussion

The research, carried out using the ELECTRE III multi-criteria decision support method, allowed us to draw a ranking of the vehicle models that meet the expectations of users who rarely use car-sharing systems. According to the results, the best model turned out to be the V12 car model. The selected model is a mid-range electric crossover passenger car. The V11 variant took second place, and the variants V4 and V9 ex aequo were third.

When analyzing the results in detail, in terms of the vehicle size, it should be stated that the main positions were taken by the models representing the class D cars, i.e., the segment that includes middle-class passenger cars, relatively large and comfortable family and sports cars. This class includes classic passenger cars with dimensions larger than compact ones, ensuring a relatively comfortable ride for five people on longer journeys. Interestingly, the vehicles representing the smallest class of cars, i.e., A, were ranked the worst.

From the point of view of vehicle propulsion, the fully electric vehicle was classified as the highest in the ranking. Second place was also taken by a car with this type of drive. In turn, the third and fourth places are represented by cars with hybrid drives. Interestingly, the last places in the ranking were also taken by electric cars (variants V3 and V10). Such results indicate that, for the respondents, it was especially for not the fact that the vehicles had alternative propulsion but the detailed parameters characterizing individual vehicles.

When analyzing the results obtained from the point of view of the importance of individual criteria for users, it should be mentioned that the most important issues were engine power, boot capacity, rental cost, battery charging/refueling time, and safety equipment. This may prove that people who rarely use car-sharing vehicles want to use relatively large, spacious, and comfortable vehicles equipped with large luggage spaces, in which it will be possible to charge the battery or refuel the car in the shortest possible time. In addition, safety equipment issues were also important. Therefore, placing vehicles such as the V10 and V8 variants in the last places is because these cars represent class A, have a small load space, and have low engine performance. Interestingly, the size of the car, its capacity, and the performance of the engine weighed heavily on the cost of renting a car. Factors that came in the last positions deserve special attention. These were issues, such as the warranty period that were deliberately included in the analysis, as it is usually associated with high-quality vehicles. Factors such as NCAP safety and the number of doors in the vehicle were equally low rated. Such factors may indicate that the respondents treat car-sharing vehicles as an additional, occasional means of getting around, which they usually use alone or with one additional passenger, disregarding the facilities needed by families, such as more doors. It is also worth paying attention to the safety issues that did not turn out to be of key importance to the respondents, perhaps because the vehicles are not used by them frequently.

5. Conclusions

In conclusion, the research conducted allowed us to achieve the goal of the work, which was to select vehicles for car-sharing systems from the point of view of users who rarely use the services. The research showed that the V12 model representing the D vehicle class, equipped with a high-performance electric motor, was the best solution. Furthermore, it should be emphasized that the vehicles with alternative drives were placed in the highest rankings. By taking into account the detailed expectations of users about the fleet, it should be noted that the most important criteria include engine power, boot capacity, rental cost, battery charging/time of refueling, and safety equipment. Therefore, the fleet preferred by users who rarely use car-sharing systems is relatively large, spacious, and comfortable, with vehicles equipped with high-performance engines. By comparing the results obtained with real business practices, it should be noted that the V12 variant car, which leads in the ranking, is the main model used in the German car-sharing model, WeShare, in Berlin or Hamburg [58]. Therefore, these vehicles are successfully used in urban conditions, as evidenced by their use in large metropolitan car-sharing.

An interesting finding was that small city vehicles were ranked the lowest. It should be mentioned that the idea of car-sharing services assumed that the vehicles used in the systems would be small city cars, whose task would be to free public space [59]. However, the results obtained show that this type of vehicle will not be the first choice among users who rarely use car-sharing systems. This is a valuable note for car-sharing service operators who, when planning to diversify their fleet, should pay attention to the real needs of their users. Of course, from the point of view of public space, small city cars will be the best solution due to their dimensions, but then it is worth undertaking detailed research on the needs of users. The analyzed example shows that in the case of users rarely using car-sharing systems, small vehicles would not be rented, and as a result, a large number of them would remain unrolled in the city, becoming unprofitable for the operator and occupying public space. Taking into account the users' specific expectations, it can be assumed that if small vehicles were equipped with high-power engines and fast-charging batteries, they could significantly make gains in the final classification in the ranking. Furthermore, when considering the modernization of the vehicle fleet in the category of future rentals by rare users, the aspects of warranty, NCAP safety, or the number of doors should not be crucial. These considerations should provide important guidance to operators in their willingness to select other vehicles for their fleet than the models covered in this article.

This article has limitations. The main limitation was that the research only covered the Polish market. Moreover, they were devoted exclusively to a group of people who rarely use car-sharing systems. As there is no literature dedicated directly to the selection of the fleet of vehicles for car-sharing systems, the author did not refer to the research conducted by other authors in discussing the results.

In future work, the author plans to analyze other user groups to obtain the full range of user approaches to the vehicle fleet. In addition, the author plans to conduct research for countries other than Poland to compare users' preferences in terms of the vehicle fleet.

Funding: Publication supported under the rector's pro-quality grant. Silesian University of Technology, 12/010/RGJ22/1041.

Institutional Review Board Statement: According to our University Ethical Statement, following: the following shall be regarded as research requiring a favorable opinion from the Ethic Commission in the case of human research (based on document in polish: https://prawo.polsl.pl/Lists/Monitor/Attachments/7291/M.2021.501.Z.107.pdf (accessed on 21 March 2022): research in which persons with limited capacity to give informed or research on persons whose capacity to give informed or free consent to participate in research and who have a limited ability to refuse research before or during their implementation, in particular: children and adolescents under 12 years of age, persons with intellectual disabilities, persons whose consent to participate in the research may not be fully voluntary, prisoners, soldiers, police officers, employees of companies (when the survey is conducted at their workplace), persons who agree to participate in the research on the basis of false information about the purpose and course of the research (masking instruction, i.e., deception) or do not know at all that they are subjects (in so-called natural experiments); research in which persons particularly susceptible to psychological trauma and mental health disorders are to participate, mental health, in particular: mentally ill persons, victims of disasters, war trauma, etc., patients receiving treatment for psychotic disorders, family members of terminally or chronically ill patients; research involving active interference with human behavior aimed at changing it, research involving active intervention in human behavior aimed at changing that behavior without direct intervention in the functioning of the brain, e.g., cognitive training, psychotherapy psychocorrection, etc. (this also applies if the intended intervention is intended to benefit (this also applies when the intended intervention is to benefit the subject (e.g., to improve his/her memory); research concerning controversial issues (e.g., abortion, in vitro fertilization, death penalty) or requiring particular delicacy and caution (e.g., concerning religious beliefs or attitudes towards minority groups) minority groups); research that is prolonged, tiring, physically or mentally exhausting. Our research is not conducted on people meeting the mentioned condition. Any of the researched people, where any of them had limited capacity to be informed or any of them had been susceptible to psychological trauma and mental health disorders; the research did not concern the mentioned-above controversial issues; the research was not prolonged, tiring, physically or mentally exhausting.

Informed Consent Statement: Informed consent was obtained from all subjects involved in the study.

Data Availability Statement: The data presented in this study are available on request from the author.

Conflicts of Interest: The author declares no conflict of interest.

References

1. Cantelmo, G.; Amini, R.E.; Monteiro, M.M.; Frenkel, A.; Lerner, O.; Tavory, S.S.; Galtzur, A.; Kamargianni, M.; Shiftan, Y.; Behrischi, C.; et al. Aligning Users' and Stakeholders' Needs: How Incentives Can Reshape the Carsharing Market. *Transp. Policy* **2022**, *126*, S0967070X22001901. [CrossRef]
2. Jochem, P.; Frankenhauser, D.; Ewald, L.; Ensslen, A.; Fromm, H. Does Free-Floating Carsharing Reduce Private Vehicle Ownership? The Case of SHARE NOW in European Cities. *Transp. Res. Part A Policy Pr.* **2020**, *141*, 373–395. [CrossRef]
3. Global Market Insights. Car Sharing Market Size by Model (P2P, Station-Based, Free-Floating), by Business Model (Round Trip, One Way), by Application (Business, Private), COVID-19 Impact Analysis, Regional Outlook, Application Potential, Price Trend, Competitive Market Share & Forecast, 2021–2027. Available online: https://www.gminsights.com/industry-analysis/carsharing-market (accessed on 12 July 2022).
4. Globe Newswire. Car Sharing Market Trends 2021—Regional Statistics and Forecasts 2024 | Europe, North America & APAC: Graphical Research. Available online: https://www.globenewswire.com/news-release/2021/02/03/2168780/0/en/Car-Sharing-Market-Trends-2021-Regional-Statistics-and-Forecasts-2024-Europe-North-America-APAC-Graphical-Research.html (accessed on 9 August 2022).
5. Del Mar Alonso-Almeida, M. To Use or Not Use Car Sharing Mobility in the Ongoing COVID-19 Pandemic? Identifying Sharing Mobility Behaviour in Times of Crisis. *Int. J. Environ. Res. Public Health* **2022**, *19*, 3127. [CrossRef]
6. Yao, Z.; Gendreau, M.; Li, M.; Ran, L.; Wang, Z. Service Operations of Electric Vehicle Carsharing Systems from the Perspectives of Supply and Demand: A Literature Review. *Transp. Res. Part C Emerg. Technol.* **2022**, *140*, 103702. [CrossRef]
7. Aguilera-García, Á.; Gomez, J.; Antoniou, C.; Vassallo, J.M. Behavioral Factors Impacting Adoption and Frequency of Use of Carsharing: A Tale of Two European Cities. *Transp. Policy* **2022**, *123*, 55–72. [CrossRef]
8. Vanheusden, W.; van Dalen, J.; Mingardo, G. Governance and Business Policy Impact on Carsharing Diffusion in European Cities. *Transp. Res. Part D Transp. Environ.* **2022**, *108*, 103312. [CrossRef]
9. Friesen, M.; Mingardo, G. Is Parking in Europe Ready for Dynamic Pricing? A Reality Check for the Private Sector. *Sustainability* **2020**, *12*, 2732. [CrossRef]
10. Ciari, F.; Bock, B.; Balmer, M. *Modeling Station-Based and Free-Floating Carsharing Demand: A Test Case Study for Berlin, Germany;* Emerging and Innovative Public Transport and Technologies, Transportation Research Board of the National Academies: Washington, DC, USA, 2014.
11. Ferrero, F.; Perboli, G.; Rosano, M.; Vesco, A. Car-sharing services: An annotated review. *Sustain. Cities Soc.* **2018**, *37*, 501–518. [CrossRef]
12. Nourinejad, M.; Roorda, M. Carsharing operations policies: A comparison between one-way and two-way systems. *Transportation* **2015**, *42*, 97–518. [CrossRef]

13. Shaheen, S.; Chan, N.; Bansal, A.; Cohen, A. *Shared Mobility—A Sustainability & Technologies Workshop, Definitions, Industry Developments and Early Understanding*; University of California, Transportation Sustainability Research Center: Barkeley, CA, USA, 2015; pp. 1–30.
14. Sprei, F.; Habibi, S.; Englund, C.; Pettersson, S.; Voronov, A.; Wedlin, J. Free-Floating Car-Sharing Electrification and Mode Displacement: Travel Time and Usage Patterns from 12 Cities in Europe and the United States. *Transp. Res. Part D Transp. Environ.* **2019**, *71*, 127–140. [CrossRef]
15. Changaival, B.; Lavangnananda, K.; Danoy, G.; Kliazovich, D.; Guinand, F.; Brust, M.; Musial, J.; Bouvry, P. Optimization of Carsharing Fleet Placement in Round-Trip Carsharing Service. *Appl. Sci.* **2021**, *11*, 11393. [CrossRef]
16. Ströhle, P.; Flath, C.M.; Gärttner, J. Leveraging Customer Flexibility for Car-Sharing Fleet Optimization. *Transp. Sci.* **2019**, *53*, 42–61. [CrossRef]
17. Monteiro, C.M.; Machado, C.A.S.; de Oliveira Lage, M.; Berssaneti, F.T.; Davis, C.A.; Quintanilha, J.A. Optimization of Carsharing Fleet Size to Maximize the Number of Clients Served. *Comput. Environ. Urban Syst.* **2021**, *87*, 101623. [CrossRef]
18. Lemme, R.F.F.; Arruda, E.F.; Bahiense, L. Optimization Model to Assess Electric Vehicles as an Alternative for Fleet Composition in Station-Based Car Sharing Systems. *Transp. Res. Part D Transp. Environ.* **2019**, *67*, 173–196. [CrossRef]
19. Carlier, A.; Munier-Kordon, A.; Klaudel, W. Optimization of a one-way carsharing system with relocation operations. In Proceedings of the 10th International Conference on Modeling, Optimization and SIMulation MOSIM 2014, Nancy, France, 5–7 November 2014.
20. Nourinejad, M.; Roorda, M.J. A dynamic carsharing decision support system. *Transp. Res. Part E Logist. Transp. Rev.* **2014**, *66*, 36–50. [CrossRef]
21. Barrios, J.A.; Godier, J.D. Fleet Sizing for Flexible Carsharing Systems: Simulation-Based Approach. *Transp. Res. Rec.* **2014**, *2416*, 1–9. [CrossRef]
22. Lu, M.; Chen, Z.; Shen, S. Optimizing the Profitability and Quality of Service in Carshare Systems Under Demand Uncertainty. *Manuf. Serv. Oper. Manag.* **2018**, *20*, 162–180. [CrossRef]
23. Hui, Y.; Wang, Y.; Sun, Q.; Tang, L. The Impact of Car-Sharing on the Willingness to Postpone a Car Purchase: A Case Study in Hangzhou, China. *J. Adv. Transp.* **2019**, *2019*, 1–11. [CrossRef]
24. Jain, T.; Rose, G.; Johnson, M. Changes in Private Car Ownership Associated with Car Sharing: Gauging Differences by Residential Location and Car Share Typology. *Transportation* **2022**, *49*, 503–527. [CrossRef]
25. Liao, F.; Molin, E.; Timmermans, H.; van Wee, B. Carsharing: The Impact of System Characteristics on Its Potential to Replace Private Car Trips and Reduce Car Ownership. *Transportation* **2020**, *47*, 935–970. [CrossRef]
26. Szałek, A.; Pielecha, I.; Cieslik, W. Fuel Cell Electric Vehicle (FCEV) Energy Flow Analysis in Real Driving Conditions (RDC). *Energies* **2021**, *14*, 5018. [CrossRef]
27. Gschwendtner, C.; Krauss, K. Coupling Transport and Electricity: How Can Vehicle-to-Grid Boost the Attractiveness of Carsharing? *Transp. Res. Part D Transp. Environ.* **2022**, *106*, 103261. [CrossRef]
28. Cieslik, W.; Szwajca, F.; Rosolski, S.; Rutkowski, M.; Pietrzak, K.; Wójtowicz, J. Historical Buildings Potential to Power Urban Electromobility: State-of-the-Art and Future Challenges for Nearly Zero Energy Buildings (nZEB) Microgrids. *Energies* **2022**, *15*, 6296. [CrossRef]
29. Wielinski, G.; Trépanier, M.; Morency, C. Electric and Hybrid Car Use in a Free-Floating Carsharing System. *Int. J. Sustain. Transp.* **2017**, *11*, 161–169. [CrossRef]
30. Pielecha, I.; Cieślik, W.; Szałek, A. Energy Recovery Potential through Regenerative Braking for a Hybrid Electric Vehicle in a Urban Conditions. *IOP Conf. Ser. Earth Environ. Sci.* **2019**, *214*, 012013. [CrossRef]
31. Migliore, M.; D'Orso, G.; Caminiti, D. The environmental benefits of carsharing: The case study of Palermo. *Transp. Res. Procedia* **2020**, *48*, 2127–2139. [CrossRef]
32. Shaheen, S.; Martin, E.; Totte, H. Zero-Emission Vehicle Exposure within U.S. Carsharing Fleets and Impacts on Sentiment toward Electric-Drive Vehicles. *Transp. Policy* **2020**, *85*, A23–A32. [CrossRef]
33. Liao, F.; Correia, G. Electric Carsharing and Micromobility: A Literature Review on Their Usage Pattern, Demand, and Potential Impacts. *Int. J. Sustain. Transp.* **2022**, *16*, 269–286. [CrossRef]
34. Turoń, K.; Kubik, A.; Łazarz, B.; Czech, P.; Stanik, Z. Car-sharing in the context of car operation. *IOP Conf. Ser. Mater. Sci. Eng.* **2018**, *421*, 032027. [CrossRef]
35. Turoń, K.; Kubik, A.; Chen, F. Operational Aspects of Electric Vehicles from Car-Sharing Systems. *Energies* **2019**, *12*, 4614. [CrossRef]
36. Turoń, K.; Kubik, A.; Chen, F. What Car for Car-Sharing? Conventional, Electric, Hybrid or Hydrogen Fleet? Analysis of the Vehicle Selection Criteria for Car-Sharing Systems. *Energies* **2022**, *15*, 4344. [CrossRef]
37. Puzio, E. The development of shared mobility in Poland using the example of a city bike system. *Res. Pap. Wrocław. Univ. Econ.* **2020**, *64*, 162–170. [CrossRef]
38. Statista. Forecast Revenues from Carsharing Services in Poland from 2019 to 2025. Available online: https://www.statista.com/statistics/1059362/poland-carsharing-revenues/ (accessed on 5 June 2022).
39. Roy, B. *How Outranking Relation Halps Multiple Criteria Decision Making*; University of South Carolina Press: Columbia, SC, USA, 1973.

40. Geneletti, D. Multi-Criteria Analysis. LIAISE Toolbox. Available online: http://beta.liaise-toolbox.eu/ia-methods/multi-criteria-analysis (accessed on 15 July 2022).
41. Shaheen, S.; Cohen, A. Carsharing and Personal Vehicle Services: Worldwide Market Developments and Emerging Trends. *Int. J. Sustain. Transp.* **2013**, *7*, 5–34. [CrossRef]
42. Shaheen, S.; Cohen, A. Innovative Mobility Carsharing Outlook Winter 2016: Carsharing Market Overview, Analysis, and Trends Innovative Mobility Carsharing Outlook—Winter 2016. Available online: http://innovativemobility.org/?project=innovative-mobility-carsharing-outlook-winter-2016 (accessed on 17 March 2018).
43. Istanbul Metropolitan Municipality & Japan International Cooperation Agency, the Study on Integrated Urban Transport Master Plan for Istanbul Metropolitan Area in the Republic of Turkey. Available online: https://openjicareport.jica.go.jp/pdf/11965720 _01.pdf (accessed on 5 June 2022).
44. Li, W.; Li, Y.; Fan, J.; Deng, H. Siting of Carsharing Stations Based on Spatial Multi-Criteria Evaluation: A Case Study of Shanghai EVCARD. *Sustainability* **2017**, *9*, 152. [CrossRef]
45. Jahan, A.; Edwards, K.L. Multi-criteria Decision-Making for Materials Selection. In *Multi-Criteria Decision Analysis for Supporting the Selection of Engineering Materials in Product Design*; Butterworth-Heinemann: Oxford, UK, 2013; pp. 31–41. [CrossRef]
46. Awasthi, A.; Breuil, D.; Chauhan, S.S.; Parent, M.; Reveillere, T. A Multicriteria Decision Making Approach for Carsharing Stations Selection. *J. Decis. Syst.* **2007**, *16*, 57–78. [CrossRef]
47. Saaty, T. How to make decision: The analytic hierarchy process. *Eur. J. Oper. Res.* **1990**, *48*, 9–26. [CrossRef]
48. Kobryń, A. *Wielokrotne Wspomaganie Decyzji w Gospodarowaniu Przestrzenią*; Difin: Warszaw, Poland, 2014.
49. Zlaugotne, B.; Zihare, L.; Balode, L.; Kalnbalkite, A.; Khabdullin, A.; Blumberga, D. Multi-Criteria Decision Analysis Methods Comparison. *Environ. Clim. Technol.* **2020**, *24*, 454–471. [CrossRef]
50. Figueira, J.R.; Greco, S.; Roy, B.; Słowiński, R. ELECTRE methods: Main features and recent developments. In *Handbook of Multicriteria Analysis*; Springer: Berlin/Heidelberg, Germany, 2010; pp. 51–89.
51. Norese, M.F. ELECTRE III as a support for participatory decision-making on the localisation of waste-treatment plants. *Land Use Policy* **2006**, *23*, 76–85. [CrossRef]
52. La Scalia, G.; Micale, R.; Certa, A.; Enea, M. Ranking of shelf life models based on smart logistic unit using the ELECTRE III method. *Int. J. Appl. Eng. Res.* **2015**, *10*, 38009–38015.
53. Saaty, R.W. The Analytic Hierarchy Process—What It Is and How It Is Used. *Math. Model.* **1987**, *9*, 161–176. [CrossRef]
54. Battisti, F. ELECTRE III for Strategic Environmental Assessment: A "Phantom" Approach. *Sustainability* **2022**, *14*, 6221. [CrossRef]
55. López, J.C.L.; Solares, E.; Figueira, J.R. An Evolutionary Approach for Inferring the Model Parameters of the Hierarchical Electre III Method. *Inf. Sci.* **2022**, *607*, 705–726. [CrossRef]
56. Auto Magazine. Top 20: The Most Popular Models of the 20 Brands in Europe. Available online: https://magazynauto.pl/ wiadomosci/top-20-najpopularniejsze-modele-20-marek-w-europie,aid,1335 (accessed on 10 July 2022).
57. Auto World Portal. The Most Popular Car Classes in Poland. Available online: https://www.auto-swiat.pl/wiadomosci/ aktualnosci/najpopularniejsze-klasy-samochodow-w-polsce/gmkcf8w (accessed on 23 August 2022).
58. WeShare Car-Sharing Operator. Available online: https://www.we-share.io/en (accessed on 24 August 2022).
59. Glotz-Richter, M. Car-Sharing—"Car-on-call" for reclaiming street space. *Procedia—Soc. Behav. Sci.* **2012**, *48*, 1454–1463. [CrossRef]

Article

Linear Quadratic Regulator and Fuzzy Control for Grid-Connected Photovoltaic Systems

Azamat Mukhatov [1], Nguyen Gia Minh Thao [2,*] and Ton Duc Do [1,*]

1   Department of Robotics and Mechatronics, School of Engineering and Digital Sciences, Nazarbayev University, Nur-Sultan 010000, Kazakhstan; azamat.mukhatov@nu.edu.kz
2   Graduate School of Engineering, Toyota Technological Institute, Nagoya 468-8511, Japan
*   Correspondence: nguyen.thao@toyota-ti.ac.jp (N.G.M.T.); doduc.ton@nu.edu.kz (T.D.D.)

Abstract: This work presents a control scheme to control a grid-connected single-phase photovoltaic (PV) system. The considered system has four 250 W solar panels, a non-inverting buck-boost DC-DC converter, and a DC-AC inverter with an inductor-capacitor-inductor (LCL) filter. The control system aims to track and operate at the maximum power point (MPP) of the PV panels, regulate the voltage of the DC link, and supply the grid with a unity power factor. To achieve these goals, the proposed control system consists of three parts: an MPP tracking controller module with a fuzzy-based modified incremental conductance (INC) algorithm, a DC-link voltage regulator with a hybrid fuzzy proportional-integral (PI) controller, and a current controller module using a linear quadratic regulator (LQR) for grid-connected power. Based on fuzzy control and an LQR, this work introduces a full control solution for grid-connected single-phase PV systems. The key novelty of this research is to analyze and prove that the newly proposed method is more successful in numerous aspects by comparing and evaluating previous and present control methods. The designed control system settles quickly, which is critical for output stability. In addition, as compared to the backstepping approach used in our past study, the LQR technique is more resistant to sudden changes and disturbances. Furthermore, the backstepping method produces a larger overshoot, which has a detrimental impact on efficiency. Simulation findings under various weather conditions were compared to theoretical ones to indicate that the system can deal with variations in weather parameters.

Keywords: fuzzy control; grid-connected PV system; incremental conductance algorithm; linear quadratic regulator; maximum power point tracking; unity power factor

Citation: Mukhatov, A.; Thao, N.G.M.; Do, T.D. Linear Quadratic Regulator and Fuzzy Control for Grid-Connected Photovoltaic Systems. *Energies* **2022**, *15*, 1286. https://doi.org/10.3390/en15041286

Academic Editors: Marco Pasetti and Wojciech Cieslik

Received: 2 December 2021
Accepted: 1 February 2022
Published: 10 February 2022

Publisher's Note: MDPI stays neutral with regard to jurisdictional claims in published maps and institutional affiliations.

1. Introduction

Renewable energy is emerging as one of the main sources of energy for the future. The key reason for this is the depletion and pollution of fossil fuels. Renewable energy sources are available, clean, eco-friendly, and cost-effective. There are various types of renewable energy sources, of which solar and wind energy systems have become more and more popular in many countries. According to [1,2], harmonic resonances, which often occur in grid-connected wind power farms, cause negative effects on the power quality of the grid.

Nowadays, solar energy is widely used around the world and demonstrates impressive results. To effectively obtain electricity from solar energy, photovoltaic (PV) systems should be installed. The system efficiency is strongly affected by two major factors, as follow [3,4]:

(a)   Weather factors such as the temperature and solar radiation;
(b)   Hardware factors such as power electronic devices and system loads.

While the prior factor is uncontrollable, the second one depends on the designer, system operator and electric grid. To improve the efficiency of the power electronic parts, appropriate converter topologies together with efficient control schemes are required. From [5–8], there are two modes of operation for the PV systems, which are:

•   Stand-alone mode;

- Grid-connected mode.

Between these two modes of operation, the grid-connected mode is preferable, as it can avoid the issues of storage systems in the stand-alone mode. For grid-connected systems, the following two problems need to be solved simultaneously [7–9]:

- Management of several combined systems;
- Regulation of each power stage or system.

Solving the second problem often requires the following tasks:

- tracking the maximum power point (MPP);
- minimizing the harmonics, which usually cause negative effects on the power grid and devices;
- maintaining the DC-link voltage within a desired range;
- keeping the unity power factor (PF) at the output of the filter [10].

One of the most important parts of this research is the MPP tracking part, which is mainly used to find and keep the output power of a PV panel at its maximum value [11–13]. The MPP tracking (MPPT) technique can be divided into two main categories: the perturb and observe (PO) technique and the incremental conductance (INC) algorithm. The negative side of the PO algorithm is its high computational complexity, but it leads to high efficiency. It measures voltage and current values to periodically estimate the power of the solar panel and compares it with the previous power. If the power of the PV module has increased $(dP/dV > 0)$, the system will start adjustments in that direction; otherwise, it will adjust in the opposite way. These operations continue until system finds the MPP. In fact, the technique depends on perturbations of the voltage, so if the perturbations are high, the speed of the technique is fast. The advantages of this method are simplicity without interest in the previous PV characteristics; however, the main drawback is that oscillations happening near the MPP, which may lead to power losses in varying weather conditions [10]. The INC type is more advantageous in terms of accuracy in finding and tracking the MPP compared to the second type; therefore, in this paper, the INC algorithm is improved by fuzzy control and then implemented in the grid-connected PV system.

Considering current controller strategies, generally, they can be divided into two main categories: on/off controllers and pulse width modulation (PWM)-based control techniques [14]. The first group has two subdivisions, which are hysteresis control and predictive control. Hysteresis control has high dynamics and fast response; however, its major drawbacks are variations of the switching frequency and high complexity of the system. Predictive control has positive sides such as less computation time, better regulation, and a decrease in offset error. On the other hand, it requires identifying a proper model for the system and the installation cost of the system is high. The second group can be divided into linear and non-linear control [14]. Proportional-integral (PI) controller and its different update versions such as multiple generalized integral (MGI) [15] and Manta Ray Foraging Optimization [16] control are the well-known classical control techniques, along with their improved versions, which can be easily designed to control the current. However, the key disadvantage of this controller is its poor compensation of lower-order harmonics and presence of steady-state errors [17]. The proportional-resonant (PR) controlled can compensate for the harmonics. Moreover, this type of controller has high dynamics, is less complex, and can reach a high gain at the resonance frequency. However, this controller has problems reaching the power factor control, which means that the system is not able to control the losses in the system [18].

Generally, the power factor is a ratio between working power and apparent power. Thus, if there is no control/maintenance of a high-power factor, the system efficiency is consequently low. On the other hand, predictive deadbeat control has a high level of harmonic compensation and rapid fast-tracking performance. The disadvantage of this controller is that it requires a lot of computation effort [19]. Harmonic compensation and steady-state error can also be done by repetitive controllers; their slow tracking response is their main drawback [20]. In various dynamic conditions, the effectiveness of the third order complex filter (TOCF) control algorithm may be demonstrated. This controller serves as a distribution static compensator, enhances grid power quality by reducing harmonics, and balances grid currents while maintaining a unity power factor. However, the controller's biggest disadvantage is its enormous computational load [21]. Another method which possibly could be proposed for mitigation of the issue is the model predictive control (MPC) method [22] and its improved versions [23,24]. However, these methods present oscillations when the load varies. In addition, the THD of MPC is higher compared to other methods, which makes the technique unsuitable for implementation in this case. There exist different kinds of adaptive filtering techniques, such as leaky least mean mixed norm (LLMMN), least mean mixed norm (LMMN), and least mean square (LMS). However, all of the above-mentioned adaptive approaches have the drawback of being unable to maintain acceptable performance in the presence of DC-offset [25]. The combined affine projection sign algorithm (CAPSA) has high stability and tracking performance, but output has the same problem as the previous mentioned method, which is the high THD [26]. Artificial neural networks (ANN) also can be implemented as controllers in this system, but the main disadvantage are their low output power compared to fuzzy logic [27]. Fuzzy logic controllers (FLC) [14] are one of the popular intelligent control techniques. They are extensively used in renewable energy systems due to their efficiency and ease of use. They are also robust and applicable to a wide range of the dynamics systems, from linear to nonlinear systems. Moreover, this type of controller can perform complex estimations, which are not possible with conventional methods [11]. The linear quadratic regulator (LQR) is an effective control method that is applicable for both linear and nonlinear systems. In this method, the control gain is designed to minimize a quadratic cost function by selection of appropriate weighting matrices. In our study case, the cost function is the quadratic function of the tracking error between the current and its reference and the control efforts. This technique was chosen to be implemented due to its properties such as stability, robustness, and ease in application. Moreover, the computational complexity of the LQR controller is not high, which means that it is fairly simple to implement.

This paper proposes a complete control solution for grid-connected single-phase PV systems based on fuzzy control and an LQR. Our past related research on this topic was conducted with a different type of controller, namely the backstepping approach. The present study is a significant extension and improvement of our former research in terms of enhancing the quality of the control method. The proposed technique is the LQR in appropriate combination with fuzzy control and improved INC algorithms for grid-connected photovoltaic systems; furthermore, detailed explanations on developing the fuzzy association rules of the designed fuzzy logic controllers are newly presented in this study. The main originality of this paper is to show and prove by comparison of our former and present control methods that the newly suggested method is more effective in various aspects. Specific details of the PV system and controllers can be found in our past work in [28], which was used as the basis for this paper. The major advantage of the LQR method is its ability to react in a rapid manner to changes of the system, namely, changes in the module temperature or solar irradiation. In other words, the system can reach its settling time faster, which is important to stabilize the behavior of output. Moreover, it can be said that the LQR technique is robust when faced with different disturbances and changes compared to the backstepping technique and its improved fault tolerant version [29]. In addition, the backstepping method has a higher overshoot, which significantly impacts efficiency in a negative way. As was mentioned above, the speed of the LQR is faster, which makes this kind of controller preferable. These are the key contributions of this study compared to our past research in [28]. Simulation results under different weather conditions show that the proposed control system can cope with changes in weather parameters effectively, and were compared with theoretical ones. Moreover, it was shown that variations in weather parameters do not significantly affect the performance of the proposed control system.

The remains of the paper are organized as follows. Section 2 shows the modeling of the grid-connected PV system, which includes the system description, PV panel model, and modeling of power converters. The control system design is depicted in Section 3, which consists of the MPPT control module, DC-link voltage regulator module, and current controller module. In addition, Section 4 provides simulation results in MATLAB, in which the first test case is with a fixed module temperature, and the second test case is with an unchanged solar irradiation. A detailed comparison and assessment of efficacy between the LQR control method in this study and the backstepping approach in our past work [28] is presented in Section 5; brief comparisons between this research and other related works are also shown in this section. The conclusions are described in the last section.

2. PV Grid-Connected System Modeling

2.1. System Description

This paper considers a grid-connected PV system consisting of two stages of power conversion. The nominal power of the system is 1 kW. Figure 1 shows the circuitry of the system; the power generated from the PV array is directed to the non-inverting buck-boost DC-DC converter. After that, to supply the grid, the obtained result is converted to AC via the single-phase DC-AC inverter. To remove unwanted noises and disturbances injected to the grid, the LCL output filter was used [30,31].

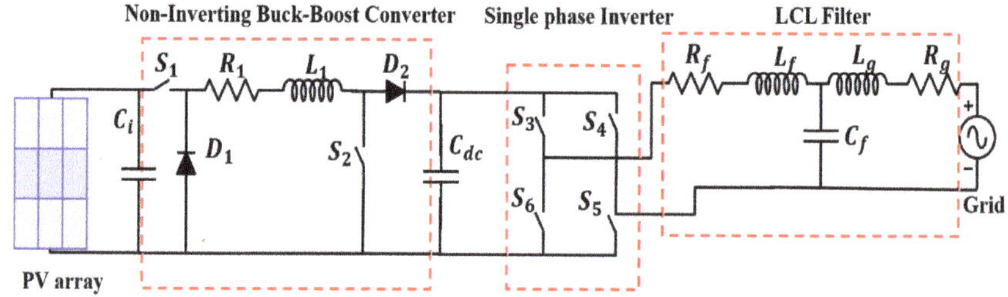

Figure 1. PV single-phase grid-connected system.

2.2. PV Panel Model

The PV panels used in this paper have characteristics as presented in [28]. The provided data is applicable when the module temperature is 25 °C and the solar radiation is 1000 W/m^2. In total, the PV array consists of four panels, where the nominal power of each panel is 250 W. Figure 2 shows the impacts of module temperature and solar radiation on the power and voltage of the PV panel, respectively. Table 1 represents MPPs of the PV panel and array in terms of power and voltage.

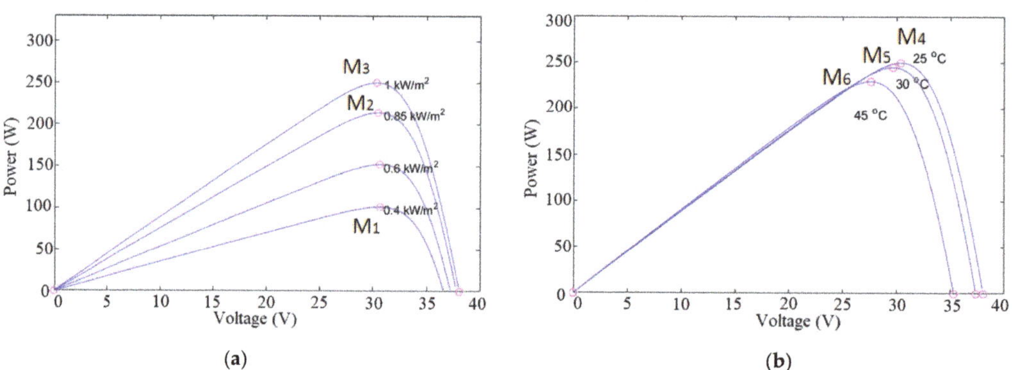

Figure 2. Power obtained from PV panels for different (**a**) solar radiation values and (**b**) module temperature values.

Table 1. Power and Voltage at Maximum Power Points.

MPPs	M1	M2	M3/M4	M5	M6
V_p (V)	29.3	30.32	30.4	29.8	28.04
$P_{P,panel}$ (W)	95.9	211.7	250	244.8	225.7
$P_{P,array}$ (W)	383.6	846.8	1000	979.2	912.8

2.3. Modeling of Converters

Figure 1 illustrates all components of the system including the single-phase inverter and the non-inverting buck-boost converter [7]. The input control signals of the non-inverting buck-boost converter and the single-phase inverter are α_p and β_p, respectively.

$$\alpha_p = \begin{cases} 0; & S_1 \text{ and } S_2 \text{ are OFF} \\ 1; & S_1 \text{ and } S_2 \text{ are ON} \end{cases} \quad \beta_p = \begin{cases} 1; & S_3 \text{ and } S_5 \text{ are ON, } S_4 \text{ and } S_6 \text{ are OFF} \\ 0; & S_3, S_4, S_5 \text{ and } S_6 \text{ are OFF} \\ -1; & S_3 \text{ and } S_5 \text{ are OFF, } S_4 \text{ and } S_6 \text{ are ON} \end{cases}$$

The modeling technique, specifically averaging, and Kirchhoff's laws were used to estimate a mathematical model for the two converters. Equation (1) and Table 2 demonstrate details of the previously mentioned procedure

$$\begin{cases} \dot{x}_1 = \frac{1}{C_i}\bar{I}_P - \alpha\frac{1}{C_i}x_2 \\ \dot{x}_2 = \alpha\frac{1}{L_1}x_1 - \frac{R_1}{L_1}x_2 + (\alpha - 1)\frac{1}{L_1}x_3 \\ \dot{x}_3 = (1 - \alpha)\frac{1}{C_{DC}}x_2 - \beta\frac{1}{C_{DC}}x_4 \\ \dot{x}_4 = \beta\frac{1}{L_f}x_3 - \frac{R_f}{L_f}x_4 - \frac{1}{L_f}x_5 \\ \dot{x}_5 = \frac{1}{C_f}x_4 - \frac{1}{C_f}x_6 \\ \dot{x}_6 = \frac{1}{L_g}x_5 - \frac{R_g}{L_g}x_6 - \frac{1}{L_g}V_g \end{cases} \tag{1}$$

Table 2. Variables.

Variable	Symbol in Figure 1	Averaged Variable in (1)
PV array voltage	V_p	x_1
Current through the inductor L_1	i_{L1}	x_2
DC link voltage	V_{DC}	x_3
Input current of the LCL filter	i_f	x_4
Voltage on the capacitor C_f	V_{Cf}	x_5
RMS value of the electric grid current	i_g	x_6
Control signal of the non-inverting buck-boost DC-DC converter	α_p {0,1}	α (0,1)
Control signal of the single-phase DC-AC inverter	β_p {−1,0,1}	β [−1,1]
PV array current	I_p	I_p
RMS value of the electric grid voltage	V_g	V_g

3. Control System Design

The design of the control system considered in this study is shown in Figure 3 and includes three main parts: the MPPT controller, the DC link voltage regulator, and the current controller. In this paper, detailed explanations on developing the fuzzy rules of the two designed fuzzy logic controllers are presented, which are useful as references for designing other fuzzy controllers.

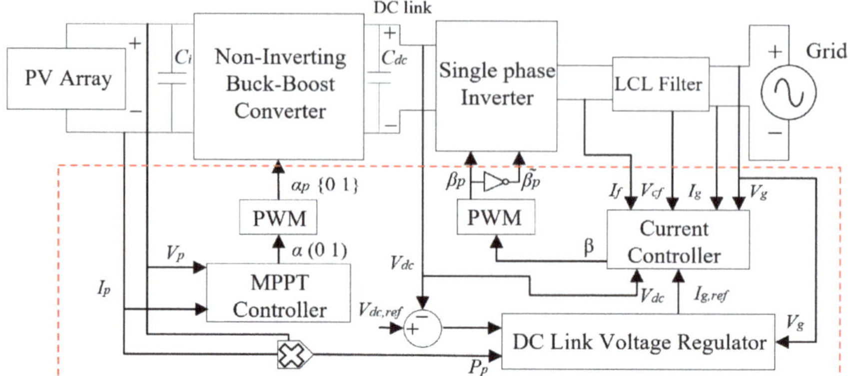

Figure 3. System schematics in detail.

3.1. The MPPT Controller Module

The PV array produces its optimal power despite varying weather with the help of the designed MPPT controller. According to Figure 4, this controller has two parts: the first fuzzy logic controller (FLC-1) and the proportional-integral (PI-1) controller.

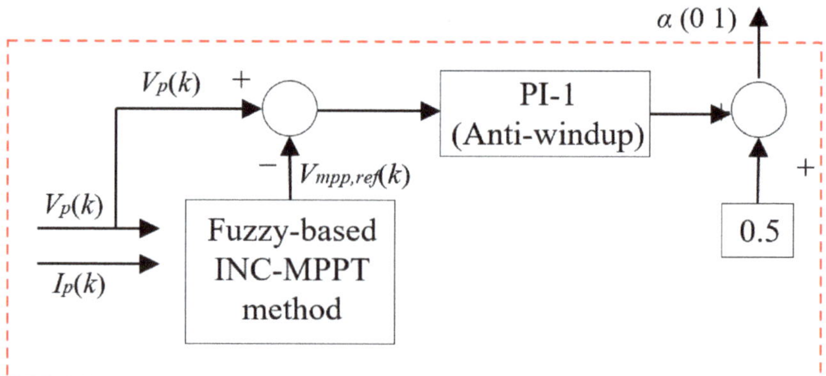

Figure 4. MPPT controller module schematics.

3.1.1. FLC-1

The main idea of this sub-controller is to improve the conventional INC-MPPT algorithm in terms of response time and efficiency by combining it with a fuzzy logic controller (FLC-1). According to Figure 5, the FLC-1 has two inputs and one output. The first input can be one of the following two kinds:

- $|A_p(k)|$—the absolute value of a modified slope of the power–voltage (P-V) curve as expressed in (2). This equation also includes a pre-scaling module $G_p(k)$, as shown in (3);
- $|dI_p(k)|$—the change in the current of PV panels in absolute value.

$$A_P(k) = G_P(k)[S_P(k)] = G_P(k)\left[I_p(k) + V_p(k)\frac{dI_P(k)}{dV_P(k)}\right] \tag{2}$$

$$G_P(k) = \frac{1}{1 + g_1\left[\frac{P_P(k)}{P_{P,total}^{max}}\right]} \tag{3}$$

where $P_{P,total}^{max}$ = 1000 W is the maximum power of PV panels and g_1 is a positive coefficient. The second input is the INC algorithm's prior step-size $\Delta V(k-1)$. Figure 6 shows the detailed flowchart of the proposed method. In addition, the aforementioned scaling module $G_p(k)$ is used to suitably increase the sensitivity of slope $S_p(k)$ as given in Figure 7.

$$\lim_{Pp(k)\to 0} G_P(k) = 1, \quad \lim_{Pp(k)\to P^{max}_{P,total}} G_P(k) = \frac{1}{1+g1}$$

$$P_P(k) = V_P(k) \times I_P(k) \tag{4}$$

$$dI_P(k) = I_P(k) - I_P(k-1) \tag{5}$$

$$dV_P(k) = V_P(k) - V_P(k-1) \tag{6}$$

To avoid significant changes in the step-size and instability of the PV output power, a switching module is implemented as described in Figure 5. According to the first input, namely, $|A_p(k)|$ or $|dI_p(k)|$, the system will use the appropriate output coefficient g_2 as shown in Table 3.

Table 3. Switching module operation.

If the Input Is $|A_P(k)|$		If the Input Is $|dI_P(k)|$
$\|A_P(k)\| > 0.1$ $g_2 = 0.25$	$\|A_P(k)\| \leq 0.1$ $g_2 = 0.1$	$g_2 = 1$ (where every value of $\|dI_P(k)\|$)

As is known from previous parts of this paper, the inputs are in the range of [0,1]. It should be noted that all the inputs have the same number of linguistic variables, specifically five linguistic variables: VS—Very Small, SM—Small, ME—Medium, LA—Large, VL—Very Large. The output has nine linguistic variables in a range of $[-1;1]$; in detail, NL—Negative Large, NM—Negative Medium, NS—Negative Small, NZ—Negative Zero, ZE—Zero, PZ—Positive Zero, PS—Positive Small, PM—Positive Medium, PL—Positive Large. As a result, there are 49 fuzzy rules associated in the FLC-1.

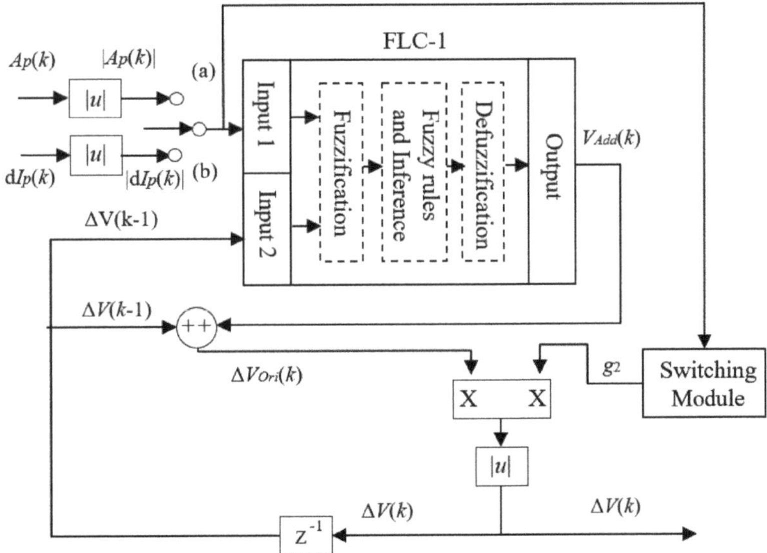

Figure 5. FLC-1 structure.

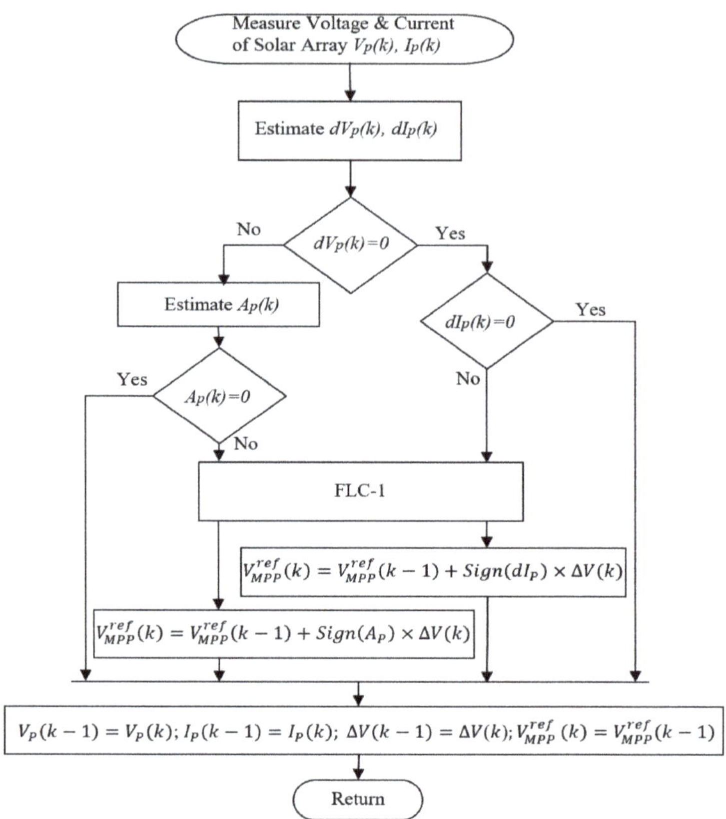

Figure 6. INC-MPPT algorithm with fuzzy logic.

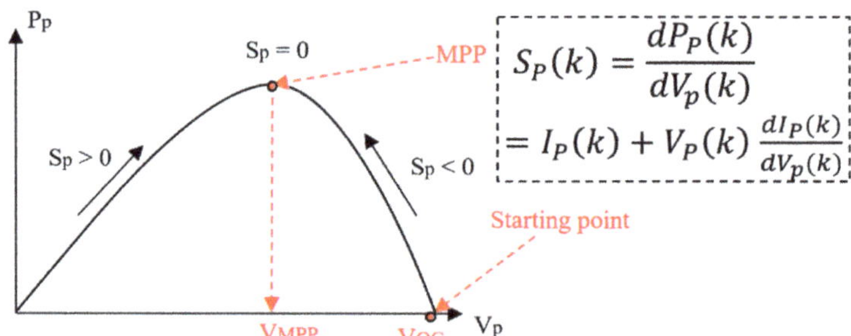

$$S_P(k) = \frac{dP_P(k)}{dV_p(k)}$$
$$= I_P(k) + V_P(k)\frac{dI_P(k)}{dV_p(k)}$$

Figure 7. PV panel power–voltage (P-V) curve.

All the association rules of the FLC-1 are shown in Table 4, while the membership functions of the inputs and output can be referred to in [28]. To explain the fuzzy rules in Table 4, we can analyze several sample cases as follows. In the first case, when the two inputs $\Delta V(k-1)$ and $|dI_p(k)|$ are VS, that means that the PV system is close to the MPP and the step voltage is also very small; thus, the output of the FLC-1 as the additional voltage $V_{add}(k)$ should be ZE to avoid fluctuations in the PV voltage at the steady state. Whereas, in another case when $\Delta V(k-1)$ is LA and $|dI_p(k)|$ is ME, the additional voltage $V_{add}(k)$ will be NZ because the tendency of the PV system is automatically approaching the MPP. On the other hand, when $\Delta V(k-1)$ is vs. and $|dI_p(k)|$ is VL, it means that the PV system is far from the MPP; therefore, the output $\Delta V(k-1)$ should be PL to force the PV system to quickly move to the MPP. Furthermore, when $|dI_p(k)|$ is VL and $\Delta V(k-1)$ is VL, the output $\Delta V(k-1)$ can be chosen as either PZ or ZE for the PV system to automatically move to the MPP; in this study, we want to increase the speed for searching the MPP, so the output $\Delta V(k-1)$ is set as PZ in this case. In general, the other fuzzy rules in Table 4 can be suitably interpreted with the same deductive method.

Table 4. Fuzzy association rules for FLC-1.

$V_{add}(k)$		$|A_p(k)|$ or $|dI_p(k)|$				
		VS	**SM**	**ME**	**LA**	**VL**
	VS	ZE	PZ	PS	PL	PL
	SM	NZ	ZE	PZ	PM	PL
$\Delta V(k-1)$	ME	NS	NZ	ZE	PS	PM
	LA	NM	NS	NZ	PZ	PS
	VL	NL	NM	NS	ZE	PZ

3.1.2. PI-1 Controller

The PI-1 controller with an anti-windup block (refer to [32]) serves as the second sub-controller of the system. Figures 4 and 8 show detailed schematics of the controller.

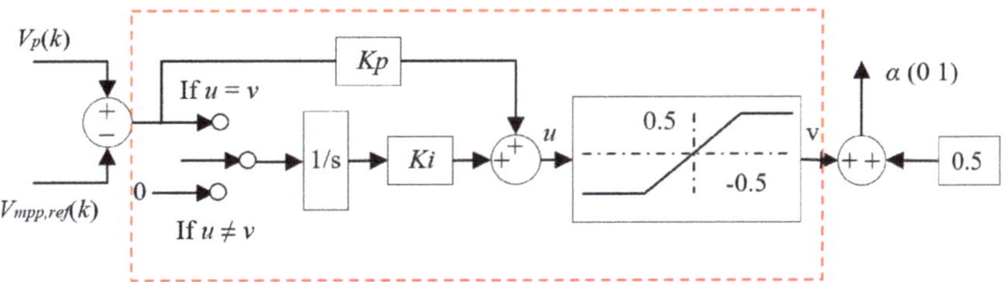

Figure 8. PI-1 controller in detail.

3.2. DC Link Voltage Regulator Module

According to Figure 3, the objective of the DC link voltage regulator is to determine an appropriate value of the reference grid current I_g^{ref} used for the current controller module. We note that the ultimate goal here is to make the DC link voltage V_{DC} reach its desired value V_{DC}^{ref} once the actual grid current I_g is well regulated to its reference I_g^{ref} [7,8] by the proposed current controller module using the LQR technique, which will be shown in detail in Section 3.3. In existing studies, a conventional PI controller has often been used to generate the reference grid current I_g^{ref} from the DC link voltage difference $e_{Vdc} = V_{DC}^{ref} - V_{DC}$, as shown in the upper left part of Figure 9. Nevertheless, it is difficult to manually choose and tune optimal values for the coefficients of the PI-2 controller due to the high nonlinearity of the grid-connected PV system, including a LCL output filter. Furthermore, the response of the conventional PI-2 controller usually has fairly large overshoot in the transient state and achieves the steady state in a relatively slow manner. Thus, this paper proposes a novel hybrid control scheme for the DC link voltage regulator module using another FLC (named FLC-2) as depicted in Figure 9 to overcome the above-mentioned drawbacks of the traditional PI-2 controller and remarkably enhance the response time.

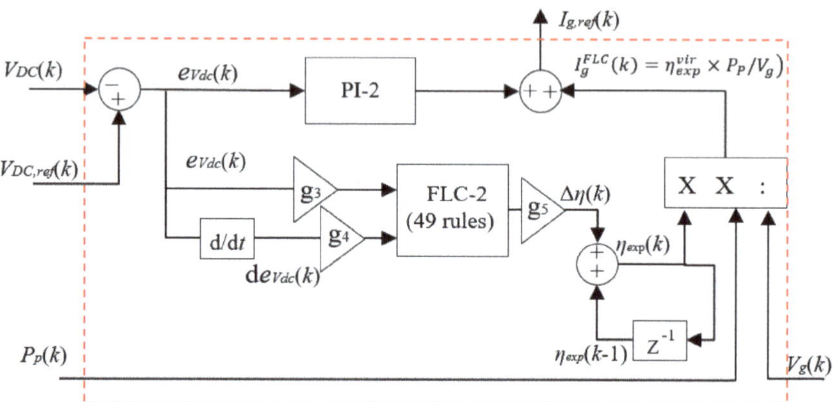

Figure 9. Proposed PI-fuzzy hybrid control scheme, where g_3, g_4 and g_5 are design coefficients.

Firstly, the theoretical relation between the output power of the DC-AC inverter supplied to the grid P_{AC} and the output power of the PV array P_P can be expressed as follows:

$$P_{AC} = \eta_{DC-AC} \times P_{DC-link} = \eta_{DC-AC} \times (\eta_{DC-DC} \times P_P) = \eta_{Exp} \times P_P \qquad (7)$$

where:

η_{DC-DC}—the efficiency of the buck-boost DC-DC converter can be estimated theoretically as a ratio value of the DC link power $P_{DC-link}$ and the output power of PV array P_P;
η_{DC-AC}—the efficiency of the DC-AC inverter can be estimated theoretically as a ratio value of the output power of the inverter P_{AC} and the DC link power $P_{DC-link}$;
$\eta_{Exp} = \eta_{DC-DC} \times \eta_{DC-AC}$—the overall efficiency of the grid-connected PV system.

Equation (7) can be written as:

$$V_g I_g (\cos\theta_g) = \eta_{Exp} P_P \qquad (8)$$

here $\cos\theta_g$ is the PF of the PV system, V_g is the rms value of the grid voltage, and I_g is the rms value of the grid current. In the normal operation of the grid, the rms value of the grid voltage is often larger than zero, meaning that $V_g > 0$ V.
Hence,

$$I_g = \eta_{Exp} \frac{P_P}{V_g(\cos\theta_g)} \qquad (9)$$

When the PF = 1, the grid current will reach the following value.

$$I_g = \eta_{Exp} P_P / V_g \qquad (10)$$

However, the actual overall efficiency η_{Exp} depends not only on the PF, but also on other component parameters of the DC-DC buck-boost converter and the DC-AC inverter, as well as the operating conditions of the PV system (e.g., the PV power, temperature, input voltage, and so forth); as a result, it is difficult to accurately estimate a particular value for η_{Exp}. We note that Equations (7)–(10) are only explanations of the theoretical relations among the parameters η_{Exp}, P_P, P_{AC}, I_g, and V_g, which are used as the reference basis for introducing and calculating a new "virtual efficiency" η_{exp}^{vir} in our proposed hybrid control scheme for the DC link voltage regulator module, as depicted in Figure 9. Hence, the calculated value of the "virtual efficiency" η_{exp}^{vir} in this figure and (11) is not the actual value of the overall efficiency of the PV system η_{Exp} in (7). In this study, we calculate the "virtual efficiency" value η_{exp}^{vir} instead of estimating the actual overall efficiency η_{Exp}.

To fulfill the above goal, the FLC-2 is designed to frequently update a suitable value for the "virtual efficiency" $\eta_{exp}^{vir}(k)$ in real-time, as described in (11) and Figure 9. Then, from (10), an additional value I_g^{FLC} for adjusting the reference grid current I_g^{ref} can be computed as $I_g^{FLC}(k) = \eta_{exp}^{vir}(k) \times P_P(k)/V_g(k)$, as shown in the right part of Figure 9. Finally, this computed additional value $I_g^{FLC}(k)$ is used to effectively compensate for the output of the conventional PI-2 controller $I_g^{PI}(k)$ to appropriately determine the reference grid current $I_g^{ref}(k)$, as presented in the upper right part of Figure 9; in detail, $I_g^{ref}(k) = I_g^{PI}(k) + I_g^{FLC}(k)$. The key aims of $I_g^{FLC}(k)$ generated by our proposed control scheme using the FLC-2 are to significantly improve the response time for updating a suitable value for the reference current $I_g^{ref}(k)$ and to elevate the effectiveness of the conventional PI-2 controller against effects caused by the high nonlinearity of the PV system.

In fact, using the proposed current controller module (refer to Figure 3 and Section 3.3), when the grid current $I_g(k)$ is regulated to its reference $I_g^{ref}(k)$ suitably generated by the designed PI-Fuzzy hybrid control scheme (see Figure 9), the DC link voltage V_{DC} achieves its desired value V_{DC}^{ref} [7,8]. This means that both the error values of the DC link voltage (e_{Vdc} in Figure 9) and the grid current in (14) and (15) are considered and regulated by the proposed complete control system, as given in the lower part of Figure 3.

The designed FLC-2 has two inputs and one output.

The two inputs are:

$e_{Vdc}(k)$—error between desired and present DC link voltage;
$de_{Vdc}(k)$—change in error.

The output is:

$\Delta\eta(k)$—step in efficiency, added to the "virtual efficiency" η_{exp}^{vir} to reach the desired value:

$$\eta_{exp}^{vir}(k) = \eta_{exp}^{vir}(k-1) + \Delta\eta(k) \tag{11}$$

The two inputs are:

$$e_{Vdc}(k) = V_{DC}^{ref} - V_{DC}(k) \tag{12}$$

$$de_{Vdc}(k) = e_{Vdc}(k) - e_{Vdc}(k-1) \tag{13}$$

All the inputs have the same number of linguistic variables, specifically seven; the range is $[-20;20]$: NL—Negative Large, NM—Negative Medium, NS—Negative Small, ZE—Zero, PS—Positive Small, PM—Positive Medium, PL—Positive Large

The output $\Delta\eta(k)$ has nine linguistic variables, and they range from $[-1;1]$: NL—Negative Large, NM—Negative Medium, NS—Negative Small, NZ—Negative Zero, ZE—Zero, PZ—Positive Zero, PS—Positive Small, PM—Positive Medium, PL—Positive Large. As a result, there are 49 fuzzy rules formed in the FLC-2.

All the association rules of the FLC-2 are presented in Table 5, while the membership functions of the inputs and output can be referred to in [28]. To interpret the fuzzy rules in Table 5, we can analyze and evaluate some sample cases as follows. Firstly, when $de_{Vdc}(k)$ is NL and $e_{Vdc}(k)$ is PL, the output of the fuzzy controller $\Delta\eta(k)$ should be ZE since the tendency of the DC-link voltage $V_{dc}(k)$ is automatically approaching its reference value. On the other hand, when $de_{Vdc}(k)$ is ZE and $e_{Vdc}(k)$ is NL, it means that $V_{dc}(k)$ is much smaller than its reference value; thus, the output $\Delta\eta(k)$ should be PL to force $V_{dc}(k)$ to rapidly move to the desired value. Furthermore, when $de_{Vdc}(k)$ is ZE and $e_{Vdc}(k)$ is PS, it means that $V_{dc}(k)$ is marginally larger than its reference value; hence, the output $\Delta\eta(k)$ should be NS to slightly decrease $V_{dc}(k)$ to its desired value without oscillation at the steady state. In general, the other fuzzy rules in Table 5 can be appropriately explained with a similar deductive technique.

Table 5. Fuzzy association rules for FLC-2.

$\Delta\eta(k)$		$e_{Vdc}(k)$						
		NL	**NM**	**NS**	**ZE**	**PS**	**PM**	**PL**
	NL	PU	PU	PL	PL	PM	PS	ZE
	NM	PU	PU	PL	PM	PS	ZE	NS
	NS	PU	PL	PM	PS	ZE	NS	NM
$de_{Vdc}(k)$	ZE	PL	PM	PS	ZE	NS	NM	NL
	PS	PM	PS	ZE	NS	NM	NL	NU
	PM	PS	ZE	NS	NM	NL	NU	NU
	PL	ZE	NS	NM	NL	NL	NU	NU

3.3. Current Controller Module

In this section, the current controller is designed by an optimal control method. Firstly, from (2), we have the following dynamic model,

$$\begin{cases} \dot{x}_4 = -\frac{R}{L}x_4 - \frac{1}{L}x_5 + \frac{u}{L} \\ \dot{x}_5 = \frac{1}{C}x_4 - \frac{1}{C}x_6 \\ \dot{x}_6 = \frac{1}{L_g}x_5 - \frac{R_g}{L_g}x_6 - \frac{1}{L_g}V_g \end{cases} \tag{14}$$

The main purpose of the current controller is to make the grid current i_g (i.e., x_6) converge to its reference x_{6ref}. Then, from the third equation of (14), the error dynamics of x_6 and the reference for x_5 (i.e., x_{5ref}) can be derived as,

$$\dot{x}_6 - \dot{x}_{6ref} = \frac{1}{L_g}\left(\left(x_5 - x_{5ref}\right) + x_{5ref}\right) - \frac{R_g}{L_g}\left(\left(x_6 - x_{6ref}\right) + x_{6ref}\right) - \frac{1}{L_g}V_g \tag{15}$$

Thus, we have

$$\dot{\tilde{x}}_6 = \frac{1}{L_g}\tilde{x}_5 - \frac{R_g}{L_g}\tilde{x}_6 - \frac{1}{L_g}x_{5ref} \tag{16}$$

where x_{5ref} is determined by

$$x_{5ref} = R_g x_{6ref} + v_g + \dot{x}_{6ref}L_g \tag{17}$$

Similarly, with x_{5ref} achieved from (17), combined with the second equation of (14),

$$\dot{x}_5 - \dot{x}_{5ref} = \frac{1}{C}\left(\left(x_4 - x_{4ref}\right) + x_{4ref}\right) - \frac{1}{C}\left(\left(x_6 - x_{6ref}\right) + x_{6ref}\right) \tag{18}$$

then

$$\dot{\tilde{x}}_5 = \frac{1}{C}\tilde{x}_4 - \frac{1}{C}\tilde{x}_6 \tag{19}$$

where

$$x_{4ref} = x_{6ref} + \dot{x}_{5ref}C_f \tag{20}$$

From x_{5ref} and x_{4ref}, obtained in (17) and (20), respectively, the first equation of (14) can be rewritten as

$$\dot{\tilde{x}}_4 = -\frac{R}{L}\tilde{x}_4 - \frac{R}{L}x_{4ref} - \frac{1}{L}\tilde{x}_5 - \frac{1}{L}x_{5ref} - \dot{x}_{4ref} + \frac{1}{L}u_1 + \frac{1}{L}u_2 \tag{21}$$

Hence, we have

$$\dot{x}_4 = -\frac{R}{L}\tilde{x}_4 - \frac{1}{L}\tilde{x}_5 + \frac{1}{L}u_1 \tag{22}$$

where

$$u_2 = Rx_{4ref} + x_{5ref} + L\dot{x}_{4ref} \tag{23}$$

Here, we decompose the control input u into two terms: u_1 and u_2; in detail, u_1 is used for feedback control to stabilize the error dynamics, whereas u_2 is the compensating term used to compensate for the offset in the reference tracking problem. Finally, the error dynamics of (15) are achieved by combining (22), (19), and (16), as follows:

$$\begin{bmatrix} x_4 \\ x_5 \\ x_6 \end{bmatrix} = \begin{bmatrix} -\frac{R_g}{L_g} & \frac{1}{L_g} & 0 \\ -\frac{1}{L} & 0 & \frac{1}{C} \\ 0 & -\frac{1}{L} & -\frac{R}{L} \end{bmatrix} + \begin{bmatrix} 0 \\ 0 \\ \frac{1}{L} \end{bmatrix} u_1 \tag{24}$$

Equation (24) is rewritten in the following form:

$$\dot{x} = Ax + Bu_1 \tag{25}$$

Consider the following cost function:

$$J(x,u) = \int_0^\infty x^T Q x + u_1^T Q u_1 \tag{26}$$

where $Q \geq 0$ and $R > 0$ are the weighting matrices with appropriate dimensions; that is, 3×3 and a scalar, respectively. After that, this cost function is minimized by the following control law:

$$u_1 = -Kx = -R^{-1}B^T P x \tag{27}$$

where K is the controller gain matrix, and P is the positive definite solution of the algebraic Riccati equation as follows

$$PA + A^T P - PBR^{-1}B^T P + Q = 0 \tag{28}$$

Typically, Q is chosen to be diagonal:

$$Q = \begin{bmatrix} q_1 & 0 & 0 \\ 0 & q_2 & 0 \\ 0 & 0 & q_3 \end{bmatrix} \tag{29}$$

where its elements and R can be selected by the following criteria,

$$q_i = \frac{1}{t_{si}(x_{imax})^2}, \ R = \frac{1}{(u_{1max})^2}, \ p > 0 \tag{30}$$

In (30), x_{imax} is the $|x_i|$ constraint, u_{imax} is the $|u_i|$ constraint, and t_{si} is the required settling time of x_i.

4. Simulation Results

The simulation performed in MATLAB/Simulink and all related parameters of the considered PV system can be referred to in [28]. The results with the designed LQR control are illustrated in Figures 10–13, in which the time unit in the horizontal axis is second.

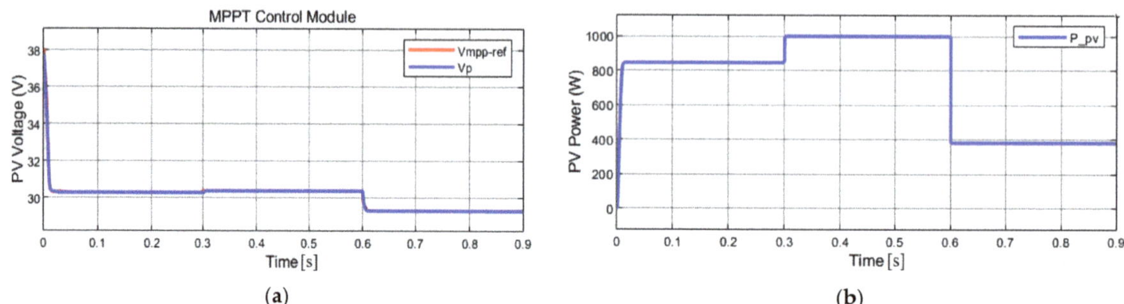

Figure 10. MPPT module performance for constant temperature: (**a**) PV voltage; (**b**) PV output power.

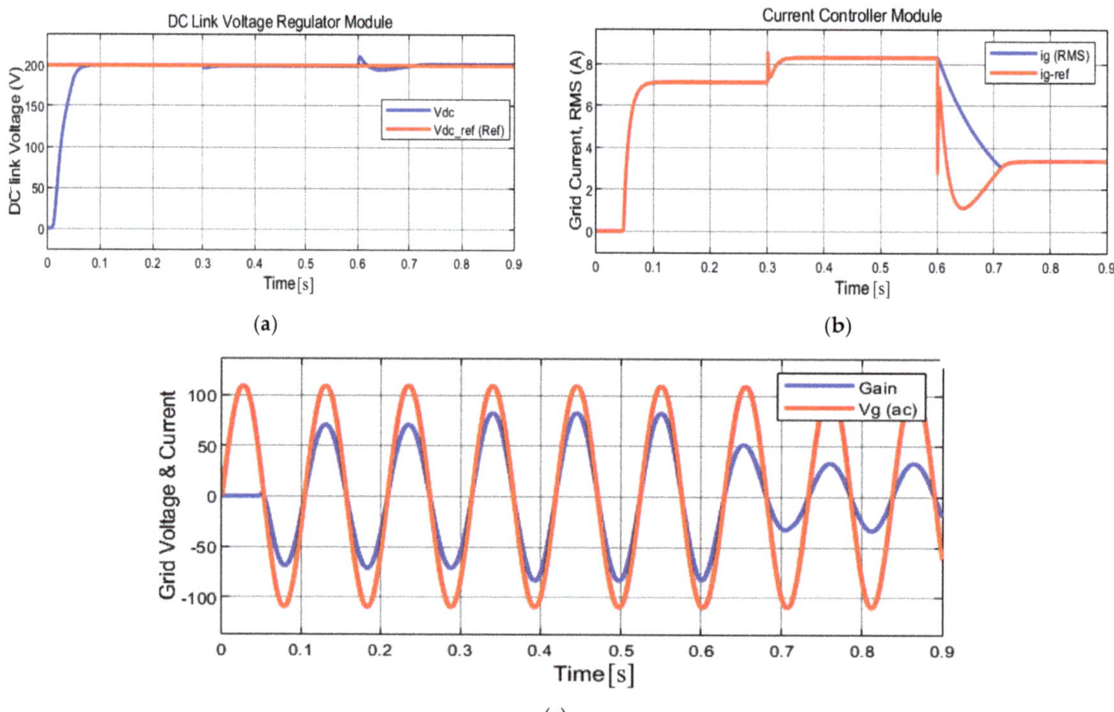

Figure 11. DC link voltage regulator module and current controller module performances with the LQR method for constant temperature case: (**a**) DC link voltage; (**b**) grid current magnitude; (**c**) grid voltage waveform V_g (V) and current waveform where Gain is $10 \times Ig$ (A).

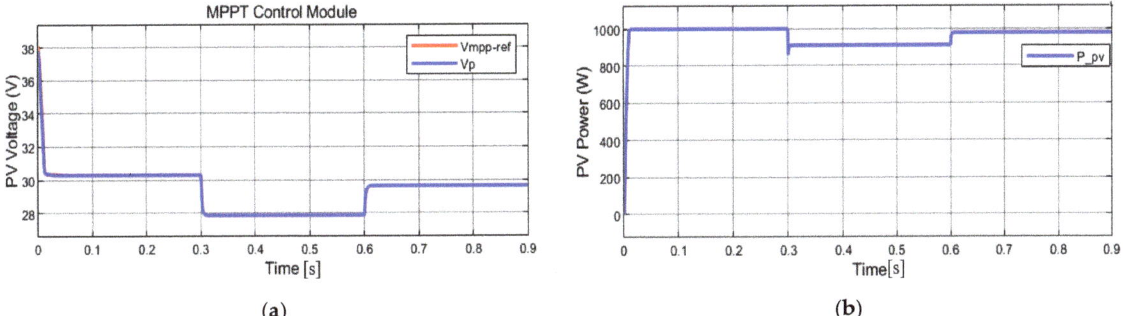

Figure 12. MPPT module performance for constant solar irradiation: (**a**) PV voltage; (**b**) PV output power.

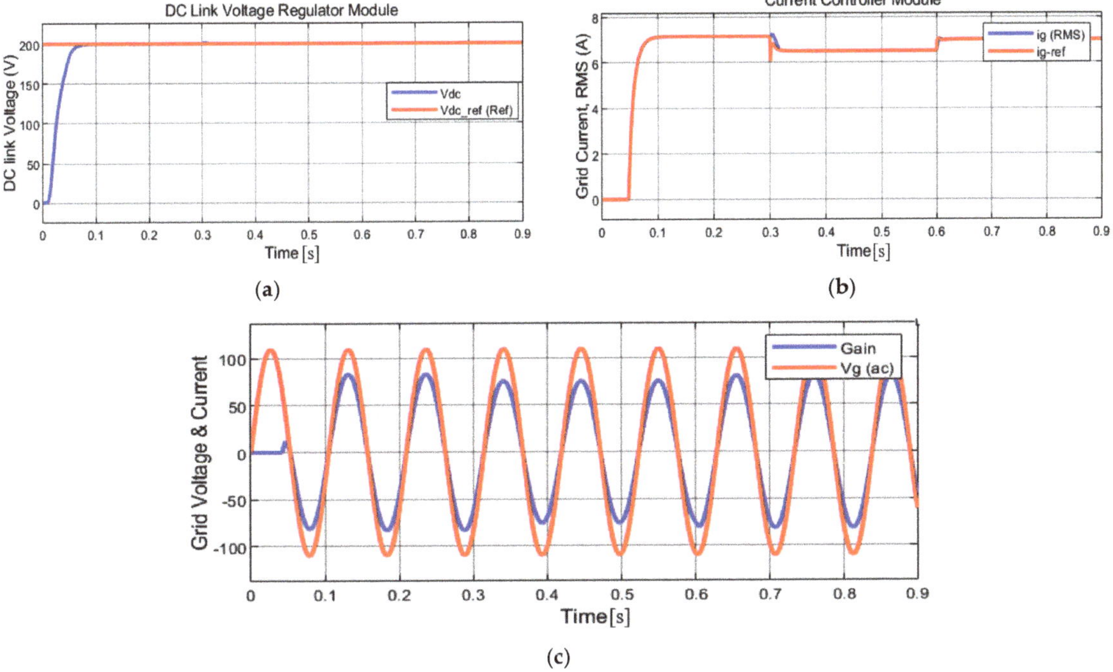

Figure 13. DC link voltage regulator module and current controller module performances with the LQR for constant solar irradiation case: (**a**) DC link voltage; (**b**) grid current magnitude; (**c**) grid voltage waveform V_g (V) and current waveform where gain is $10 \times Ig$ (A).

4.1. Simulation 1: Constant Module Temperature

This case considers when the PV module temperature is constant at 25 °C. Irradiation starts from 850 W/m^2 at a time from 0 s to 0.3 s, then it becomes 1000 W/m^2 from 0.3 s to 0.6 s and finally becomes 400 W/m^2 from 0.6 s to 0.9 s. Figure 10 shows that the results of the voltage V_p of the PV array are close to the reference values for the MPPT. The obtained output powers of the PV array are 847 W, 998 W, and 385 W, which match to the reference data provided in Table 1. Thus, this means that the power loss is small in this test.

The DC-link voltages correspond to each other in Figure 11. Furthermore, the grid current is equal to the reference values. Finally, it was shown and proven that the voltage and current of the grid are in phase, which means that the power factor of the grid-connected PV system is nearly unity.

4.2. Simulation 2: Constant Solar Irradiation

In the second case, solar irradiation is constant at 1000 W/m^2, but the module temperature is varying. From $t = 0$ s to 0.3 s temperature is 25 °C, next, at 0.3 s the temperature is 45 °C, and lastly, at 0.3 s the temperature is 30 °C. According to Figure 12, the performance of the panel is 30.38 V/1000 W, 27.92 V/912.1 W, and 29.76 V/978.8 W, which is highly close to the values represented in Table 1. Despite the module temperature change, V_{DC} matched its reference at all times. In addition, the RMS value of I_g is maintained according to the reference trend, as presented in Figure 13. The phases of the grid voltage and current match, which means that the system's power factor is unity.

5. Comparison between LQR and Backstepping Approaches

This research suggests the suitable combination of an LQR and fuzzy control for grid-connected PV systems. To show the effectiveness of the provided technique, it is important to make comparisons between some other methods, such as photovoltaic grid-connected systems using fuzzy logic and backstepping approaches [28] (see this reference paper for the specific details of simulations). Figures 14–17 present the simulation results of fuzzy control and the backstepping approach for a grid-connected photovoltaic system with the module temperature (Figures 14 and 15), and then with constant solar irradiation (Figures 16 and 17), in which the time unit in the horizontal axis is seconds. The obtained simulation results should be compared to those of the above-mentioned method. Specifically, the results in Figures 11 and 13 should be compared with those in Figures 15 and 17, respectively. We can see that both the control methods have good results.

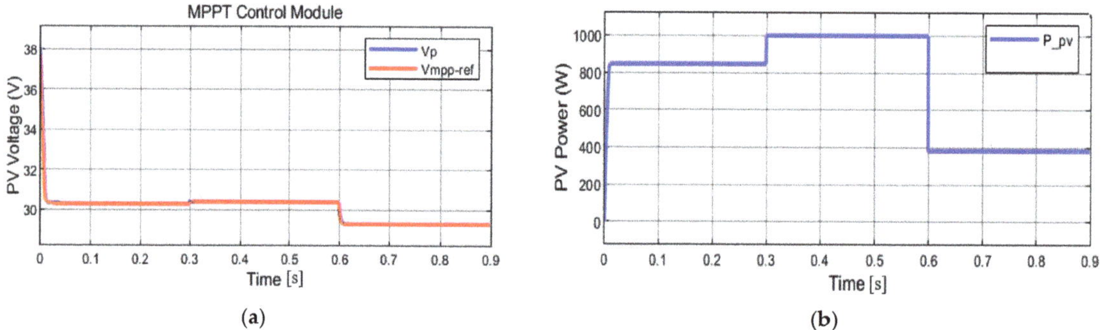

Figure 14. MPPT module performance in case with the backstepping approach for constant temperature: (**a**) PV voltage; (**b**) PV output power.

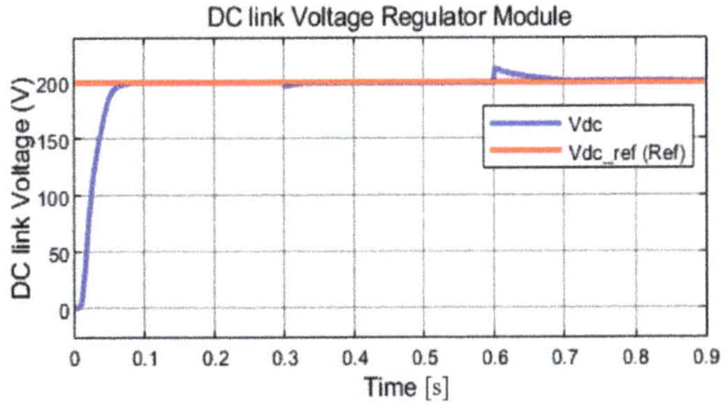

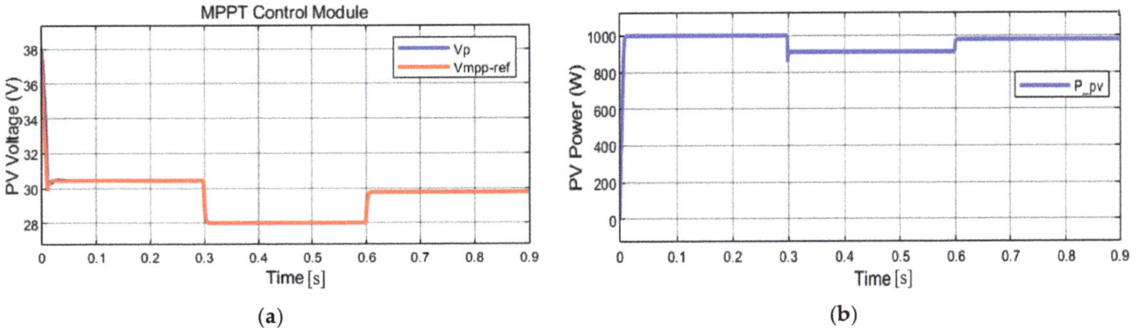

Figure 15. DC link voltage regulator module and current controller module performances with backstepping approach for constant temperature case: (**a**) DC link voltage; (**b**) grid current magnitude; (**c**) grid voltage waveform V_g (V) and current waveform where gain is $10 \times Ig$ (A).

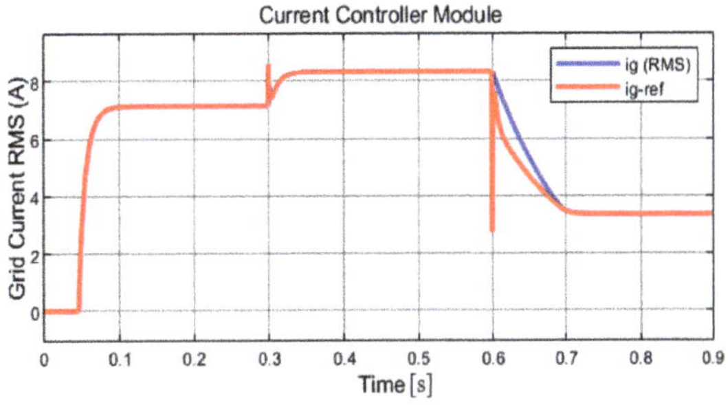

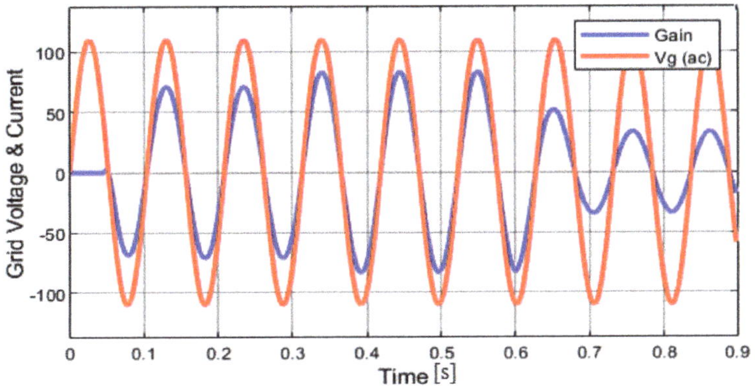

Figure 16. MPPT module performance in case with the backstepping approach for constant solar irradiation: (**a**) PV voltage; (**b**) PV output power.

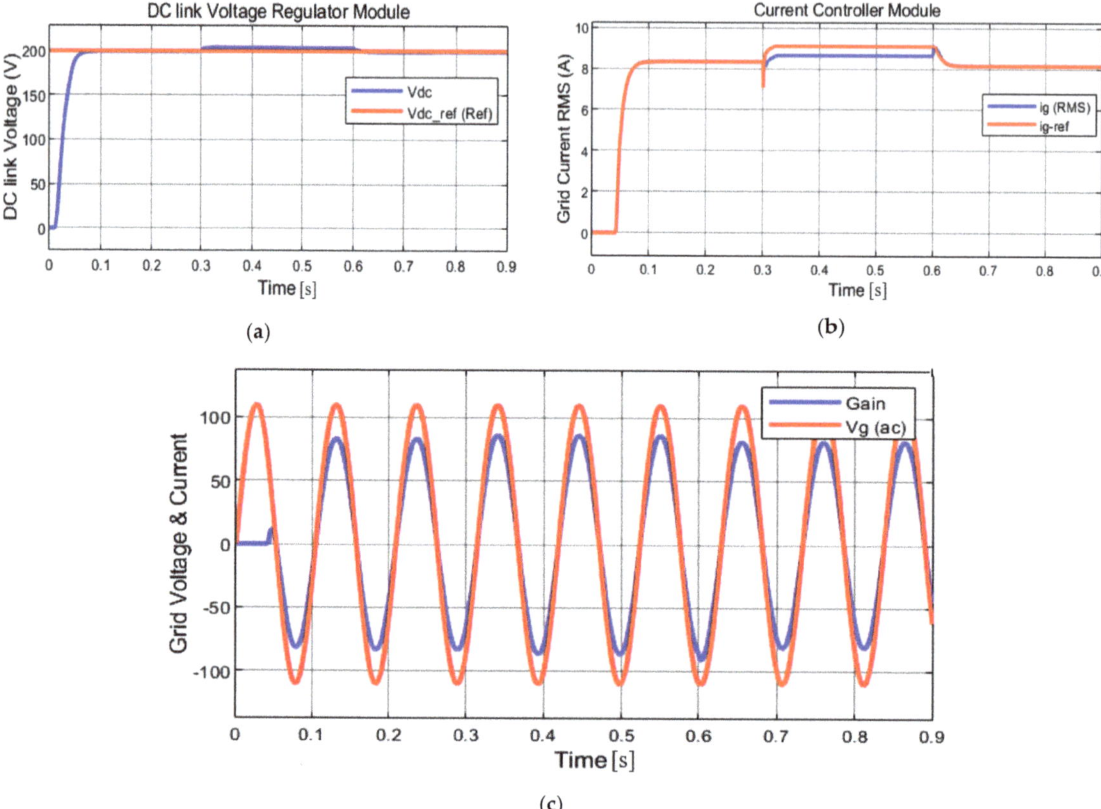

Figure 17. DC link voltage regulator module and current controller module performances with backstepping approach for constant solar irradiation case: (**a**) DC-link voltage; (**b**) grid current magnitude; (**c**) grid voltage waveform V_g (V) and current waveform where gain is $10 \times Ig$ (A).

5.1. Simulation 1: Constant Module Temperature

It is obvious that the simulation results of the MPPT parts in both the cases are the same since the main changes were not related to the MPPT controller, but to the LQR. Thus, the behaviors shown in Figures 10 and 14 are same; the performances presented in Figures 12 and 16 are also similar. Comparing the DC link voltage regulator module and the current controller module in both the cases, it can be clearly seen that the LQR case reacts to the changes in the module temperature and irradiation faster; in other words, the settling time of the LQR technique is lower compared to the backstepping method. Moreover, the LQR is robust when faced with different temperature and irradiation changes, which makes this technique preferable. In addition, in the case of the backstepping method, overshooting of the signal was observed, which significantly degrades the output and overall efficiency of the considered PV system. Furthermore, the response speed of the designed LQR is faster; consequently the rise time and peak time of the LQR are lower than those of the backstepping approach.

5.2. Simulation 2: Constant Solar Irradiation

In addition, the fuzzy-based INC-MPPT controller (see Figures 3–7 and Table 4) in this paper is improved from our past method [33], with significant modifications in the pre-scaling module $G_p(k)$ for increasing the sensitivity of the slope $S_p(k)$, in the fuzzy association rules for boosting the speed of searching the MPP, and in the control parameter values, which are designed and tuned to be more suitable for grid-connected PV systems with the LCL output filter. The detailed analysis and evaluation of the effectiveness of the prior related MPPT method for a small stand-alone PV system in various simulations and experiments can be referred to in Sections 4 and 5 of our past work [33], in which cases of partial shadow on the PV array were also investigated in experiments. Moreover, the influence of the module temperature on the performance of the stand-alone PV system and MPPT controller can be found in Sections 2.2 and 4.3 of the prior study [33]. In fact, our present paper proposes a complete control solution for grid-connected single-phase PV systems including the LCL output filter based on fuzzy control and LQR techniques with multiple objectives as described in Sections 1 and 3 above; thus, it is noted that the separate assessment of the MPPT controller compared to other MPPT systems is out the scope of this study.

Due to substantial differences in configurations of considered PV systems and control objectives between this study and other existing works, it seems inappropriate and difficult to directly compare the detailed effectiveness of the proposed control scheme in this research to that of other studies. Therefore, we have only performed the detailed comparison of the current controller module using the LQR method proposed in the present paper with our past work using the backstepping approach [28] for evaluation, as described in Section 5 above. Moreover, for further reference, the current controller module using the LQR technique in this paper is briefly compared with the PID-fuzzy hybrid controller introduced in our other study [34], as represented in Table 6.

Table 6. Comparison between the proposed current controller module and the previously introduced PID-fuzzy controller in [34].

Control Methods	The Current Controller Module Using the LQR in This Paper	The Introduced PID-Fuzzy Hybrid Controller in [34]
Design and computational complexity	- Fairly simple design based on an optimal control method using the algebraic Riccati equation (see Section 3.3)	- Includes three separate FLCs for controlling gains K_p, K_I, and K_D, in which each FLC consists of 49 association rules.
Field of application in power systems	- For a complete single-phase grid-connected PV system including the LCL output filter with the consideration of DC link voltage regulation.	- For a simple single-phase grid-connected PV system including the L output filter without the consideration of DC link voltage regulation.
Control objectives, performance, and effectiveness	- Can well regulate the grid current with a unity power factor and DC link voltage under different operating conditions of the PV system. - Very fast response, and small overshoot. - Has good efficacy and is robust faced with large variations in solar radiation and PV module temperature. - Theoretical analysis of stability of the LQR controller has been confirmed.	- Can control the active and reactive power supplied to the grid with the step change of reference signals for the two powers. - Quick response and relatively small overshoot. - Has not yet been checked with large variations in solar radiation and PV module temperature. - Theoretical analysis of stability of the hybrid controller has not yet been performed.

In this grid-connected PV system using a DC-AC inverter and LCL output filter, due to parasitic capacitance and grounding resistance, the issues of common mode (CM) voltage and leakage current may become significant if the design and control of the inverter and LCL filter are not appropriate [35,36]. As presented in other existing studies [37,38], the issues of CM voltage and leakage current in grid-connected PV systems can be effectively investigated and reduced using various techniques such as improved PWM methods, modified topologies of PV inverters with complementary switches [37], design and implementation of CM filters [38,39], active damping control approaches [40], and so forth. On the other hand, our paper focuses on developing a complete control scheme for grid-connected single-phase PV systems including an LCL filter based on fuzzy logic and an

LQR method with three key goals comprising the MPPT, DC link voltage regulation, and injection of the PV power into the grid with a unity PF. Hence, it should be noted that the assessment and reduction of the CM voltage and leakage current are beyond the scope of the present paper; these issues will be thoroughly considered in our future work.

6. Conclusions

Based on fuzzy control and an LQR, this study provides a comprehensive control solution for grid-connected single-phase PV systems. In terms of improving the quality of controller methods, this work represents a substantial extension and enhancement of our past research. For grid-connected solar systems, the suggested approach is an LQR suitably combined with fuzzy control, in which the design procedures of all the controllers are also described in detail. The major novelty of this study is to demonstrate and verify that the newly proposed approach is more successful in different aspects by comparing our past and present control methods. The LQR technique's major benefit is its ability to react quickly to unexpected changes in the system, such as changes in module temperature and solar irradiation. In other words, the systems achieve their settling period sooner, which is necessary to steady the output behavior.

Furthermore, as compared to the backstepping approach, the LQR method is more resistant to various changes in weather conditions. The backstepping approach also has the greater overrun, which has a detrimental effect on the efficiency of the investigated PV system. As previously stated, the LQR has the quicker response speed, making this type of controller more desirable. These are the major contributions of our present work as compared to the earlier research. Moreover, the results of simulations under different weather circumstances were compared to theoretical ones, indicating that the proposed system can cope with variations in weather parameters well. It was also demonstrated that abrupt changes in weather factors had no significant effects on the proposed control system's performance.

In future work, intelligent models based on fuzzy control for effectively predicting PV power and load demand will be thoroughly studied and implemented to improve the effectiveness and quality of grid-connected solar energy systems, especially under adverse conditions such shaded solar PV modules.

Author Contributions: A.M. worked on all tasks; T.D.D. and N.G.M.T. worked on the methodology and supervised. All the authors participated in writing, editing, and review. All authors have read and agreed to the published version of the manuscript.

Funding: This work was supported by Nazarbayev University under the Faculty Development Competitive Research Grant Program (FDCRGP), Grant No. 11022021FD2924.

Data Availability Statement: Not applicable.

Conflicts of Interest: The authors declare no conflict of interest.

References

1. Torquato, R.; Arguello, A.; Freitas, W. Practical Chart for Harmonic Resonance Assessment of DFIG-Based Wind Parks. *IEEE Trans. Power Deliv.* **2020**, *35*, 2233–2242. [CrossRef]
2. Thao, N.G.M.; Uchida, K.; Kofuji, K.; Jintsugawa, T.; Nakazawa, C. A comprehensive analysis study about harmonic resonances in megawatt grid-connected wind farms. In Proceedings of the 2014 International Conference on Renewable Energy Research and Application (ICRERA), Milwaukee, WI, USA, 19–22 October 2014; pp. 387–394.
3. Alajmi, B.N.; Ahmed, K.H.; Finney, S.J.; Williams, B.W. Fuzzy-Logic-Control Approach of a Modified Hill-Climbing Method for Maximum Power Point in Microgrid Standalone Photovoltaic System. *IEEE Trans. Power Electron.* **2011**, *26*, 1022–1030. [CrossRef]
4. Kottas, T.L.; Boutalis, Y.S.; Karlis, A.D. New Maximum Power Point Tracker for PV Arrays Using Fuzzy Controller in Close Cooperation with Fuzzy Cognitive Networks. *IEEE Trans. Energy Convers.* **2006**, *21*, 793–803. [CrossRef]
5. Hamidia, F.; Abbadi, A.; Boucherit, M.S. Maximum Power Point Tracking Control of Photovoltaic Generation Based on Fuzzy Logic. In *International Conference in Artificial Intelligence in Renewable Energetic Systems*; Springer: Cham, Switzerland, 2018; pp. 197–205. [CrossRef]
6. Menniti, D.; Pinnarelli, A.; Brusco, G. Implementation of a novel fuzzy-logic based MPPT for grid-connected photovoltaic generation system. In Proceedings of the 2011 IEEE Trondheim PowerTech, Trondheim, Norway, 19–23 June 2011. [CrossRef]

7. Yang, B.; Li, W.; Zhao, Y.; He, X. Design and Analysis of a Grid-Connected Photovoltaic Power System. *IEEE Trans. Power Electron.* **2009**, *25*, 992–1000. [CrossRef]

8. El Fadil, H.; Giri, F.; Guerrero, J. Grid-connected of photovoltaic module using nonlinear control. In Proceedings of the 2012 3rd IEEE International Symposium on Power Electronics for Distributed Generation Systems (PEDG), Aalborg, Denmark, 25–28 June 2012; pp. 119–124. [CrossRef]

9. Wai, R.-J.; Wang, W.-H. Grid-Connected Photovoltaic Generation System. *IEEE Trans. Circuits Syst. I Regul. Pap.* **2008**, *55*, 953–964. [CrossRef]

10. Reddy, D.; Ramasamy, S. A fuzzy logic MPPT controller based three phase grid-tied solar PV system with improved CPI voltage. In Proceedings of the 2017 Innovations in Power and Advanced Computing Technologies (I-PACT), Vellore, India, 21–22 April 2017; pp. 1–6. [CrossRef]

11. Kumar, J.; Rathor, B.; Bahrani, P. Fuzzy and P amp; O MPPT techniques for stabilized the efficiency of solar PV system. In Proceedings of the 2018 International Conference on Computing, Power and Communication Technologies (GUCON), Greater Noida, India, 28–29 September 2018; pp. 259–264. [CrossRef]

12. Kumar, B.P.; Winston, D.P.; Christabel, S.C.; Venkatanarayanan, S. Implementation of a switched PV technique for rooftop 2 kW solar PV to enhance power during unavoidable partial shading conditions. *J. Power Electron.* **2017**, *17*, 1600–1610. [CrossRef]

13. Andrew-Cotter, J.; Uddin, M.N.; Amin, I.K. Particle swarm optimization based adaptive neuro-fuzzy inference system for MPPT control of a three-phase grid-connected photovoltaic system. In Proceedings of the 2019 IEEE International Electric Machines & Drives Conference (IEMDC), San Diego, CA, USA, 12–15 May 2019; pp. 2089–2094. [CrossRef]

14. Arulkumar, K.; Palanisamy, K.; Vijayakumar, D. Recent advances and control techniques in grid connected Pv system—A review. *Int. J. Renew. Energy Res.* **2016**, *6*, 1037–1049.

15. Mukundan, N.; Singh, Y.; Naqvi, S.B.Q.; Singh, B.; Pychadathil, J. Multi-Objective Solar Power Conversion System with MGI Control for Grid Integration at Adverse Operating Conditions. *IEEE Trans. Sustain. Energy* **2020**, *11*, 2901–2910.

16. Alturki, F.A.; Omotoso, H.O.; Al-Shamma'A, A.A.; Farh, H.M.H.; Alsharabi, K. Novel Manta Rays Foraging Optimization Algorithm Based Optimal Control for Grid-Connected PV Energy System. *IEEE Access* **2020**, *8*, 187276–187290. [CrossRef]

17. Blaabjerg, F.; Teodorescu, R.; Liserre, M.; Timbus, A.V. Overview of Control and Grid Synchronization for Distributed Power Generation Systems. *IEEE Trans. Ind. Electron.* **2006**, *53*, 1398–1409. [CrossRef]

18. Teodorescu, R.; Blaabjerg, F.; Liserre, M.; Loh, P.C. Proportional-resonant controllers and filters for grid-connected voltage-source converters. *IET Proc. Electr. Power Appl.* **2006**, *153*, 750–762. [CrossRef]

19. Wu, G.R.; Dewan, R.; Slemon, S.B. Analysis of a PWM AC to DC Voltage Source Converter under the Predicted Current Control with a Fixed Switching Frequency. *IEEE Trans. Ind. Appl.* **1991**, *27*, 756–764. [CrossRef]

20. Zhang, K.; Kang, Y.; Xiong, J.; Chen, J. Direct repetitive control of SPWM inverter for UPS purpose. *IEEE Trans. Power Electron.* **2003**, *18*, 784–792. [CrossRef]

21. Singh, B.; Jain, V. TOCF Based Control for Optimum Operation of a Grid Tied Solar PV System. *IEEE Trans. Energy Convers.* **2020**, *35*, 1171–1181. [CrossRef]

22. Shan, Y.; Hu, J.; Guerrero, J.M. A Model Predictive Power Control Method for PV and Energy Storage Systems with Voltage Support Capability. *IEEE Trans. Smart Grid* **2019**, *11*, 1018–1029. [CrossRef]

23. Manoharan, M.S.; Ahmed, A.; Park, J.-H. An Improved Model Predictive Controller for 27-Level Asymmetric Cascaded Inverter Applicable in High-Power PV Grid-Connected Systems. *IEEE J. Emerg. Sel. Top. Power Electron.* **2019**, *8*, 4395–4405. [CrossRef]

24. Mahfuz-Ur-Rahman, A.M.; Islam, R.; Muttaqi, K.M.; Sutanto, D. Model Predictive Control for a New Magnetic Linked Multilevel Inverter to Integrate Solar Photovoltaic Systems with the Power Grids. *IEEE Trans. Ind. Appl.* **2020**, *56*, 7145–7155. [CrossRef]

25. Mansour, A.M.; Arafa, O.M.; Marei, M.I.; Abdelsalam, I.; Aziz, G.A.A.; Sattar, A.A. Hardware-in-the-Loop Testing of Seamless Interactions of Multi-Purpose Grid-Tied PV Inverter Based on SFT-PLL Control Strategy. *IEEE Access* **2021**, *9*, 123465–123483. [CrossRef]

26. Verma, A.; Singh, B. CAPSA Based Control for Power Quality Correction in PV Array Integrated EVCS Operating in Standalone and Grid Connected Modes. *IEEE Trans. Ind. Appl.* **2020**, *57*, 1789–1800. [CrossRef]

27. Sarita, K.; Kumar, S.; Vardhan, A.S.S.; Elavarasan, R.M.; Saket, R.K.; Shafiullah, G.M.; Hossain, E. Power Enhancement with Grid Stabilization of Renewable Energy-Based Generation System Using UPQC-FLC-EVA Technique. *IEEE Access* **2020**, *8*, 207443–207464. [CrossRef]

28. Thao, N.G.M.; Uchida, K. Control the photovoltaic grid-connected system using fuzzy logic and backstepping approach. In Proceedings of the 2013 9th Asian Control Conference ASCC, Istanbul, Turkey, 23–26 June 2013; pp. 1–8. [CrossRef]

29. Katir, H.; Abouloifa, A.; Noussi, K.; Lachkar, I.; El Aroudi, A.; Aourir, M.; El Otmani, F.; Giri, F. Fault Tolerant Backstepping Control for Double-Stage Grid-Connected Photovoltaic Systems Using Cascaded H-Bridge Multilevel Inverters. *IEEE Control Syst. Lett.* **2021**, *6*, 1406–1411. [CrossRef]

30. Liserre, M.; Blaabjerg, F.; Hansen, S. Design and Control of an LCL-Filter-Based Three-Phase Active Rectifier. *IEEE Trans. Ind. Appl.* **2005**, *41*, 1281–1291. [CrossRef]

31. Sun, W.; Chen, Z.; Wu, X. Intelligent optimize design of LCL filter for three-phase voltage-source PWM rectifier. In Proceedings of the 2009 IEEE 6th International Power Electronics and Motion Control Conference, Wuhan, China, 17–20 May 2009; pp. 970–974. [CrossRef]

32. Li, X.-L.; Park, J.-G.; Shin, H.-B. Comparison and Evaluation of Anti-Windup PI Controllers. *J. Power Electron.* **2011**, *11*, 45–50. [CrossRef]
33. Thao, N.G.M.; Uchida, K.; Nguyen-Quang, N. An Improved Incremental Conductance-Maximum Power Point Tracking Algorithm Based on Fuzzy Logic for Photovoltaic Systems. *SICE J. Control. Meas. Syst. Integr.* **2014**, *7*, 122–131. [CrossRef]
34. Thao, N.G.M.; Dat, M.T.; Binh, T.C.; Phuc, N.H. PID-fuzzy logic hybrid controller for grid-connected photovoltaic inverters. In Proceedings of the International Forum on Strategic Technology, Ulsan, Korea, 13–15 October 2010; pp. 140–144. [CrossRef]
35. Zeb, K.; Khan, I.; Uddin, W.; Khan, M.A.; Sathishkumar, P.; Busarello, T.D.C.; Ahmad, I.; Kim, H.J. A Review on Recent Advances and Future Trends of Transformerless Inverter Structures for Single-Phase Grid-Connected Photovoltaic Systems. *Energies* **2018**, *11*, 1968. [CrossRef]
36. Figueredo, R.S.; De Carvalho, K.C.M.; Ama, N.R.N.; Matakas, L. Leakage current minimization techniques for single-phase transformerless grid-connected PV inverters—An overview. In Proceedings of the 2013 Brazilian Power Electronics Conference, Gramado, Brazil, 27–31 October 2013; pp. 517–524. [CrossRef]
37. Hedayati, M.H.; John, V. EMI and ground leakage current reduction in single-phase grid-connected power converter. *IET Power Electron.* **2017**, *10*, 938–944. [CrossRef]
38. Figueredo, R.S.; Matakas, L. Integrated Common and Differential Mode Filter With Capacitor-Voltage Feedforward Active Damping for Single-Phase Transformerless PV Inverters. *IEEE Trans. Power Electron.* **2019**, *35*, 7058–7072. [CrossRef]
39. Liu, Y.; Lai, C.-M. LCL Filter Design with EMI Noise Consideration for Grid-Connected Inverter. *Energies* **2018**, *11*, 1646. [CrossRef]
40. Niyomsatian, K.; Vanassche, P.; Gyselinck, J.J.C.; Sabariego, R.V. Active-Damping Virtual Circuit Control for Grid-Tied Converters with Differential-Mode and Common-Mode Output Filters. *IEEE Trans. Power Electron.* **2020**, *35*, 7583–7595. [CrossRef]

Article

Plug-in Hybrid Ecological Category in Real Driving Emissions

Kinga Skobiej * and **Jacek Pielecha**

Faculty of Civil and Transport Engineering, Poznan University of Technology, pl. M. Sklodowskiej-Curie 5, 60-965 Poznan, Poland; jacek.pielecha@put.poznan.pl
* Correspondence: kinga.d.skobiej@doctorate.put.poznan.pl; Tel.: +48-61-665-2239

Abstract: Transportation, as one of the most growing industries, is problematic due to environmental pollution. A solution to reduce the environmental burden is stricter emission standards and homologation tests that correspond to the actual conditions of vehicle use. Another solution is the widespread introduction of hybrid vehicles—especially the plug-in type. Due to exhaust emission tests in RDE (real driving emissions) tests, it is possible to determine the real ecological aspects of these vehicles. The authors of this paper used RDE testing of the exhaust emissions of plug-in hybrid vehicles and on this basis evaluated various hybrid vehicles from an ecological point of view. An innovative solution proposed by the authors is to define classes of plug-in hybrid vehicles (classes from A to C) due to exhaust emissions. An innovative way is to determine the extreme results of exhaust gas emission within the range of acceptable scatter of the obtained results. By valuating vehicles, it will be possible in the future to determine the guidelines useful in designing more environmentally friendly power units in plug-in hybrid vehicles.

Keywords: exhaust emission; energy consumption; real driving emissions test

Citation: Skobiej, K.; Pielecha, J. Plug-in Hybrid Ecological Category in Real Driving Emissions. *Energies* **2021**, *14*, 2340. https://doi.org/10.3390/en14082340

Academic Editor: Constantine D. Rakopoulos

Received: 1 March 2021
Accepted: 17 April 2021
Published: 20 April 2021

Publisher's Note: MDPI stays neutral with regard to jurisdictional claims in published maps and institutional affiliations.

1. Introduction

The main problem of the ever-growing industry is its negative impact on the natural environment. One of the most dynamically changing sectors of industry is transport, which significantly contributes to air pollution. In order to reduce the impact of vehicles on the environment, increasingly stringent emission standards are introduced and solutions are sought that could allow minimization of the exhaust emissions from vehicles. One of the solutions proposed by the carmakers aiming at a global reduction of exhaust emissions is to replace as many conventional vehicles as possible with electric ones. Yet, due to the high cost of batteries, low vehicle range, and the lack of infrastructure, fully electric vehicles cannot fully replace the conventional ones. Manufacturers of conventional vehicles still struggle with the increasingly stringent exhaust emission standards.

The emission standards are set forth worldwide to control the pollutants emitted from vehicles to the atmosphere. The exhaust emissions are measured under the conditions of a predefined type of approval test. This part of the vehicle certification is responsible for its ecological properties and is the same for all passenger cars. The course of the test reflects the most probable road conditions and its performance, identical for all vehicles, entitles a comparison of the emission results among all the tested vehicles. These days, however, increased attention has been drawn to the performance of road tests, i.e., performed under actual traffic conditions. Currently, these tests are specified in the EU regulations as RDE (real driving emissions) [1–6]. They are carried out to most accurately reproduce the actual traffic conditions in the environmental aspect. Such tests must be performed in compliance with certain precisely defined requirements while at the same time allowing a relative arbitrariness, which significantly spreads the obtained exhaust emission results despite meeting the RDE requirements. The performed qualitative and quantitative analyses of the exhaust emissions in different tests was the subject of [7]. The authors proved that the values of relative road emissions depend more on the distance covered during the test than the duration of the test. The exhaust road emission values determined in different tests depend mainly on the type of test and are greater in shorter tests compared to the RDE test. The analysis of the investigations has confirmed that it is possible to shorten the tests by approximately 20% without a significant change in the exhaust emission results. This is confirmation of the increasing importance being placed on road test results, which must be verified for not only static parameters, but also, to a large extent, for dynamic parameters.

The detailed requirements of the RDE tests and the possibilities of optimization of a combustion engine were investigated by Pielecha and Skobiej [8]. The analysis of the emission level of individual exhaust components allowed the authors to show that the exhaust emission level may be lower by 26% to 81% compared to the road test performed under actual traffic conditions. The performed tests indicate that already at the stage of design, the engines can be optimized in terms of their exhaust emissions. Roadside emission tests have also been shown to be the most reliable vehicle performance information vehicle. The literature presented in this paper intends to indicate the actions and research aimed at increasing the environmental aspects of hybrid vehicles. These are both solutions involving better management of the energy stored in the batteries and technical solutions affecting the development of this type of vehicle.

A partial solution to the problem of the limited range of electric vehicles as well as its dependence on the traffic conditions is the introduction of a plug-in hybrid [9,10]. A plug-in hybrid combines the advantages of hybrid and electric vehicles. Compared to typical hybrids, the batteries used in plug-in hybrid electric vehicles (PHEVs) have greater capacity and range and can be charged from an external power supply. Such vehicles are more economical in terms of fuel consumption and more ecological in terms of exhaust emissions. Due to the fact that approximately 30–50 km [11] can be covered using electrical energy, PHEVs are more ecological than the conventional ones.

Cieslik et al. [12] investigated an electric vehicle under varied vehicle operating conditions, particularly the influence of the weather on the energy consumption by the vehicle. Under actual traffic conditions in sub-zero temperatures, the energy consumption in such a vehicle is greater by 14% compared to warm weather. Similar investigations were carried out by Yi et al. [13], indicating that energy consumption in the electric vehicle (EV) may vary drastically depending on the driving conditions, which is extremely impactful on the vehicle range.

Li et al. in [14], observing the growing popularity of electric and plug-in hybrid vehicles, proposed a methodology of assessment of the energy distribution within a vehicle. Plug-in hybrids allow charging of the batteries while driving, which reduces the demand for energy from the external sources.

Pielecha et al. [15] compared vehicles of different powertrains: conventional, plug-in hybrid, and electric, in tests under actual traffic conditions. They showed that the plug-in hybrid consumes 20% less energy than the conventional vehicle.

As per the IEA, the Global EV Outlook 2020 reported [16] that since 2010 (last 10 years), the number of electric vehicles has increased and today their share in the market amounts to 2.5% (1 in 40 new vehicles is fitted with an electric powertrain), including 74% fully electric ones and 26% plug-in hybrids. The increase in the number of plug-in hybrid vehicles is presented in Figure 1, where the 2010–2019 phase is compared. In 2012, when the sales of plug-in hybrids were initiated, already 100,000 vehicles were located in the US. In 2017, the number of these vehicles exceeded 1 million, for which the greatest share was in Europe. The sales of plug-in hybrid vehicles reached a level of 2.4 million worldwide in 2019 and the highest number of these vehicles was recorded in Europe.

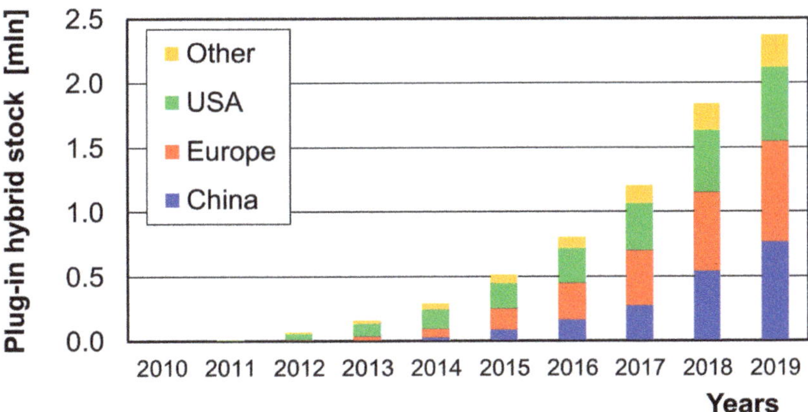

Figure 1. Increasing the number of plug-in hybrid vehicles between 2010 and 2019 globally (based on [16]).

Due to the fact that this type of vehicle is equipped with two sources of propulsion—a conventional internal combustion engine and an electric motor—it is necessary to optimally develop an energy management strategy and calibrate its parameters; the new system is designed to reduce emissions on the one hand and minimize energy consumption on the other.

2. Aim of the Paper

The addressing of the topic discussed in the paper is a result of social expectations, related to fair provision of vehicle emission-related information. Accusations were made that emission tests (particularly the type of approval tests) do not reflect the actual vehicle emissions, and, at the same time, the provided fuel economy information significantly diverges from the values obtained in daily vehicle operation. A similar topic, but in terms of a very broad scope of research, was addressed by the Initial Green Vehicle Index Roadmap project [17], where the assumption was to develop a basis for the comprehensive assessment of vehicles in terms of exhaust emissions, fuel consumption, and energy. The project, divided into phases between 2019 and 2030, assumed the study of exhaust emissions in type approval tests (NEDC, WLTC, and RDE) and, on this basis, the classification of vehicles on a 10-point scale of two indices: the clean air index and energy efficiency index. The clean air index compared on-road exhaust emissions with the permissible values—the lower the exhaust emissions, the higher the index value. The second index, the energy efficiency index, assessed the energy consumption of vehicles represented in kWh/100 km. Determination of the two indices and their respective values made it possible to assign a rating from 0 to 5 stars to the vehicle.

Therefore, the authors of this paper see the need for an objective approach to the assessment of the fuel consumption and exhaust emissions under actual operation of a vehicle. Such a particularization is necessary in the times of valuation of much data influencing, not only the vehicle operation, but also the choices of the end-users. A very good example is the valuation of the energy consumption of electric vehicles (though, conditioning of this parameter on external conditions should be required). The categorization of exhaust emission assessments conditional on the probability distribution of the results obtained in road tests is a novel approach. A novel method of determining the above will be the estimation of the extreme values of the exhaust emissions for a given vehicle within the admissible spreads of the road test results. Such a scenario, in the first stage, forced the application of a mathematical optimization apparatus in order to determine the minimum and the maximum values of the exhaust emissions and fuel consumption and, in the second stage, the need to practically validate the possibilities of obtaining such results from vehicles fitted with given powertrains (plug-in hybrid vehicles). The realization of such a research plan resulted in the development of a new tool that may be used in the environment-related assessment of motor vehicles under actual traffic conditions. In relation to this, the discipline of transport would gain another solution, serving the purpose of protecting the natural environment. Currently, the certificate of type approval provides the final values related to the conformity of a vehicle to a given ecological category without energy- and emission-related valuations in the said category. Currently, sold vehicles in the emission category of Euro 6 are not classified in terms of adequacy of the actual emission and fuel consumption with the values stated in the certificate of type approval. The aim of the paper is the development of a procedure, according to which one could assess whether the road emission test performed on a plug-in hybrid is reliable and, at the same time, indicate an interval of probability of meeting the requirements for tests carried out in the actual traffic. Such a task remains in line with the optimization tasks, in which one needs to determine the extreme values and indexes, to which the obtained emission result is to be compared. In the paper, the authors introduced valuation of vehicles in road tests (classes from A to C), which facilitates the assessment of the vehicle emission category of a PHEV (not only in the type approval but also in road exhaust emission tests).

The effect of the paper will be the possibility of assessing the adequacy of the exhaust emissions under actual traffic conditions based on the vehicle emission category and determining the index (valuation), directly translating the results of a type approval test to the road tests. By evaluating plug-in hybrids in environmental tests, it will potentially be possible not only to assess their actual environmental impact, but also to establish guidelines that will contribute to the design of more environmentally friendly powertrains in the future. The research presented in this paper is an alternative approach to the issues of environmental assessment of vehicles, although limiting oneself only to the emission of exhaust compounds—in the case of PHEVs—is not a complete solution. The next challenge, which is a continuation of the research, will be to develop a methodology for evaluating the energy consumption of such vehicles. The consequence of these actions will be the determination of the ecological assessment of conventional vehicles (evaluation of exhaust emissions), hybrid vehicles (evaluation of exhaust emissions and energy consumption), and electric vehicles (only energy consumption).

This procedure is motivated by the following reasons:

1. The proposal to categories vehicles was not developed on a large scale (with the exceptions quoted above), particularly in road emission tests. The projects so far have been mainly based on the exhaust emission results of homologation tests, in which the only criterion was the maximum value of exhaust emission. There is a research niche in the approach proposed in the paper—the authors propose two extreme values: maximum (used so far), but they also outline the methodology for determining the minimum value of exhaust emission.

2. The information defining the categorization of vehicles in road emissions tests in terms of, among other things, exhaust emissions reflects the latest literature data well:

   - Electrified vehicles (ZEV and PHEV) will have a significantly large share (about 90%) of the European market in 2030, of which more than 70% of such vehicles will be equipped with internal combustion engines [18].
   - The environmental impact of hybrid combustion vehicles powered, e.g., by hydrogen is significantly lower than that of cars powered only with electric motors [19].

3. The proposal to categorize vehicles is not required by any legislation; however, the proposed method of exhaust emission assessment may be an indication of the direction of the development of regulations. Using the authors' methodology, the trend of evaluating the exhaust gas emission in the road tests could be maintained, e.g., with partial elimination of very complex type-approval tests.

4. Increase public awareness of the environmental aspects of vehicles, which have a local and direct impact on human health (exhaust emissions), as opposed to energy consumption, which can be produced in a very distant area or from renewable sources.

3. Methodology

3.1. Requirements for the Tests under Actual Traffic Conditions

From 2017 onwards, the approval process for a new type of passenger car in the European Union includes a procedure for measuring emissions under real traffic conditions. The European Union Regulation (715/2007/EC [1] and 692/2008 [2]) on RDE tests is a response to the results of tests concerning the increased emission of nitrogen oxides from cars equipped with compression ignition engines, despite the fact that such vehicles met the acceptable standards in laboratory conditions. According to the RDE rules (Package 1–4), for all new approvals from September 2020, the emission of nitrogen oxides measured under road conditions will not exceed 1.43 times (CF—conformity factors) the maximum limit (for Euro 6d-Temp is 60 mg/km), i.e., 86 mg/km (Table 1). The parameters of the road tests are not arbitrary, and the moving average windows method (MAV) (also referred to in the literature as EMROAD, developed by the JRC) is used to determine emissions.

Table 1. Requirements for RDE testing in Europe [3–6].

2015	2016	2017	2018	2019	2020	2021	2022
	Euro 6b			Euro 6d-Temp		Euro 6d	
	NEDC			WLTC			
research and concept phase				Conformity Factor (CF)			
			$CF_{NOx} = 2.1$	$CF_{NOx} = 1.5$		$CF_{NOx} = 1.43$	
				$CF_{PN} = 1.5$		$CF_{PN} = 1.5$	

The route shall be chosen in such a way that the test is carried out without interruption, the data is recorded continuously, and the duration of the test is between 90 and 120 min. The electrical energy for the PEMS (portable emission measurement system) shall be supplied by an external power supply and not by a source that draws its energy directly or indirectly from the engine of the vehicle under test. The installation of the PEMS system shall be carried out in such a way that it affects vehicle emissions, performance, or both as little as possible. Care shall be taken to minimize the mass of the installed equipment and potential aerodynamic changes in the test vehicle. RDE tests shall be conducted on working days, and on paved roads and streets (e.g., off-road driving is not permitted). Prolonged idling after the first ignition of the combustion engine at the start of the emission test shall be avoided (Table 2).

Table 2. Specific requirements of the RDE test [3–6].

Parameter	Requirement
ambient temperature (t_a)	normal range: $0\,°C \leq t_a \leq 30\,°C$ lower extended range: $-7\,°C \leq t_a < 0\,°C$ (emission corrective factor 1/1.6) upper extended range: $30\,°C < t_a \leq 35\,°C$ (emission corrective factor 1/1.6)
driving test altitude (h)	normal range: $h \leq 700$ m a.s.l.; extended range: $700 < h \leq 1300$ m a.s.l.
impact evaluation of ambient weather and road conditions as well as the driving style	total altitude increase: less than 1200 m/100 km; relative positive acceleration (RPA): move than RPA_{min} (for all road conditions); the product of velocity and acceleration ($V \cdot a_+$): less than $V \cdot a_{+\,max}$ (for all road conditions)

Table 2. *Cont.*

Parameter	Requirement
cold start	duration of the cold start period is defined from engine start to first of 5 min or coolant temp ≥ 70 °C; max velocity during cold start ≤ 60 km/h; the average speed (including stops) shall be between 15 km/h and 40 km/h; total stop time during cold start <90 s; idling after ignition <15 s
any vehicle stop	no longer than 180 s
vehicle aftertreatment systems operation	a single regeneration of the particulate filter justifies repeating the RDE test; the occurrence of two filter regenerations is to be included in the results of the RDE test
driving comfort systems operation	regular use as intended by the manufacturer (for example: use of the air conditioning)
vehicle load	vehicle mass: driver (and passenger) along with the PEMS equipment; maximum load $<90\%$ of the sum of the mass of the passengers and the vehicle curb weight
test requirements	duration 90 min–120 min
urban test phase requirements	29–44% share of the whole test time; distance: more than 16 km; vehicle speed: up to 60 km/h; average speed: 15 km/h–40 km/h; vehicle stop: 6–30% of the urban phase of the test time
rural test phase requirements	23–43% share of the whole test time; distance: more than 16 km; vehicle speed (V): 60 km/h $< V \leq 90$ km/h
motorway test phase requirements	23–43% share of the whole test time; distance: more than 16 km; vehicle speed: more than 90 km/h; vehicle speed over 100 km/h for at least 5 min; vehicle speed over 145 km/h no more than 3% of the test phase time

3.2. Research Equipment

In order to measure the concentration of toxic compounds in the engine exhaust gas, mobile exhaust gas analyzers were used in stationary tests. The concentration of gaseous compounds was measured with the use of a Semtech DS analyser by Sensors. It enables measurement of the concentration of carbon monoxide, hydrocarbons, nitrogen oxides, and carbon dioxide, and, on the basis of oxygen concentration, the coefficient of excess air is determined.

The main purpose of the Semtech DS analyser is to measure the concentration of gaseous compounds from automotive vehicles. In this version, it can be used to test engines powered by different fuels, whose composition should be taken into account in the final data treatment (post processing). It is a representative of a group of PEMS-type measuring devices. It therefore meets the ISO 1065 standard for testing exhaust emissions with mobile systems. In addition to the possibility of using the analyser for in-service vehicle tests, it can be used as a measuring device for stationary tests, e.g., on an engine dynamometer. The Semtech DS analyser consists of the following measurement modules:

- Flame ionization detector (FID), which uses the change in electrical potential resulting from ionization of the particles in the flame; it is used to determine the total concentration of hydrocarbons;
- NDUV (non-dispersive ultra violet)-type analyser, using ultraviolet radiation to measure concentrations of nitrogen oxide and nitrogen dioxide;
- NDIR (non-dispersive infrared) analyser, using infrared radiation to measure concentrations of carbon monoxide and carbon dioxide;
- Electrochemical analyser to determine the oxygen concentration in the exhaust gas.

The Semtech DS analyser, in cooperation with a suitable flow meter, enables measurement of the exhaust mass flow rate. An important aspect is the appropriate thermal condition of the equipment, which is necessary to ensure stable indications. The time needed to obtain the appropriate temperature of the analyzer is 60 min. The measurement starts with the introduction of the exhaust sample into the analyser through a measuring probe that maintains a temperature of 191 °C. The exhaust gas sample is then filtered from the particulate matter. The filtered sample is then subjected to a hydrocarbon concentration measurement. The next step is to cool the exhaust sample to 4°C and start measuring the concentration of nitrogen oxides, carbon monoxide, carbon dioxide, and oxygen. Table 3 presents the values of measurement uncertainty and the measurement range of particular modules of the Semtech DS analyser.

Table 3. Uncertainty of indications of individual measurement modules of the Semtech DS analyzer.

Uncertainty	NDIR Analyser Indications		NDUV Analyser Indications	
Component	Carbon Monoxide	Carbon Dioxide	Nitrogen Oxide	Nitrogen Dioxide
measuring range	0–8%	0–20%	0–2500 ppm	0–500 ppm
extended uncertainty of measurement	±3% reading (or 50 ppm CO)	±3% of reading (or 0.1% CO_2)	±3% reading (or 15 ppm)	±3% reading (or 10 ppm)
		whichever is greater		

For the measurement of particle diameters, a TSI Incorporated analyser—EEPS 3090 (Engine Exhaust Particle Sizer™ Spectrometer) was used. It allowed measurement of the discrete range of particle diameters (from 5.6 nm to 560 nm) based on their different electrical mobility. Exhaust fumes are directed to the device through a dilution system and a system that maintains the required temperature. A pre-filter stops particles larger than 1 μm in diameter that are outside the measuring range of the device. After passing through the neutralizer, the particles receive a positive electrical charge depending on their diameter. The particles deflected by the high-voltage inner electrode enter the ring slot. In the space between the inner electrode (having a positive electrical charge) and the outer cylinder (built as a stack of isolated electrodes arranged in rings), the electric charge of the collected particles (on the outer electrodes) is read by the processing system. Technical data of the TSI 3090 EEPS analyser are presented in Table 4.

Table 4. Spectrometer characteristics of EEPS TSI 3090.

Parameter	Value
range of measured particle diameters	5.6 nm–560 nm
measurement resolution (number of channels)	16 channels per decade (total 32)
reading frequency	10 Hz
airflow	40 dm^3/min
flue gas sample flow	10 dm^3/min
sample temperature range	10 °C–52 °C
operating temperature of the device	0 °C–40 °C
analyser weight	32 kg

3.3. Characteristics of the Research Objects

The characteristic features of the research objects are presented in Table 5. All vehicles were plug-in hybrids. They varied in terms of the displacement of the fitted combustion engine, maximum power output, and torque. The vehicles were selected so as to most efficiently diversify and compare vehicles of different environmental performances. Vehicle A is characterized by the highest power output of the combustion engine and the electric motor and the highest battery capacity (13.6 kW·h). Vehicle B is distinguished by a continuously variable transmission, the lowest curb weight, and an average battery capacity (8.8 kW·h). Vehicle C is characterized by the highest engine capacity but has the lowest battery capacity (3.3 kW·h). Depending on the battery capacity, the vehicles were referred to as large battery (vehicle A), medium battery (vehicle B), and small battery (Vehicle C).

Table 5. Technical parameters of plug-in vehicles.

Technical Parameters	Vehicle A Large Battery	Vehicle B Medium Battery	Vehicle C Small Battery
engine	gasoline, Turbo, R4	gasoline, R4	gasoline, R4
fuel system	direct injection	multipoint injection	direct injection
engine displacement	1395 cm^3	1798 cm^3	1999 cm^3
max. power	115 kW + 85 kW (electric)	72 kW + 53 kW (electric)	113 kW + 50 kW (electric)
max. torque	250 Nm/1500–3500 rpm + 330 Nm (electric)	142 Nm/3600 rpm + 163 Nm (electric)	189 Nm/5000 rpm + 205 Nm (electric)
gearbox	automatic, 6 gears	automatic infinitely variable e-CVT	automatic, 6 gears
size (length/width/height)	4869/1864/1503 mm	4540/1760/1490 mm	4855/1860/1470 mm
curb weight	1655 kg	1375 kg	1740 kg
average CO$_2$ emissions	31–42 g/km (WLTP)	28–35g/km (WLTP)	33 g/km (WLTP)
euro standard	Euro 6d-Temp	Euro 6-Temp	Euro 6
model year	2020	2020	2019
battery	13.6 kW·h	8.8 kW·h	3.3 kW·h

3.4. Adopted Method to Search for the Function Minimum

In search for the lowest road (specific) emission in the RDE test, a task was applied consisting in seeking the lowest value of the function of many variables while fulfilling the imposed conditions. Such a task was formulated in the form:

$$\min \mathbf{f(x)} \tag{1}$$

fulfilling the limitations:

$$\mathbf{h(x)} = 0, \tag{2}$$

$$\mathbf{g(x)} \leq 0 \tag{3}$$

where: **f**—objective function of the optimization task, **h(x)**—equality limitations vector, and **g(x)**—inequality limitation vector.

The above-presented general form of the objective function and the function of limitations requires an introduction of the modification of the objective function and limitations to enable an application of the general reduced gradient method. The general reduced gradient method belongs to a group of methods searching for the function minimums of many variables without limitations. The equality and inequality limitations can be considered by including them in the objective function:

$$K(\mathbf{x}) = \ \mathbf{f(x)} \ \sum_i h_i^2(\mathbf{x}) + 1/(2\mu) \ \sum_i W(\mathbf{g}) + 1/(2\mu), \tag{4}$$

where: $W(\mathbf{g}) = g(\mathbf{x}) - g^{(min)}$ for $g \leq g^{(min)}$ and μ—par parameter modified in the optimization process.

The generalized reduced gradient (GNG) algorithm implemented in the Excel add-in Solver was used to perform the analysis. The principle of operation of the GNG algorithm in a shortened version is presented on the basis by Lasdon et al. [20]. Such an algorithm works very well for solving nonlinear problems; however, it is sensitive to the choice of initial data.

4. Results

4.1. Validation of the Tests for Compliance with the Requirements

In order to be able to compare the exhaust emissions in the first place, the nature of the test drives had to be compared. The performed tests were validated for their compliance with the RDE procedure, which requires three phases: urban, rural, and motorway. First, a formal check of the test drives was performed, and the detailed data are shown in Table A1 in Appendix A. All tests were repeated 5 times; tests in which extreme values were reached were rejected. For the purposes of this paper, the test that did not deviate by more than 10% from the average value in terms of the emissions of each exhaust constituent was assumed to be representative.

Despite meeting all the formal requirements for the tests, the most vital parameters were also compared that could have impact on the different emission results of the investigated vehicles. The course of the test route clearly indicates the three phases of the test (Figure 2a). We can distinguish the urban phase 20–25 km, the rural phase—25 km to 50 km, and the motorway phase—60 to 90 km. All three phases were rather similar in terms of their average speeds (Figure 2b). The spread of the average speed for the urban phase did not exceed 1 km/h; for the rural phase, 2 km/h; and for the motorway phase (the greatest), 5 km/h. The average value of the speed in the entire test was similar in each drive and amounted to 52, 56, and 54 km/h for vehicles A, B, and C, respectively.

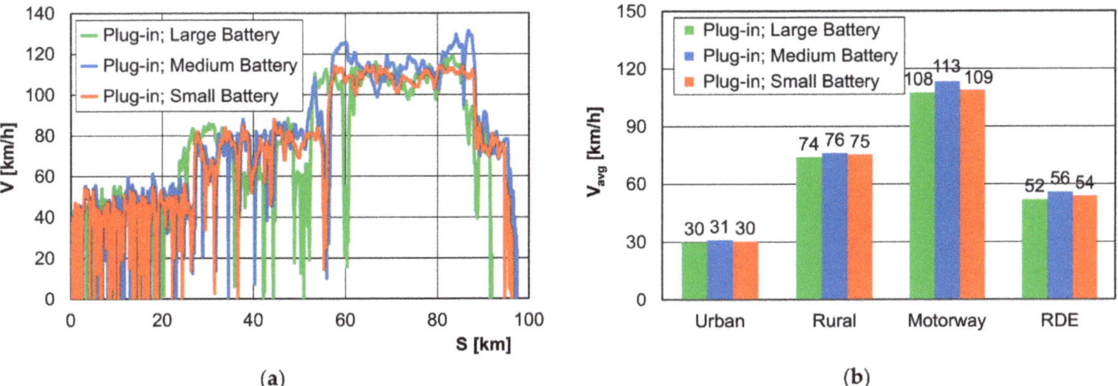

Figure 2. The tracings of the speed in the tests performed on the plug-in hybrids of different battery capacities (**a**) and average values of the speed in each phase of the test and in the entire RDE test (**b**).

In a second step, the authors compared the contribution of each phase of the RDE test for individual vehicles. The formal requirement is that the urban phase fills from 29% to 44%, the rural phase 23% to 43%, and the motorway phase 23% to 43% of the entire RDE test. The obtained results for the urban phase were 36.7%, 33.1%, and 34.1% for vehicles A, B, and C, respectively (Figure 3). This means that the data variation described with the coefficient of variation CoV (the ratio of the standard deviation to the mean) is 1.5%. This is a very small value and indicates a very high probability of the obtained results. For the rural phase of the test, the obtained results are 27.5%, 31.6%, and 32.7%, which gives a CoV = 7.4%. This value is several times higher than that from the urban phase. This phase of the test is characterized by a greater variability of vehicle speed and a greater share of unpredictable conditions. For the motorway phase, the shares in the entire test were 35.5%, 35.3%, and 33.2%, which gives a value of CoV = 3.3%. Such a comparison clearly confirms the similarity of the phase shares in the RDE test. It is noteworthy that, despite the allowable high variability in phase shares, the authors tried to keep the test drives very similar throughout the study, so that the emission results were only affected by traffic parameters (which the authors had no control over).

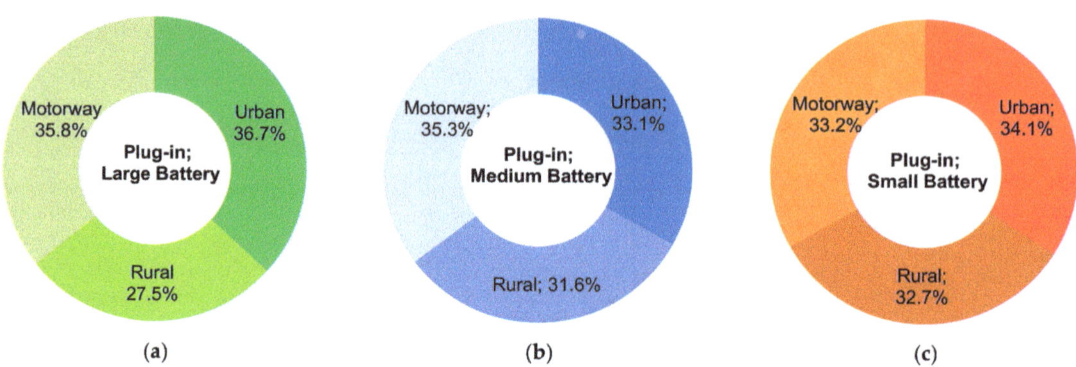

Figure 3. Comparison of the share of individual phases of the RDE test for the plug-in hybrids with a large (**a**), medium (**b**), and small (**c**) battery capacity.

The third and last stage of the comparison was determining the parameters characterizing the dynamic states of the vehicle operation. The first parameter was the 95% centile of the product of speed and positive acceleration (Figure 4a), which, in each phase of the test, should be lower than the predefined maximum. The requirement is to make sure that the test drive is not excessively dynamic, and the vehicle accelerations do not drastically increase the dynamics, resulting in increased exhaust emissions. The values of this parameter for the urban part, shown in Figure 4a, are in the range 9–12 m^2/s^3 and are similar for all test drives. For the rural phase, the obtained values of this parameter in individual drives have a greater spread (14 m^2/s^3 to 18 m^2/s^3). For the motorway phase, these values are the most similar and amount to approximately 14 m^2/s^3. It is noteworthy that in each phase of the test, the admissible value is not exceeded (marked with a continuous line). This suggests the correctness of the test realization in terms of driving dynamics. At the same time, it indicates a significant similarity, which will constitute a basis for the comparison of the exhaust emissions from the investigated vehicles.

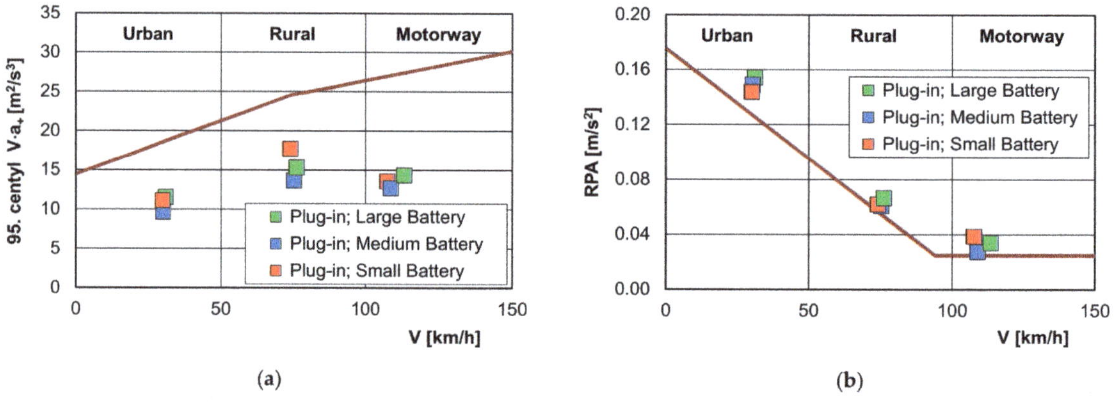

Figure 4. Comparison of the 95% centile of the product of vehicle speed and positive acceleration (**a**) and the relative positive acceleration (**b**) in each phase of the test performed on the plug-in hybrids.

The other parameter describing the dynamic conditions was relative positive acceleration (Figure 4b). This parameter described the minimum dynamic conditions for the test drive not to be a steady state and to eliminate driving with the use of cruise control. Unfortunately, this parameter is determined on such a level that the performed tests are characterized by only slightly higher values than the minimum ones. For the urban part of the test, for all the vehicles, the relative positive acceleration fell in the range 0.14 m/s^2 to 0.16 m/s^2 and was approximately 10% greater than the minimum (13 m/s^2) defined for the obtained average speed in this phase (approximately 30 km/h). In the rural phase of the test, the values were even closer to one another, and fell in the range of approximately 0.6–0.7 m/s^2 and were greater than the minimum by approximately 5% (for the average speed of 75 km/h). In the motorway phase of the test, the relative positive acceleration was the lowest. It amounted to 0.3 $\pm$ 0.04 m/s^2 and was slightly higher than the minimum defined on the level of 0.25 m/s^2.

The presented characteristics of the tests as well as the steady state and dynamic properties of the test drives prove that the formal requirements (presented in a concise form) for each of the test drives and each phase of the test were met. Such a situation implies the possibility of moving on to the next stage of the investigations, consisting in determining the exhaust emission values. Similar test conditions and similar dynamic parameters obtained for all the vehicles tested provide grounds for concluding that the exhaust gas emission results are not burdened with inaccuracy resulting from differences in the test course. At the same time, the similarity of the test runs indicate differences resulting only from individual vehicle characteristics, such as the engine used or battery capacity.

4.2. Exhaust Emission Results

Due to the fact that the engine was not in the initial phase of the RDE test, the characteristic aspect of the emission of individual exhaust components is the initial flat period of the relation. It results from the fact that the initial phase of the RDE test is performed only with the electric motor activated and the cold phase of the test is moved from the urban phase to the rural one (Figure 5a). This is particularly visible when we analyze the emission of carbon dioxide (shown in increments), from which we conclude that the vehicle of the lowest battery capacity drove approximately 40 km on the electric motor. The other vehicles used the electric motor for the distance of over 50 km. The advantage of this situation is that the engine warm-up time is reduced. This is due to the fact that there is a higher engine speed and load during the rural phase of the test. This results in much faster catalytic converter firing and a reduced cold start time. The downside, however, is the fact that the cold engine operation at higher loads and speeds results in higher emissions of individual exhaust components. This is particularly visible in the analysis of the emission of carbon monoxide (Figure 5b), where the start of a cold engine (vehicle A) in a very short time resulted in emission of almost 50% of the entire emission of this component (approximately 30 mg/km for the distance of approximately 50 km and for the entire test this value was approximately 60 mg/km for the distance of approximately 100 km).

A separate comment is required related to the nature of the changes in the emission of carbon monoxide for the medium-battery plug-in hybrid, for which this emission drops as the test continues (from 70 km onwards). This is caused by the lowest increment of the mass of carbon monoxide compared to the distance covered by the vehicle. In the outstanding two cases, the increase in the mass of carbon monoxide was greater than the increment in the covered distance. For each investigated vehicle, the emission of carbon monoxide was lower than the admissible one (1000 mg/km Euro 6d-Temp). For individual vehicles, the obtained values of the emission of this component, defined as the total mass of the emitted component against the total covered distance, are: 63, 74, and 109 mg/km, for vehicle C, B, and A, respectively.

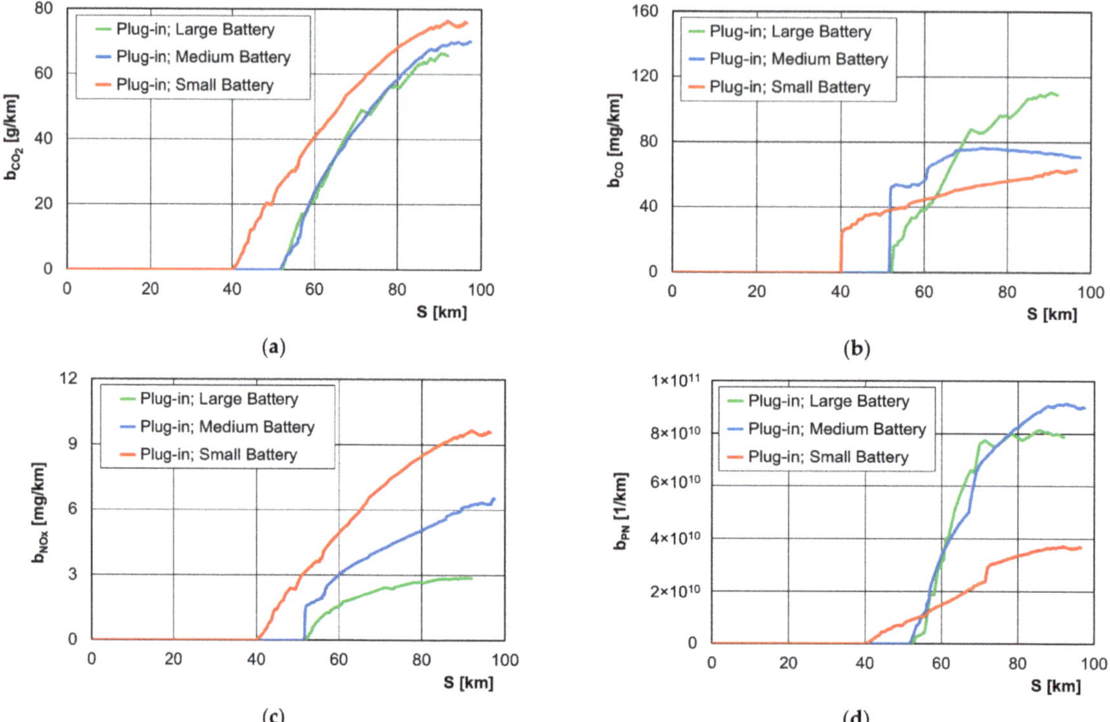

Figure 5. Dependence of the emission of carbon dioxide (**a**), carbon monoxide (**b**), nitrogen oxides (**c**), and particle number (**d**) on the distance covered during the road tests for individual investigated vehicles; the curves were made using the emission intensity results obtained in each second of the test.

The nature of the changes in the emission of nitrogen oxides (Figure 5c) was heavily dependent on the engine displacement and the vehicle weight. Vehicle C (of the highest curb weight and engine displacement) was characterized by the highest emission of nitrogen oxides and, at the same time, had a constant growing trend in the rural and motorway cycles. The final value of the emission of nitrogen oxides obtained using all the data related to the emission intensity in the entire RDE test was 9.5 mg/km, which is an approximately 50% higher value compared to vehicle B (6.4 mg/km) and more than 3 times higher than the results obtained for vehicle A (approximately 3 mg/km). In the final described case, vehicle A had the engine of the lowest displacement, but the engine was turbocharged. The catalytic converter significantly reduces the concentration of nitrogen oxides in the exhaust system, which is particularly visible in modern turbocharged engines. The obtained values of the emission of nitrogen oxides do not exceed the limits prescribed in the Euro 6d-Temp standard, which amounts to 60 mg/km.

A different tracing from the previous one was obtained for particle number (Figure 5d). The smallest increase was observed for vehicle C, for which the greatest emission of nitrogen oxides was recorded, which confirms the inversely proportional relation between these exhaust components. When comparing the obtained results of all the investigated vehicles, we know that the particle number is lower than the prescribed limit, which amounts to $6 \cdot 10^{11}$ 1/km for the direct-injected engines. Determining the total exhaust emissions in the tests requires an application of an averaging algorithm of the emissions in the measurement windows that are determined based on the emission of carbon dioxide in individual phases of the WLTC test (red dots in Figure 5).

The value of the road emission of carbon dioxide in the RDE test, determined individually for each of the vehicles, in the majority of cases falls between the lower ($Tol_{low} = 0$) and the upper ($Tol_{high} = +45\%$) limit of tolerance. In all the investigations of the plug-in hybrids, one can see a similarity of the distribution of the average emission of carbon dioxide in the measurement windows (Figure 6): for the speed in the range up to 75 km/h, the observed values do not exceed 100 g/km. In the range from 75 km/h to 125 km/h, the emission values of carbon dioxide are from 100 g/km to 200 g/km for vehicle A (Figure 6a); for vehicle B, up to 220 g/km (Figure 6b); and for vehicle C, up to 150 g/km (Figure 6c).

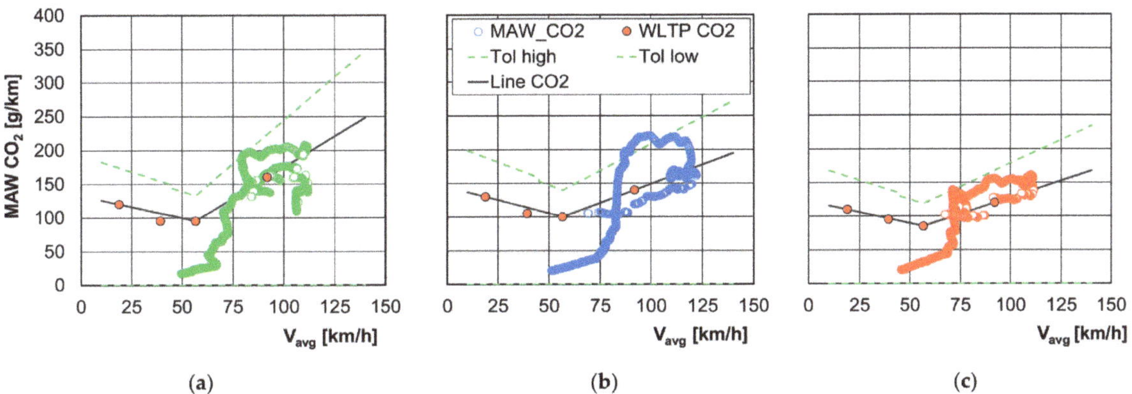

Figure 6. Characteristic curves determining the relation between the emissions of carbon dioxide in the measurement windows that are the basis for the determination of the emission of individual exhaust components in the RDE test for: Plug-in, Large Battery (**a**), Plug-in, Medium Battery (**b**), Plug-in, Small Battery (**c**).

Analyzing the carbon dioxide emissions for the different phases of the RDE test reveals zero emissions of this component in the urban phase for all plug-in hybrid vehicles tested (Figure 7a). The differences are visible only in the subsequent phases of the test: in the rural phase, the greatest emission occurs for the vehicle with the smallest battery capacity (49 g/km) and the lowest emission occurs for vehicle B with the medium battery capacity (32 g/km). The greatest values of the emission of carbon dioxide were recorded in the motorway phase of the test: 3–4 times higher compared to the rural phase. The greatest emission of carbon dioxide (159 g/km) is for the vehicle with the lowest engine displacement, which suggests the load of the powertrain. At the same time, it is confirmation of the fact that downsized engines fitted in vehicles of relatively high curb weight do not ensure the expected environmental results. The final result of the road emission of carbon dioxide for all the vehicles is 63 ± 2 g/km, which means that none of the values differed from one another by more than 3.2%. This is a very similar result confirmed by the gas mileage of 2.6, 2.8, and 3.0 dm^3/100 km for vehicles A, B, and C, respectively. This is caused by the fact that the electric mode was used in each phase of the test by each vehicle and its share was 60%, 53%, and 48% for vehicles A, B, and C, respectively.

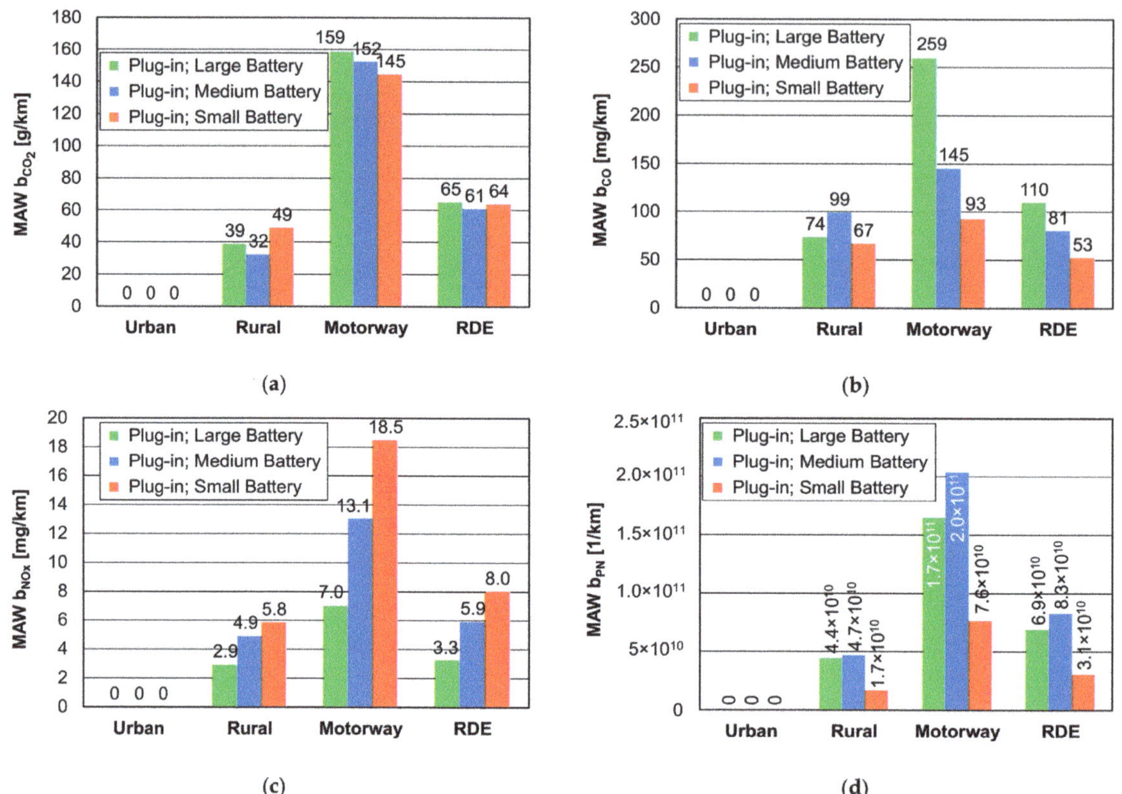

Figure 7. Road emission of carbon dioxide (**a**), carbon monoxide (**b**), nitrogen oxides (**c**), and particle number (**d**) for each phase of the test and in the entire RDE test for individual plug-in hybrid vehicles; the values were determined using an algorithm allowing for averaging in the measurement windows.

Much greater differences in individual phases of the test were recorded for the emission of carbon monoxide (Figure 7b). Similar to the previous case, the urban phase was characterized with zero emission of this component and in the rural phase, the differences were significant and amounted to 33% (for vehicle B, the emission of this component was 99 mg/km and for vehicle C, 67 mg/km). Vehicle A reached an intermediate value of 74 mg/km. An even greater divergence was recorded in the motorway phase of the test, where, similar to carbon dioxide, the highest values (259 mg/km) were recorded for vehicle A (downsizing) and the smallest (93 mg/km) for vehicle C (the greatest displacement). This translated into the final values that, for vehicle A (110 mg/km), were twice as high compared to vehicle C (53 mg/km). It is noteworthy that these values are 10–20 times lower than the admissible ones prescribed in the Euro 6d-Temp standard, where the limit is 1000 mg/km.

The road emission of nitrogen oxides was dependent on the engine displacement: the greater the engine displacement, the higher the emission of this component in individual phases of the RDE test (Figure 7c). The road emission of nitrogen oxides in the rural phase for all the tested vehicles was the result of the engine cold start in this phase. The recorded values of the road emission in the rural phase were 2.9 g/km, 4.9 mg/km, and 5.8 mg/km for vehicles A, B, and C, respectively. Vehicle C generated twice the mass of nitrogen oxides compared to vehicle A (assuming the same distance in the rural phase of the test). The values of this parameter in the motorway phase were even more varied: vehicle C generated 2.5 times the mass of nitrogen oxides compared to vehicle A, which could have been caused by the mileage of this vehicle in the first place (MY 2019) and the reduced efficiency of the catalytic converter. The consequence is the final result of the emission of nitrogen oxides for vehicle A of 3.3 mg/km and vehicle C of twice as much (8 mg/km). These values are more than 10 times lower that the ones prescribed in the RDE standard: 60 mg/km (b_{NOx} Euro 6d-Temp) × 1.43 (CF_{NOx}) = 86 mg/km. The road emission of the particle number was quite the opposite compared to nitrogen oxides. The highest values in each phase of the test and in the entire RDE test were recorded for vehicle B (Figure 7d). These values were approximately 10–20% higher compared to vehicle A, whose values were 0, $4.4 \cdot 10^{10}$ 1/km, $1.7 \cdot 10^{11}$ 1/km, and $6.9 \cdot 10^{10}$ 1/km in the urban, rural, motorway phase, and the entire test, respectively. The lowest emission of the particle number was vehicle C, for which the value in the entire RDE test was $3.1 \cdot 10^{10}$ 1/km. All the obtained values were several times lower than the admissible limit of this component prescribed in the RDE test requirements, amounting to 1.5 times the limit of the Euro 6d-Temp standard ($6 \cdot 10^{10}$ 1/km × 1.5 = $9 \cdot 10^{11}$ 1/km).

The above-presented procedure of determination of exhaust emissions under road conditions served to assess the environmental performance of three different plug-in hybrid vehicles. On this basis, the authors identified the exhaust emissions falling in the admissible limits, i.e., quantitative information was obtained. The qualitative information, however, was not assured, i.e., the result was not obtained against the environmental capabilities of a given plug-in hybrid vehicle model. The obtained results do not contain information on the scope of variability in the position of the results against the minimum and maximum obtainable values for each vehicle.

5. Discussion

The obtained on-road emission values from each vehicle were used as input parameters to determine the limits (where the values are possible). Specifying a maximum value is not questionable because it should be a limit:

$$b_{j,max} = b_{j,Euro6d\text{-}Temp} \times CF_j \tag{5}$$

where: $b_{j,Euro6d\text{-}Temp}$ denotes the admissible value of the road emission for the j-th exhaust component ($b_{CO,Euro\ 6d\text{-}Temp}$ = 1000 mg/km, $b_{NOx,Euro\ 6d\text{-}Temp}$ = 60 mg/km, $b_{PN,Euro\ 6d\text{-}Temp}$ = $6 \cdot 10^{11}$ 1/km and CF_j—conformity factor for the j-th exhaust component (CF_{CO} = 2.1, CF_{NOx} = 1.43, CF_{PN} = 1.5).

The minimum values that vehicles can theoretically obtain were determined with constant and variable emission rates. Constant emissivity should be assumed when changes in emissivity do not appear in individual phases of the RDE test. This means that a significantly small standard deviation occurs from the average value described with the coefficient CoV < 10%. The exact values of CoV for the emission intensity of all the exhaust components from all the investigated vehicles are presented in Table 6.

Table 6. CoV coefficient for all the tested exhaust components.

Exhaust Components	Vehicle A		Vehicle B		Vehicle C	
	Rural	**Motorway**	**Rural**	**Motorway**	**Rural**	**Motorway**
CoV_{CO2} [%]	93.9	21.7	78.3	28.4	125.4	14.6
CoV_{CO} [%]	86.0	33.5	77.5	61.0	552.0	17.5
CoV_{NOx} [%]	83.2	30.6	77.5	17.2	125.9	14.6
CoV_{PN} [%]	128.2	61.8	77.2	35.5	131.7	104.3

It can be seen from Table 6 that all CoV values are greater than 10%, so the emission intensity (or road emission value) must depend on a mean value that accurately describes the nature of changes in the said emission intensity (or road emission). In this paper, the minimum value of the exhaust emissions with both methods was determined irrespective of the above.

5.1. Determining the Minimum Road Emissions

5.1.1. Constant Road Emission Intensity

For the determination of the theoretical values of the minimum exhaust emissions, the authors used a method described in Section 3.4 of this paper. As the input values, the authors used the emission intensities of a given exhaust component $E_{j,k}$, determined from the road emissions $b_{j,k}$; $j = CO_2$, CO, NO_x, PN, for the rural and motorway phases). In the urban phase, the emission intensity was 0. The variable values of the algorithm were: urban, rural, and motorway test duration (t_k) and the distance in the urban, rural, and motorway phases (S_k). The limitations were:

- The sum of the entire test duration: $t_U + t_R + t_M \in (90\text{–}120\ \text{min})$;
- Average speed in the urban phase $S_U/t_U \in (15\text{–}40\ \text{km/h})$;
- Average speed in the rural phase $S_R/t_R \in (40\text{–}80\ \text{km/h})$;
- Average speed in the motorway phase $S_M/t_M \in (80\text{–}140\ \text{km/h})$.

The initial values were: $t_U = t_R = t_M = 30$ min and $S_U = S_R = S_M = 16$ km.

The values of the emission intensity of individual exhaust components (Table 7) were obtained based on the average road emission, the time, and the distance covered in each test phase. The objective function had a form:

$$b_{j,RDE} = 0.34\ b_{j,U} + 0.33\ b_{j,R} + 0.33\ b_{j,M} \tag{6}$$

and upon including the constant emission intensity $E_{j,k}$ in each phase of the test:

$$b_{j,RDE} = 0.34\ E_{j,U}\ t_U/S_U + 0.33\ E_{j,R}\ t_R/S_R + 0.33\ E_{j,M}\ t_M/S_M. \tag{7}$$

Table 7. The value of the constant emission intensity for the phases of the RDE test (rural, motorway) as the algorithm input data.

Emission Intensity	Vehicle A		Vehicle B		Vehicle C	
	Rural	**Motorway**	**Rural**	**Motorway**	**Rural**	**Motorway**
E_{CO2} [g/s]	0.80	4.74	0.69	4.80	1.02	4.37
E_{CO} [mg/s]	1.51	7.75	2.10	4.57	1.40	2.80
E_{NOx} [mg/s]	0.06	0.21	0.10	0.41	0.12	0.56
E_{PN} [1/s]	$9.1{\cdot}10^8$	$4.9{\cdot}10^9$	$9.9{\cdot}10^8$	$6.4{\cdot}10^9$	$3.6{\cdot}10^8$	$2.3{\cdot}10^9$

Using the Solver tool (Excel MS Office^{TM}), for each of the vehicles, the duration of each of the test phases (t_k), the distance covered in each of the test phases (S_k), and the share of each test phase in the entire RDE test (u_k) were determined. Detailed data are presented in Tables A2–A4 (Appendix A), for vehicles A–C, respectively. From the comparison of the data in the tables, the theoretical minimum value of the exhaust emissions is greater than zero, which means that the plug-in hybrid vehicle in the RDE test will always use a combustion engine. For vehicle A, the minimum value of the road emission of carbon dioxide was 49 g/km; for carbon monoxide, 83 mg/km; for nitrogen oxides, 2.5 mg/km; and for the particle number, $5.2{\cdot}10^{10}$ 1/km. It should be noted that the obtained values of the parameters of time (t_k), test phase duration (S_k), and test phase share (u_k) were different for each exhaust component and the scatter of results in individual analyzed categories, as measured with the CoV coefficient, fell in the range from 1.1% to 5.7% (Table A2). For vehicle B, the following road emission values were obtained using the same pattern: b_{CO2} = 48 g/km, b_{CO} = 65 mg/km, b_{NOx} = 4.7 mg/km, a b_{PN} = $6.6{\cdot}10^{10}$ 1/km, at the coefficient of variation CoV changing from 0.9% to 20% (Table A3). For vehicle C, the values of individual parameters were as follows: b_{CO2} = 49 g/km, b_{CO} = 41 mg/km, b_{NOx} = 6.2

mg/km, a b_{PN} = 2.4·10^{10} 1/km, at the coefficient of variation CoV changing from 0.0% do 12.4% (Table A4).

The obtained average values for individual vehicles and each exhaust component are shown in Figure 8. From this figure, the road emission of carbon dioxide for each vehicle falls in the range 60–80 g/km (Figure 7a) at the minimum value of approximately 49 g/km (Figure 8a). The value was adopted obligatorily on the level of 95 g/km (this is a target value for a manufacturer's vehicle fleet; however, this value may decrease in future years). The road emission of carbon monoxide for all the investigated vehicles was in the range 50–110 mg/km (Figure 7b) at the maximum value of 2100 mg/km and the minimum one of 40–80 mg/km (Figure 8a). The road emission of nitrogen oxides fell in the range 3.3 mg/km–8.0 mg/km (Figure 7c) at the maximum value of 86 mg/km, determined according to Equation (5).

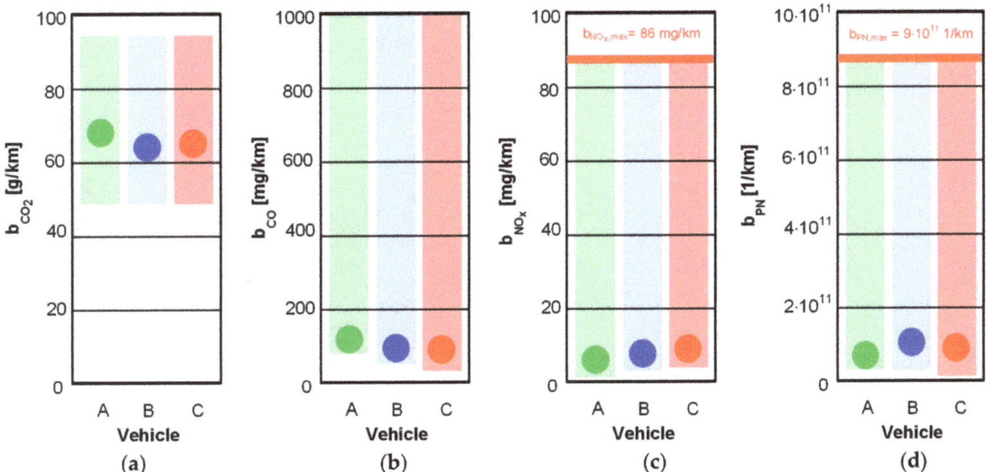

Figure 8. Values of the minimum road emission intensity of carbon dioxide (**a**), carbon monoxide (**b**), nitrogen oxides (**c**), and particle number (**d**) in the RDE test obtained according to the algorithm for a constant emission intensity; data provided in Tables A2–A4.

The theoretical minimum value of the road emission of nitrogen oxides was in the range from 2.5 mg/km to 4.7 mg/km (Figure 8c). The last investigated exhaust component, the road emission of particle number, was in the range from 3.1·10^{10} 1/km to 8.3·10^{10} 1/km (Figure 7d). For this parameter, the maximum value determined in Equation (5) is 9·10^{11} 1/km and the minimum determined value falls in the range from 2.4·10^{10} 1/km to 6.6·10^{10} 1/km (Figure 8d).

5.1.2. Variable Road Emission Intensity

The determination of the minimum road emission using constant emission intensity of a given exhaust component does have its flaws: the engine warm up and increased catalyst efficiency after light-off are not allowed for. A more generalized approach may also be the use of the mathematical description of the curves presented in Figure 5, in relation to which the general form of Equation (2) assumes a form where individual values of the road emission for each exhaust component and each test phase will be dependent on the distance covered by the vehicle:

$$b_{j,RDE} = 0.34\, b_{j,U}(S) + 0.33\, b_{j,R}(S) + 0.33\, b_{j,M}(S). \tag{8}$$

In this case, we need to apply the non-continuous function for each exhaust component that will allow for the operation of the electric motor (in the range $S_k \leq S_{EV}$) during the urban and (partially) rural phases. For the outstanding distance, each course of the road emissions was described with a square Equation (9):

$$b_{j,k}(S_k) = \begin{cases} 0 & \text{for } S_k \leq S_{EV} \\ x_{j,k}(S_k)^2 + y_{j,k}S_k + z_{j,k} & \text{for } S_k > S_{EV} \end{cases} \tag{9}$$

where: $j = CO_2$, CO, NO_x, PN; S—distance [km]; S_{EV}—distance covered by the vehicle using an electric motor [km]; k = Vehicle A, Vehicle B, Vehicle C; and $x_{j,k}$, $y_{j,k}$, $z_{j,k}$—multinomial coefficients (Table 8).

Table 8. Values of the equation indexes of the z, y, z multinomial and the coefficient of determination (R^2) for the road emission in the RDE test for each plug-in hybrid vehicle.

j	k	Vehicle A	Vehicle B	Vehicle C
S_{EV} [km]		52.2	51.6	40.0
b_{CO2} [g/km]	x	−0.0351	−0.0332	−0.0201
	y	6.6697	6.5103	4.1145
	z	−250.92	−248.93	−132.92
	R^2	0.996	0.998	0.998
b_{CO} [mg/km]	x	−0.0579	−0.0322	−0.0067
	y	10.922	5.2708	1.5469
	z	−404.81	−138.60	−24.207
	R^2	0.990	0.919	0.995
b_{NOx} [mg/km]	x	−0.0024	−0.0013	−0.0021
	y	0.4972	0.2977	0.3557
	z	−16.292	−10.467	−12.417
	R^2	0.976	0.994	0.998
b_{PN} [1/km]	x	$-5.91 \cdot 10^7$	$-9.62 \cdot 10^7$	$-6.33 \cdot 10^6$
	y	$1.08 \cdot 10^{10}$	$1.59 \cdot 10^{10}$	$1.58 \cdot 10^9$
	z	$-4.05 \cdot 10^{11}$	$-5.76 \cdot 10^{11}$	$-5.51 \cdot 10^{10}$
	R^2	0.983	0.996	0.979

The values of the multinomial coefficients ($x_{j,k}$, $y_{j,k}$, $z_{j,k}$) are presented in Table 8 and their analysis indicates an increase in the road emissions upon exceeding the distance ($S_k > S_{EV}$), which is indicated by the negative value of each coefficient $x_{j,k}$.

The positive value of coefficient $y_{j,k}$ indicates a shift of the course of the emissions to the right for the increasing distance from the start of the test and the negative value of coefficient $z_{j,k}$ confirms the assumptions of zero emission of each exhaust component during the operation of the electric motor. In Table 8, the authors also provide the coefficient of determination (R^2), which indicates a very good fit of the adopted equation to the curves, showing the road emission of each exhaust component in the RDE test.

Utilizing Equation (9) and the data contained in Table 8, the minimum road emission of each of the exhaust components was determined in individual phases of the RDE test and in the entire test. The obtained detailed results regarding the duration of each phase of the test, the distance in each phase of the tests, and the share of each phase in the test are shown in Tables A5–A7 (Appendix A). The graphical presentation of the final relations is shown in Figure 9. From the figure, it is found that the minimum value of carbon dioxide for vehicles A and B is 0 g/km, which leads to the conclusion that the vehicle can cover the entire RDE test distance using an electric motor exclusively (which is compliant with the test detailed requirements). This is possible because the range of vehicles A and B on an electric motor was approximately 52 km and the minimum distance of the RDE test is 48 km. Due to the fact that the combustion engine was off, the emission of the outstanding exhaust components from vehicles A and B was also zero. Only in the case of vehicle C with the small battery capacity was it impossible to carry out the entire RDE test exclusively using the electric motor. For this particular vehicle, the theoretical minimum values of the road emission are greater than zero for each of the exhaust components. However, when comparing the results obtained for constant and variable emission intensity, one can observe lower values for the latter method. It is probable that the theoretical minimum values of the road emission obtained using the variable emission intensity method during the road tests are closer to reality, which is why the authors recommend them for application in further research.

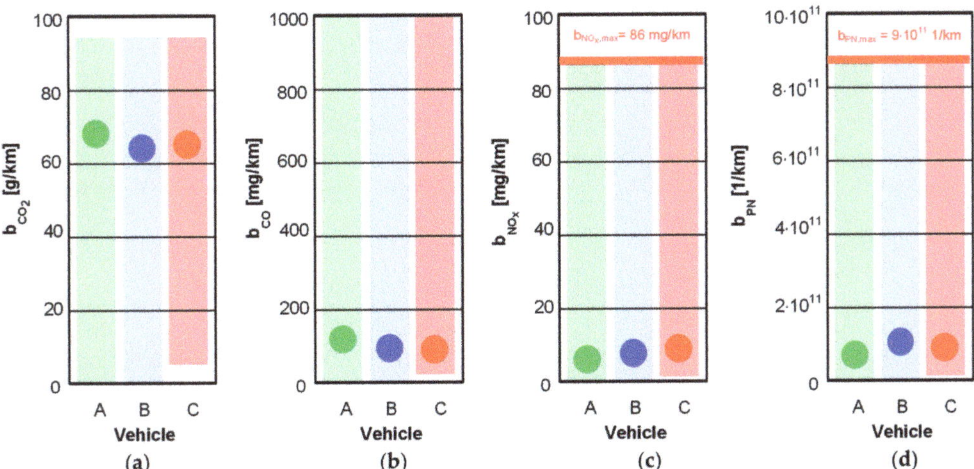

Figure 9. Value of the minimum road emission of carbon dioxide (**a**), carbon monoxide (**b**), nitrogen oxides (**c**), and particle number (**d**) obtained according to the algorithm for variable exhaust emission intensity; data provided in Tables A5–A7.

5.2. Plug-in Vehicles Emission Category

The environmental assessment of the vehicles was initiated upon the analysis of the obtained results of actual road emissions of individual exhaust components (p. 4) and upon adopting the maximum values (values adopted based on the emission standards—p. 5.1.1) and the minimum ones (values adopted based on the theoretical determination of the minimum for the measured emission intensity—p. 5.1.2). The environmental categorization (EC—ecological category) for each exhaust component was performed based on the determination of the percentage value of the obtained road emission depending on the minimum and maximum value according to the formula:

$$EC_j\ [\%] = 100\% \times (b_j\!-\!b_{j,min})/(b_{j,max}\!-\!b_{j,min}). \tag{10}$$

Equation (10) describes the process of scaling that adapts the value of any exhaust component to the new limits determined with the minimum (0%) and the maximum value (100%). Such an approach enables each road emission to be presented as a value ranging from 0–100%, as shown in Table 9. According to the results presented below, one can confirm that the highest values (EC = 64–68%) are gained for the road emission of carbon dioxide, which indicates great potential for improvement in this matter. In the case of the emission category pertaining to the road emission of carbon monoxide, the determined values are in the range from 4% to 11%. This confirms a substantial reserve of approximately 90%, which is tantamount with the fact that the analyzed vehicles generate much less of this exhaust component compared to the emission standard prescribed for the RDE test. We have a similar situation for the road emission of nitrogen oxides and particle number: the obtained values from 4–7% also confirm the rule.

Table 9. Real values of the road exhaust emissions (b_j), theoretically determined maximum value ($b_{j,\,max}$) minimum value ($b_{j,\,min}$), and the environmental assessment (EC_j) for individual plug-in hybrid vehicles.

Parameter	Vehicle A	Vehicle B	Vehicle C
b_{CO_2} [g/km]	65	61	64
b_{CO} [mg/km]	110	81	53
b_{NOx} [mg/km]	3.3	5.9	8
b_{PN} [1/km]	$6.90 \cdot 10^{10}$	$8.30 \cdot 10^{10}$	$3.10 \cdot 10^{10}$
$b_{CO2,\,max}$ [g/km]	95	95	95
$b_{CO,\,max}$ [mg/km]	1000	1000	1000
$b_{NOx,\,max}$ [mg/km]	86	86	86
$b_{PN,\,max}$ [1/km]	$9.00 \cdot 10^{11}$	$9.00 \cdot 10^{11}$	$9.00 \cdot 10^{11}$
$b_{CO2,\,min}$ [g/km]	0	0	5.99
$b_{CO,\,min}$ [mg/km]	0	0	11.43
$b_{NOx,\,min}$ [mg/km]	0	0	2.35
$b_{PN,\,min}$ [1/km]	0	0	$2.00 \cdot 10^{9}$
EC_{CO2}	68%	64%	65%
EC_{CO}	11%	8%	4%
EC_{NOx}	4%	7%	7%
EC_{PN}	8%	9%	3%

The values of emissions of individual exhaust components for individual vehicles were used for the overall environmental assessment of the analyzed vehicles. An arithmetic average relation was applied, yet the authors are aware that a valuation can be introduced in terms of the significance of each of the exhaust components. Such an action would require considering the hazard of individual exhaust components to human health and the criteria of its assessment would require further analyses. Therefore, the application of an arithmetic average allows all the discussed exhaust components to be treated equally. The obtained values, shown in Figure 10, show that all the vehicles under analysis obtain ecological values lower than 50%, which classifies them to the ecological category A.

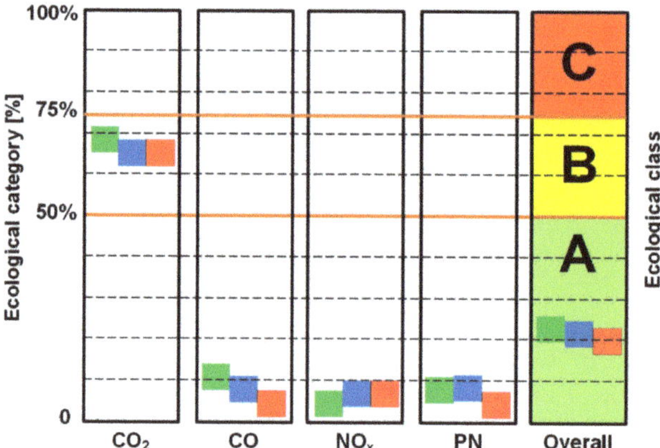

Figure 10. Value of the minimum road emission in the RDE test (■—vehicle A, ■—vehicle B, ■—vehicle C) obtained according to the algorithm for variable exhaust emission intensity; data provided in Tables A4–A6.

An explanation is required for the case in which the road emission of an exhaust component is greater than the product of the CF coefficient and the value provided in the emission standard. In such a case, the value of CF will be greater than 100%, which, for conventional vehicles, will result in the necessity to extend the range of qualification and emission categories.

6. Conclusions

In the paper, the authors presented the process of creation of a new tool that can be used in the environmental assessment of motor vehicles under actual traffic conditions. Currently, the certificate of type approval provides the final results as regards the compliance with the vehicle ecological category, yet without the emission-related valuation of the vehicle. The methodology presented in the paper and applied when investigating plug-in hybrid vehicles unveils its practical application, which is confirmation that the analyzed vehicles adapt well environmentally when it comes to the trends in electromobility, given their emission category (A). The division into emission categories of plug-in hybrid vehicles can be introduced in a different subjective way, e.g., by creating more categories from A to E.

The major conclusions of the paper can be summarized in the following points:

1. The concept of this paper was aimed at indicating the methodology of processing of the hybrid vehicles exhaust emission results, but it could also be of use for either conventional vehicles or electric ones, in which instead of analyzing the exhaust emissions one could focus on the energy consumption. The authors see no contraindications to applying the methodology in heavy-duty vehicles, in the case of which other indexes are used for the description of their emissions-related performance (referred to the performed work of a vehicle).
2. The estimated minimum exhaust emission results for the RDE test assuming constant emission rates are about 20–40% less than the actual values. However, estimating the minimum values, for variable emission rates, reveals the largest differences, as the minimum values can be zero. This is a characteristic feature of hybrid vehicles, especially PHEVs, whereby using a large electric capacity of the battery, zero emissions can be achieved in the RDE test. Therefore, the next stage of the research should be a parallel analysis consisting in the evaluation of emission aspects on the one hand, and energy consumption aspects on the other.
3. In the era of fighting for environmental protection, the additional emission class indicator of a vehicle may be an important aspect in the consumer's choice of a vehicle. Due to the campaigns of the European Union and member states, consumer awareness of the harmfulness of exhaust emissions is constantly rising. The proposed method of categorizing hybrid vehicles (the authors plan to extend the RDE study and publish an article on the energy consumption of vehicles in the future) is clear and transparent, which may support the choice of greater ecological power unit.
4. The research presented in this paper describes an alternative way to evaluate the environmental performance of vehicles, although limiting oneself only to emissions for PHEVs is not a closed solution. The next stage of work will be to develop a methodology for evaluating the energy consumption of hybrid and electric vehicles in road tests. The result of such a procedure will be the ecological evaluation of vehicles powered only by combustion engines (evaluation on the basis of exhaust emissions), hybrid vehicles—HEVs and PHEVs (evaluation on the basis of exhaust emissions and energy consumption), and vehicles powered by electricity or fuel cells (evaluation on the basis of energy consumption).

Author Contributions: Conceptualization, K.S. and J.P.; methodology, K.S. and J.P.; software, K.S.; validation, K.S. and J.P.; formal analysis, K.S; investigation, K.S. and J.P.; resources, K.S. and J.P.; data curation, K.S. and J.P.; writing—original draft preparation, K.S.; writing—review and editing, K.S. and J.P.; visualization, K.S. and J.P.; supervision, K.S. and J.P.; project administration, J.P.; funding acquisition, K.S. and J.P. All authors have read and agreed to the published version of the manuscript.

Funding: This research was funded by Poznan University of Technology, grant number 0415/SBAD/0319.

Institutional Review Board Statement: Not Applicable.

Informed Consent Statement: Not Applicable.

Data Availability Statement: The data presented in this study are available on request from the corresponding author.

Conflicts of Interest: The authors declare no conflict of interest.

Abbreviations

a	acceleration vehicle
b	road exhaust emission
CF	Conformity Factor
CoV	coefficient of variation
E	exhaust emission rate
EC	ecological category
EV	Electric Vehicle
FID	flame ionization detector
GNG	Generalized Reduced Gradient
h	driving test altitude
HEV	Hybrid Electric Vehicle
M	motorway
MAV	Moving Average Windows
NDIR	non-dispersive infrared
NDUV	non-dispersive ultra violet
NEDC	New European Driving Cycle
PEMS	Portable Emission Measurement System
PHEV	Plug-in Hybrid Electric Vehicle
R	rural
RDE	Real Driving Emissions
RPA	Relative Positive Acceleration
S	distance
t	time
u	share
U	urban
V	vehicle speed
WLTC	Worldwide-harmonized Light duty vehicles Test Cycle
WLTP	Worldwide-harmonized Light duty vehicles Test Procedure

Appendix A

Table A1. Characteristic data of the performed tests and their comparison with the admissible values.

Trip Characteristics	Vehicle A	Vehicle B	Vehicle C	Valid	Unit
Distance	91.90	97.40	96.45	–	km
Duration	106.17	104.52	107.83	90–120	min
Cold start duration	5.00	5.00	5.00	5.00	min
Urban distance	33.73	32.22	32.88	>16	km
Rural distance	25.25	30.79	31.58	>16	km
Motorway distance	32.92	34.39	31.98	>16	km
Urban distance share	36.70	33.08	34.10	29–44	%
Rural distance share	27.48	31.61	32.74	23–43	%
Motorway distance share	35.82	35.31	33.16	23–43	%
Urban average speed	30.05	31.14	30.33	15–40	km/h
Rural average speed	74.09	76.23	75.34	–	km/h
Motorway average speed	107.55	113.37	108.83	–	km/h
Total trip average speed	51.94	55.91	53.66	–	km/h
Motorway speed above 145 km/h	0.00	0.00	0.00	<3% mot. time	%

Table A1. *Cont.*

Trip Characteristics	Vehicle A	Vehicle B	Vehicle C	Valid	Unit
Motorway speed above 100 km/h	16.28	16.63	17.32	≥ 5	min
Stop share (Urban phase)	20.09	17.91	15.12	6–30	%
Initial idling duration	0.00	0.00	0.00	≤ 15	s
Cold start average speed	23.70	23.68	21.33	15–40	km/h
Cold start maximum speed	44.25	45.86	47.00	<60	km/h
Cold start stop time	0.00	31.00	18.00	≤ 90	s

Table A2. Road emissions for the RDE test (assuming constant exhaust emission intensity)—output data obtained from the algorithm and the proposed values for the generalized test (Vehicle A).

Exhaust Components	t			S			U			b
	U [s]	R [s]	M [s]	U [km]	R [km]	M [km]	U [–]	R [–]	M [–]	RDE
CO_2	4157	640	603	19.1	16.0	24.3	0.321	0.270	0.409	49.4 g/km
CO	3999	717	684	18.6	17.9	27.5	0.290	0.280	0.430	83.4 mg/km
NO_x	4106	640	654	18.9	16.0	26.3	0.309	0.261	0.430	2.51 mg/km
PN	3999	718	683	18.6	18.0	27.5	0.290	0.280	0.430	$5.2 \cdot 10^{10}$ 1/km
Average	4065	679	656	18.8	17.0	26.4	0.303	0.273	0.425	–
St. deviation	68.8	38.8	33.0	0.21	0.97	1.33	0.01	0.01	0.01	–
CoV [%]	1.7%	5.7%	5.0%	1.1%	5.7%	5.0%	4.4%	2.9%	2.1%	–

Table A3. Road emissions for the RDE test (assuming constant exhaust emission intensity)—output data obtained from the algorithm and the proposed values for the generalized test (Vehicle B).

Exhaust Components	t			S			U			b
	U [s]	R [s]	M [s]	U [km]	R [km]	M [km]	U [–]	R [–]	M [–]	RDE
CO_2	4096	878	427	16.0	21.9	17.2	0.290	0.398	0.312	48.4 g/km
CO	4013	710	677	18.4	17.8	27.2	0.290	0.280	0.430	65.2 mg/km
NO_x	4004	715	681	18.5	17.9	27.4	0.290	0.280	0.430	4.68 mg/km
PN	4040	640	720	22.4	16.0	29.0	0.333	0.237	0.430	$6.6 \cdot 10^{10}$ 1/km
Average	4038	736	626	18.8	18.4	25.2	0.301	0.299	0.400	–
St. deviation	35.8	87.2	116.4	2.31	2.18	4.69	0.02	0.06	0.05	–
CoV [%]	0.9%	11.8%	18.6%	12.3%	11.8%	18.6%	6.1%	20.0%	12.8%	–

Table A4. Road emissions for the RDE test (assuming constant exhaust emission intensity)—output data obtained from the algorithm and the proposed values for the generalized test (Vehicle C).

Exhaust Components	t			S			U			b
	U [s]	R [s]	M [s]	U [km]	R [km]	M [km]	U [–]	R [–]	M [–]	RDE
CO_2	3673	915	812	22.7	22.9	32.7	0.290	0.292	0.418	49.3 g/km
CO	4015	709	676	18.4	17.7	27.2	0.290	0.280	0.430	41.4 mg/km
NO_x	4002	716	682	18.5	17.9	27.5	0.290	0.280	0.430	6.2 mg/km
PN	3811	911	678	20.5	22.8	27.3	0.290	0.323	0.387	$2.4 \cdot 10^{10}$ 1/km
Average	3875	813	712	20.0	20.3	28.7	0.290	0.294	0.416	–
St. deviation	142	101	58	1.75	2.51	2.32	0.00	0.02	0.02	–
CoV [%]	3.7%	12.4%	8.1%	8.7%	12.4%	8.1%	0.0%	6.0%	4.2%	–

Table A5. Road emissions for the RDE test (assuming variable exhaust emission intensity)—output data obtained from the algorithm and the proposed values for the generalized test (Vehicle A).

Exhaust Components	t			S			U			b
	U [s]	R [s]	M [s]	U [km]	R [km]	M [km]	U [–]	R [–]	M [–]	RDE
CO_2	3665	1338	397	16.0	16.0	16.0	0.333	0.333	0.333	0
CO	3665	1338	397	16.0	16.0	16.0	0.333	0.333	0.333	0
NO_x	3665	1338	397	16.0	16.0	16.0	0.333	0.333	0.333	0
PN	3665	1338	397	16.0	16.0	16.0	0.333	0.333	0.333	0
Average	3665	1338	397	16.00	16.00	16.00	0.333	0.333	0.333	–
St. deviation	0.00	0.00	0.00	0.00	0.00	0.00	0.00	0.00	0.00	–
CoV [%]	0.0%	0.0%	0.0%	0.0%	0.0%	0.0%	0.0%	0.0%	0.0%	–

Table A6. Road emissions for the RDE test (assuming variable exhaust emission intensity)—output data obtained from the algorithm and the proposed values for the generalized test (Vehicle B).

Exhaust Components	t			S			U			b
	U [s]	R [s]	M [s]	U [km]	R [km]	M [km]	U [–]	R [–]	M [–]	RDE
CO_2	3602	1338	460	16.0	16.0	18.5	0.317	0.317	0.367	0
CO	3602	1338	460	16.0	16.0	18.5	0.317	0.317	0.367	0
NO_x	3602	1338	460	16.0	16.0	18.5	0.317	0.317	0.367	0
PN	3602	1338	460	16.0	16.0	18.5	0.317	0.317	0.367	0
Average	3602	1338	460	16.0	16.0	18.5	0.317	0.317	0.367	–
St. deviation	0.0	0.0	0.0	0.00	0.00	0.00	0.00	0.00	0.00	–
CoV [%]	0.0%	0.0%	0.0%	0.0%	0.0%	0.0%	0.0%	0.0%	0.0%	–

Table A7. Road emissions for the RDE test (assuming variable exhaust emission intensity)—output data obtained from the algorithm and the proposed values for the generalized test (Vehicle C).

Exhaust Components	t			S			U			b
	U [s]	R [s]	M [s]	U [km]	R [km]	M [km]	U [–]	R [–]	M [–]	RDE
CO_2	3665	1338	397	16.0	16.0	16.0	0.333	0.333	0.333	6.0 g/km
CO	3665	1338	397	16.0	16.0	16.0	0.333	0.333	0.333	11.4 mg/km
NO_x	3665	1338	397	16.0	16.0	16.0	0.333	0.333	0.333	2.3 mg/km
PN	3665	1338	397	16.0	16.0	16.0	0.333	0.333	0.333	$2.0 \cdot 10^9$ 1/km
Average	3665	1338	397	16.0	16.0	16.0	0.333	0.333	0.333	–
St. deviation	0	0	0	0.00	0.00	0.00	0.00	0.00	0.00	–
CoV [%]	0.0%	0.0%	0.0%	0.0%	0.0%	0.0%	0.0%	0.0%	0.0%	–

References

1. The European Parliament and The Council of The European Union. Commission Regulation (EC) 715/2007 of the European Parliament and of the Council of 20 June 2007 on Type Approval of Motor Vehicles with Respect to Emissions from Light Passenger and Commercial Vehicles (Euro 5 and Euro 6) and on Access to Vehicle Repair and Maintenance Information. *Off. J. Eur. Union* **2007**, *L 171*, 1–16.
2. The Commission of the European Communities. Commission Regulation (EC) 692/2008 of 18 July 2008 Implementing and Amending Regulation (EC) 715/2007 of the European Parliament and of the Council on Type-Approval of Motor Vehicles with Respect to Emissions from Light Passenger and Commercial Vehicles (Euro 5 and Euro 6) and on Access to Vehicle Repair and Maintenance Information, European Commission (EC). *Off. J. Eur. Union* **2008**, *L 199*, 1–136.
3. The European Commission. Commission Regulation (EU) 2016/427 of 10 March 2016 Amending Regulation (EC) No. 692/2008 as Regards Emissions from Light Passenger and Commercial Vehicles (Euro 6), Verifying Real Driving Emissions. *Off. J. Eur. Union* **2016**, *L 82*, 1–98.

4. The European Commission. Commission Regulation (EU) 2016/646 of 20 April 2016 Amending Regulation (EC) No. 692/2008 as Regards Emissions from Light Passenger and Commercial Vehicles (Euro 6), Verifying Real Driving Emissions. *Off. J. Eur. Union* **2016**, *L 109*, 1–22.
5. Commission Regulation (EU) 2017/1151 of 1 June 2017 Supplementing Regulation (EC) No 715/2007 of the European Parliament and of the Council on Type-Approval of Motor Vehicles with Respect to Emissions from Light Passenger and Commercial Vehicles (Euro 5 and Euro 6) and on Access to Vehicle Repair and Maintenance Information, Amending Directive 2007/46/EC of the European Parliament and of the Council, Commission Regulation (EC) No 692/2008 and Commission Regulation (EU) No 1230/2012 and Repealing Commission Regulation (EC) No 692/2008. Off. J. Eur. Union L 175. 2017, pp. 1–643. Available online: http://data.europa.eu/eli/reg/2017/1151/oj (accessed on 22 December 2020).
6. The European Commission. Commission Regulation (EU) 2017/1154 of 7 June 2017 Amending Regulation (EU) 2017/1151 Supplementing Regulation (EC) No 715/2007 of the European Parliament and of the Council on Type-Approval of Motor Vehicles with Respect to Emissions from Light Passenger and Commercial Vehicles (Euro 5 and Euro 6) and on Access to Vehicle Repair and Maintenance Information, Amending Directive 2007/46/EC of the European Parliament and of the Council, Commission Regulation (EC) No 692/2008 and Commission Regulation (EU) No 1230/2012 and Repealing Regulation (EC) No 692/2008 and Directive 2007/46/EC of the European Parliament and of the Council as Regards Real-Driving Emissions from Light Passenger and Commercial Vehicles (Euro 6). *Off. J. Eur. Union* **2017**, *L 175*, 708–732.
7. Pielecha, J.; Andrych-Zalewska, M.; Skobiej, K. The impact of using an in-cylinder catalyst on the exhaust gas emission in real driving conditions tests of a Diesel engine. *IOP Conf. Ser. Mater. Sci. Eng.* **2018**, *421*, 042064. [CrossRef]
8. Pielecha, J.; Skobiej, K. Evaluation of ecological extremes of vehicles in road emission tests. *Arch. Transp.* **2020**, *4*, 33–46. [CrossRef]
9. Katrašnik, T. Hybridization of powertrain and downsizing of IC engine—A way to reduce fuel consumption and pollutant emissions—Part 1. *Energy Convers. Manag.* **2007**, *48*, 1411–1423. [CrossRef]
10. Wirasingha, S.G.; Schofield, N.; Emadi, A. Plug-in hybrid electric vehicle developments in the US: Trends, barriers, and economic feasibility. In Proceedings of the IEEE Vehicle Power Propulsion Conference, Harbin, China, 3–5 September 2008. [CrossRef]
11. Khan, M.; Kockelma, K.M. Predicting the market potential of plug-in electric vehicles using multiday GPS data. *Energy Policy* **2012**, *46*, 225–233. [CrossRef]
12. Cieslik, W.; Szwajca, F.; Golimowski, W.; Berger, A. Experimental analysis of residential photovoltaic (PV) and battery electric vehicle (BEV) systems in terms of annual energy utilization. *Energies* **2021**, *14*, 1085. [CrossRef]
13. Yi, Z.; Bauer, P.H. Effects of environmental factors on electric vehicle energy consumption: A sensitivity analysis. *IET Electr. Syst. Transp.* **2017**, *7*, 3–13. [CrossRef]
14. Li, G.; Zhang, X.P. Modeling of plug-in hybrid electric vehicle charging demand in probabilistic power flow calculations. *IEEE Trans. Smart Grid* **2012**, *3*, 492–499. [CrossRef]
15. Pielecha, J.; Skobiej, K.; Kurtyka, K. Exhaust emissions and energy consumption analysis of conventional, hybrid, and electric vehicles in real driving cycles. *Energies* **2020**, *13*, 6423. [CrossRef]
16. IEA. Global EV Outlook 2020. 2020. Available online: https://www.iea.org/reports/global-ev-outlook-2020 (accessed on 21 October 2020).
17. Lang, M. Initial GVI Roadmap, Österreichischer Automobil Motorrad und Touring Club (ÖAMTC), GVI—Green Vehicle Index, Horizon 2020, Grant Agreement ID: 814794. 2019. Available online: https://www.gvi-project.eu/wp-content/uploads/sites/2/D4.1-Roadmap.pdf (accessed on 22 March 2021).
18. Kapus, P. Passenger Car Powertrain 4.x—Fuel Consumption, Emissions and Cost, Webinar, AVL List GmbH. 2020. Available online: https://www.avl.com/-/passenger-car-powertrain-4.x-fuel-consumption-emissions-and-cost (accessed on 2 June 2020).
19. Korn, T. The most effective technology to comply with CO_2-legislation: The new generation of hydrogen internal combustion engines. In Proceedings of the International Vienna Motor Symposium, Vienna, Austria, 22–24 April 2020.
20. Lasdon, L.S.; Waren, A.D.; Jain, A.; Ratner, M. Design and testing of a generalized reduced gradient code for nonlinear programming. *ACM Trans. Math. Softw.* **1978**, *4*, 34–50. [CrossRef]

 *energies*

Review

Challenges and Barriers of Wireless Charging Technologies for Electric Vehicles

Geetha Palani [1], Usha Sengamalai [1], Pradeep Vishnuram [1] and Benedetto Nastasi [2,*]

1   Department of EEE, SRM Institute of Science and Technology, Kattankulathur, Chennai 603203, India
2   Department of Planning, Design and Technology of Architecture, Sapienza University of Rome, Via Flaminia 72, 00196 Rome, Italy
*   Correspondence: benedetto.nastasi@outlook.com

Abstract: Electric vehicles could be a significant aid in lowering greenhouse gas emissions. Even though extensive study has been done on the features and traits of electric vehicles and the nature of their charging infrastructure, network modeling for electric vehicle manufacturing has been limited and unchanging. The necessity of wireless electric vehicle charging, based on magnetic resonance coupling, drove the primary aims for this review work. Herein, we examined the basic theoretical framework for wireless power transmission systems for EV charging and performed a software-in-the-loop analysis, in addition to carrying out a performance analysis of an EV charging system based on magnetic resonance. This study also covered power pad designs and created workable remedies for the following issues: (i) how power pad positioning affected the function of wireless charging systems and (ii) how to develop strategies to keep power efficiency at its highest level. Moreover, safety features of wireless charging systems, owing to interruption from foreign objects and/or living objects, were analyzed, and solutions were proposed to ensure such systems would operate as safely and optimally as possible.

Keywords: electric vehicle; wireless power transmission; battery system for energy storage; compensation networks; coil design; and wireless charging system

Citation: Palani, G.; Sengamalai, U.; Vishnuram, P.; Nastasi, B. Challenges and Barriers of Wireless Charging Technologies for Electric Vehicles. *Energies* **2023**, *16*, 2138. https://doi.org/10.3390/en16052138

Academic Editors: Chunhua Liu and Wojciech Cieslik

Received: 24 December 2022
Revised: 15 February 2023
Accepted: 20 February 2023
Published: 22 February 2023

1. Introduction

Electric vehicle (EV) technology has been gaining popularity due to its lower fuel emissions, and the numbers of EVs is anticipated to increase quickly. This has created a demand for ongoing improvements to charging infrastructures, especially wireless infrastructure. These must be designed for private, commercial, and public applications and be usable for both home and public charging stations. The technology of wireless power transmission can eliminate the use of wires, thus increasing the mobility, convenience, and safety of electronic devices for all users. Wireless power transfer is useful to power electrical devices where interconnecting wires are inconvenient, hazardous, or not possible. Thus, the accessibility of wireless charging stations resolves the problems with charging time, range anxiety, and charger connectivity, arguably the biggest obstacles to the widespread adoption of electric vehicles (EVs). The deployment of such effective and dependable high-power wireless charging infrastructures at close ranges would support a wider free range for EVs. However, many technical difficulties have arisen relating to the installation of EV wireless charging infrastructure. The main obstacles to wireless charging system adoption include low coupling coefficient between the transmitters and the receiver, misaligned power pads, and interruption from foreign objects like metal or live objects [1]. Electric vehicles (EVs) are a viable and feasible solution to the environmental problems the automotive industry is now experiencing. Figure 1 depicts wireless technology for charging electric vehicles.

Electrified transportation has created a paradigm shift in the transportation industry. It is regarded as being more intelligent, safe, and reliable, while also being more environmentally friendly. Less reliance on fossil fuels will result from the adoption of electrified transportation [2]. Conductive or plug-in chargers have provided trouble for EV owners struggling to meet the high voltage batteries' periodic charging needs. Electric vehicle wireless charging could be a remedy. The issues posed by conductive EV chargers would be avoided, and wireless charging would improve the EV user experience. Numerous researchers have been drawn to the idea of wireless power transmission via electromagnetic induction for use in the implementation of wireless charging of electronic devices and high-voltage batteries of electric cars. Despite the apparent advantages of EV wireless charging, substantial barriers to the economic feasibility and acceptance of wireless power transfer technology in the automotive sector have been defined. Compared to established conductive EV chargers, the main issues are high initial cost and limited power transfer efficiency.

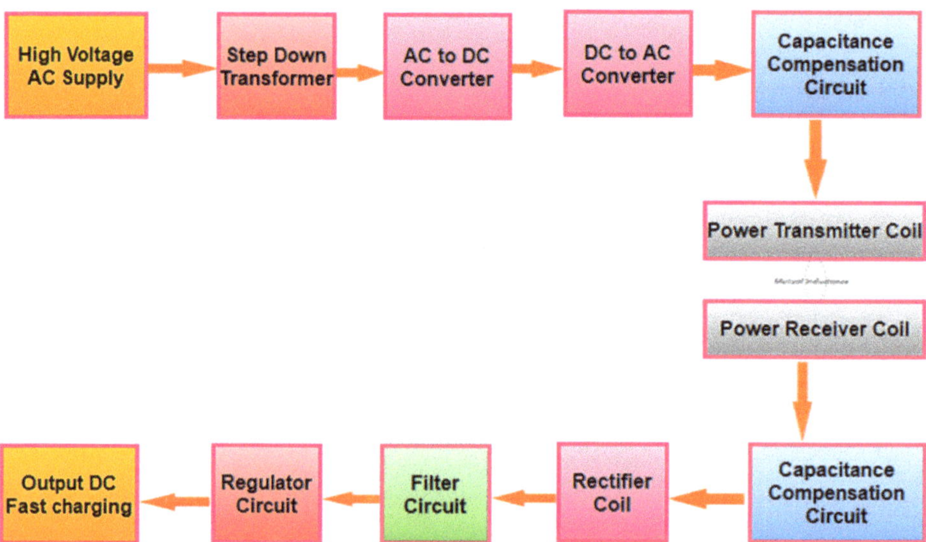

Figure 1. Wireless Electric Vehicle Charging Technology.

Additionally, significant standards and regulatory bodies have published guidance to avoid potential safety hazards. A new era of environmentally friendly and secure mobility has been made possible by burgeoning inventive research and ongoing improvements in the wireless charging technology of EVs [3].

Wireless power transmission (WPT) is a practical, cordless, trustworthy, appropriate, and all-weather power transmission or charging technology. A standard WPT system setup for charging an EV is made up of a transmitting coil buried in the ground at the charging station and a receiving coil built into the car. It consists of two separate electrical systems coupled to one another. The high-voltage EV battery can be charged wirelessly thanks to magnetic coupling. The inherent advantages of EV wireless charging performance include electrical separation, operation in harsh environments, and safety (owing to non-contact operation). However, high current is needed to charge high-voltage EV batteries, and any interruption to WPT will substantially impact the system's viability [4].

Therefore, power transfer efficiency must be close to 100% for wireless charging to be commercially feasible. Inductive charging can achieve such high efficiency provided the receiving and transmitting coils are correctly matched and there is no magnetic loss in the air. The practical constraints of transmitting and receiving coils to transfer power without loss and the necessity for precise alignment are two significant barriers to inductive coupling in wireless EV charging. MIT researchers suggested magnetic resonance coupling, with a 90% over-the-air power transmission efficiency within a few millimeters [5]. Currently, more power is lost through the process than is added to an EV's battery pack. The amount depends significantly on the electrical output and surrounding circumstances.

The standard inductive power transfer at a resonance frequency is magnetic resonance coupling. Compared to traditional inductive power transfer, where resonance aids in the maximum power transfer, the resonance coupling approach is slightly more complicated. Magnetic resonance coupling can effectively power/charge EVs wirelessly across a few millimeters [6]. Research and development must be done to maximize wireless power transfer efficiency under realistic and imperfect EV charging settings and to make wireless charging sustainable from a business standpoint. The following are the main obstacles to wireless EV charging as defined by this work.

Two crucial elements significantly impact the functionality and effectiveness of EV wireless charging systems: (i) the efficiency of the resonant frequency power electronics circuitry and (ii) the power pad in the coupling efficiency [7].

A WPT system must have at least two magnetic couplers to transmit power wirelessly; on the transmitter side, a central coupler and, on the receiver side, a secondary coupler similar to the other. For a more efficient WPT system, it is vital to have a high coupling coefficient (k) between the quality factor (Q) and the coupler. To understand the effective coupling of transmitter and receiver for wireless charging, it is crucial to investigate the effects of different coil shapes, coil structures, physical coil spacing, and coil materials, according to power transfer level, on the power transfer performance. The power electronics industry has been working hard to create efficient circuits, with significant advancements in semiconductor technology [8]. To date, there are very few standardized and optimized per-pad systems designed for commercially-feasible wireless EV charging, as the standards are still in the research stage [9]. This study examined and validated power pad design parameters for extremely effective wireless EV charging systems, keeping in mind that different vehicle segments may have different ground clearances, ranging from less than 10 cm for compact electric vehicles to roughly 30 cm for large SUVs [10]. Depending on the make and model of the car, the normalized distance between the transmission and receiver coils beneath the car (vehicle) will vary. Designing around limitations on the height and location of the coil on the vehicle body has proven difficult because it can limit other design elements, including aerodynamics, safety, and aesthetics. There is an an urgent need to design and create an EV wireless power transfer system to transfer power over the pads. The trick is to keep an appropriate power transfer rate while achieving a reasonable separation distance between the pads in the receiver and the transmitter coils. As the distance between the coils grows, the power transfer efficiency may drop quickly [11]. As the power input to the transmitting coil increases, magnetic flux leakage will simultaneously increase Therefore, a trade-off between distance, efficiency, and radiation leakage must be made for the best wireless charging system. These issues could be resolved in two ways: by creating a system with a low degree of transmitter-to-receiver mismatch, or by creating a resonant tank tailored for maximum power transfer. A transmitter to receiver pad auto-alignment strategy was established in this research work.

Regarding the ideal alignment of the coils, drivers cannot be expected to park their car precisely over the wireless charging pad. Thus, it is difficult to put into practice a purely mechanical technique for aligning the coils present in the transmitting and receiving coils. Still, alignment optimization is a creative approach with many benefits. First, complete alignment is feasible, as is maximum efficiency. Second, aligning the coils does not require a skilled driver. However, if coils are gravely misaligned, mechanical adjustment may be unavoidable [12]. Still, to facilitate a static solution, limiting such mechanical alignment methods may be desirable. An adaptive system must be developed and constructed to maintain wireless EV charging systems' best power transmission efficiency under various usage scenarios.

The impact of mounting metallic objects in EV wireless charging cannot be ignored. Therefore, substantial design attention must be paid to avoid interference from living objects and foreign object debris between the transmitting and receiving charging pads of

EVs. If these issues are not resolved, magnetic resonance coupling will be constrained and, therefore, unprofitable for wireless EV charging. This inspired the current work and this author's efforts to research and develop a method for maximizing power transfer efficiency in less-than-ideal, real-world EV charging circumstances [13].

Three alternative charging modes—static, quasi-dynamic, and dynamic charging—could be used to automate the charging of automobiles using wireless charging [14]. Static charging has several advantages, including the possibility of being deployed in convenient places like parking lots or garages and eliminating the electric shock posed by cables [15]. The QWC system enables charging for EVs when they are temporarily stationary, such as at traffic lights, extending their range and reducing the requirement for energy storage [16]. The DWC system would continually charge the EV through authorized on-road charging lanes, increasing driving range and shrinking battery size [17,18]. Utilizing wireless charging systems with an efficiency range of around 88.5%, WPT was completed by level 2 (230-V ac) powering at a rate of 7.2 kW.

Today, the most efficient forms of wireless charging are resonant CPT [19–22], utilized in dedicated lanes for dynamic charging [23–25], and resonant IPT [26–30], used in both Static WC and Dynamic WC methods. In [19], a comparison of capacitive and inductive WPT was made. IPT has allowed for successfully marketed products at low power levels for many years [31,32]. Magnetic couplers—transformers with only a few millimeters between the propagation (transmitter (Tx)) and reception (receiver (Rx)) components—have been in various phases of development for some time [33]. Many researchers have used enhanced compensation strategies to increase efficiency [34–37], air gap, and power level [38–42]. WPT distances have been extended by MIT researchers, who published a study in 2007 detailing their success in lighting a 60-W bulb at a distance of 2 m, thereby building researchers' confidence in expanding WPT to the necessary distances [43].

There have been several emerging areas of interest in research, including system design and analysis [44], component stress optimization, and compensation networks. Ongoing research has led to an increase in WPT of about 96% at a distance of 200 mm and several kilowatts of output power. The conductive method of wireless charging is currently in its stabilization phase and has produced several commercial goods and standards.

Electric vehicle (EV) charging infrastructure development and implementation are necessary for the efficient use of EVs. EVs have fewer charging stations and range-specific connectors than ICE vehicles, requiring more recharge time. To solve this refueling issue, an EV charger with high power and efficiency is required. Using a fast charger, approximately 50% of the battery may be charged in 3 min, and up to 80% may be charged in 15 min [45,46]. Fast charging methods, however, have been shown to cause high-voltage batteries to degrade. Control algorithms are necessary, based on the cost and rating of the converter used in a charger, using different microcontrollers, digital signal processors, and specialized linear integrated circuits. Ideal voltage and current ratios contribute to longer battery. To address prolonged charging issues, however, fast chargers are necessary. The conductive charging method is nearly mature [47], and established standards have been made. Inductive charging, which is still developing, has the potential to supplement conductive charging.

The structure of this paper was as follows. The definition and the benefits of wireless charging devices were covered in the first section after the introduction. EV conductive charging methods were explained in the second part. The wireless power transmission methods were described in detail in the third section. Static and dynamic wireless charging methods were presented and discussed in the fourth section. Upcoming and present standards of EV in WPT were covered in the fifth section. EV-based V2G charging methods were described in the sixth section. The quadruple power pad coil analysis for wireless EV charging was explained in the seventh section, in detail. The compositions of wireless charging were presented in the eighth section. The last and final section was the conclusion, including the final decisions determined by this paper.

2. Benefits of Wireless Charging over Wired Charging

Plug-in and wireless power transmission methods have been used to charge electric vehicles. In the plug-in technique, the electric vehicle's battery is charged at the charging station via a cord or plug. In contrast, in the wireless charging method, the battery of an electric vehicle (car) is charged utilizing wireless power transmission. The wireless charging technique is superior to wired charging in several ways. First, there is no need to carry and store cords, which could be considered the primary advantage of using wireless charging. Using this method circumvents the possibility of having wires wear out over time [48]. A wireless charging system may also reduce the size of the typical electric vehicle battery using dynamic wireless charging. Large-sized batteries are currently used in electric vehicles. However, it has been projected that similar automobile batteries will grow lighter and smaller once wireless technologies are incorporated. As a result, these two requirements will lower the overall cost of electric vehicles. Due to these advantages, significant automakers like Hyundai, Nissan, and Tesla have been investing heavily in wireless technology, primarily for electric vehicles. However, very few companies currently have wireless technology incorporated into their models. Due to all of these factors, the market for wireless EV charging is expected to undergo revenue growth.

2.1. Restrictions: Maximize the Upgrading Costs for Wireless Charging Technologies

For power transmission with a power control device (PCU), wireless charging technology for electric cars requires transmitter and receiver coils. The transmitter coil is located in the base charging pad (BCP), whereas the receiver coil is located in the vehicle charging pad (VCP). The average entire cost of a fitted commercial wireless charging system for a home is between USD 2500 and USD 3000. The cost of an electric vehicle increases when wireless charging technology is used in the vehicle. Consequently, this raises the cost of wireless electric vehicle charging [49–52].

Wireless charging technology will become more affordable in response to rising demand and the widespread manufacturing of electric vehicles. The electric vehicle industry is still in the introductory stage, regarding wireless charging technology. However, it is anticipated that most vehicle OEMs will implement this technology into their car models in the future. Therefore, it can be concluded that, given the current state of the economy and scale-induced economies of scale, the very high cost of upgrading or enriching to wireless charging technology remains a significant constraint.

2.2. Chance: Increasing the Government Funding for Wireless Charging Technology

In many nations, the advancement of wireless charging is now supported by government incentives and support for electric vehicles. The main benefits of wireless charging include full autonomy, the lack of a charging station, the decreased risk of an electric shock to the driver, and smaller battery units. The general population would be able to work for extended periods without needing to wait for their cars to charge. This increase in productive hours would also contribute to the GDP growth of a nation. The absence, or reduced requirement for, charging stations for dynamic charging is another crucial factor supporting the implementation of wireless charging in urban areas with a lack of available space [53–55].

The UK government awarded over USD 48.5 million for 12 initiatives in July 2019 to improve the experience of electric car owners and drivers. A business called Charge received USD 3.01 million from the government to install technology for wireless charging in suburban buildings. The first wireless testing technology was completed in Buckinghamshire, Marlow, in December 2021.

Successful testing for a wireless charging road occurred in Sweden in 2021, in an effort to modernize transportation processes and hasten the transition to electric mobility. The Israeli company Electron erected a dynamic wireless charging system on a 1.65-kilometer public road in Sweden's Gotland. A fully electric transport truck was charged on this intelligent road. The US state of Michigan signed an agreement that would see the construction of the world's first wireless charging road infrastructure in Detroit. In Detroit, Electron will use its dynamic charging technology to offer on-the-go charging for battery-powered automobiles. The project is anticipated to be completely operational in 2023.

2.3. Challenge: Minimizing Efficiency Loss

An electric vehicle (EV) can be charged wirelessly by parking it at the top of the base panel, i.e., without any manual connections. Compared to conventional power transfer, power loss in wireless charging is approximately 7–12% higher. Additionally, the distance over which a wireless charger may transmit using electromagnetic induction and/or magnetic resonance is constrained [56–59]. This is a significant hurdle for manufacturers, particularly in the case of SUVs and LCVs possessing a high specific ground clearance as also for automatic design [60]. The ratio of energy power efficiency to transmitter-to-receiver separation is inversely proportional. Another difficulty facing the wireless EV charging industry is safety during vehicle charging because powerful electromagnetic fields could damage the biological environment. Thus, efficiency and safety concerns have become a barrier for manufacturers in this sector. An EV electric drive system is only responsible for a 15–20% energy loss, compared to 64–75% for a gasoline engine. EVs also use regenerative braking to recapture and reuse energy that would typically be lost in braking, and waste no energy idling.

The market category with the quickest growth rate is anticipated to be 3–11-kW. The segment (by power supply range) anticipated to increase at the quickest rate over the projection period is also 3–11 kW. For very small- and medium-sized battery-powered electric vehicles (EV), 3–11-kW wireless charging devices are typically employed. Wireless charging solutions could be portable and lightweight at this power level, making them appropriate for charging at both home and work. The Nissan Leaf and Chevrolet Volt now have access to a 3.3 kW wireless charger from Plugless Power. IncWiTricity and Prodrive Technologies unveiled a wireless charging system in 2016 that was able to charge an electric vehicle up to 11 kW more efficiently than cable charging solutions. Due to its applicability in the workplace and in home charging situations where rapid charging is not a requirement, the 3–11-KW category is anticipated to dominate the market [61–64].

The 3–11-kW infrastructure in the market for wireless EV charging was anticipated to develop at the highest rate in Europe over the research period. Growing demand for home charging systems, due to rising sales of battery-powered electric cars, has been credited with driving market growth in Europe. Major OEMs have taken steps to include wireless charging in their automobiles, including BMW, Audi, and Mercedes. This could encourage market expansion in Europe. During the forecast timeline, BEVs have been projected to experience the fastest growth. By propulsion, BEVs are anticipated to experience the quickest growth during the projection period. In BEVs, as opposed to Plug-in Hybrid Electric Vehicles (PHEVs), wireless charging system adoption has been higher. In BEVs, the battery is the only source of power, and it must be charged frequently. Wireless charging technology is installable in offices, malls, public spaces, and garages. Increasing expenditures in the deployment of wireless charging technologies for battery electric vehicles have been predicted to benefit the BEV segment in the wireless EV charging market in the coming years. These countries include Sweden, Germany, Italy, and the USA, among others. The Tesla Model S, Nissan Leaf, and Jaguar I-Pace are well-known BEVs that support wireless charging. As a result, it is anticipated that, over the projected period, the BEV segment will be greater than the PHEV segment.

During the forecast years, Europe is anticipated to be the largest market for BEVs in the wireless EV charging market. This market expansion can be attributed to the existence of top auto manufacturers seeking to employ wireless charging technology in Europe. Leading players using wireless charging will persuade other significant participants in the automotive sector to use the technology for BEVs. Revenues for the BEV market in the region are anticipated to increase due to rising BEV sales in Europe, as well as the European Union's legal targets, e.g., to cut CO_2 emissions from vehicles and vans by 55% and 50%, respectively, by 2030. The number of battery-electric vehicles in Europe increased from around 1 million in 2019 to 1.8 million in 2020.

The EV market in the Asia-Pacific region is predicted to undergo the most rapid increase during the forecast period. The Asia-Pacific region includes both developed and developing countries, like South Korea and Japan, and emerging economies, like India and China. The area has become a center for the manufacture of automobiles in recent years. The Asia-Pacific region has experienced increasing demand for electric vehicles due to rising environmental concerns and the rising purchasing power of the populace. Both municipal and federal governments have shown interest in lowering carbon emissions through electrifying transportation [65–69]. As a result, the use of electric vehicles has gained familiarity in the area. Governments have concentrated on constructing robust charging infrastructures to encourage the adoption of electric vehicles. Rapid technological development in South Korea and Japan's electronic equipment manufacturing hubs is anticipated to lower the price of the wireless charging technology used in electric vehicles. Cost savings are then anticipated to fuel the expansion of the wireless EV charging market in the area. It is also anticipated that the presence of some of the top companies in the wireless EV charging trending market will aid expansion in the Asia-Pacific region. Toshiba Corporation, ZTE Corporation, Mitsubishi Electric, and Toyota Motor Corporation are a few companies active in this area.

3. EV Conductive Charging Method

A logical interconnection exists between the electrical power system and the vehicle, through an EV conductive charger. It comprises a low-frequency AC to high-frequency AC converter, including the power factor adjustment, or DC–DC converters and AC/DC rectifiers (PFC). Off-board and on-board chargers are the two categories of conductive chargers. For onboard chargers, battery current regulators and rectifiers are inside the car; for off-board chargers, they are present outside of the vehicle [70]. Conductive chargers are categorized according to the level included in their power transmission. These device are capable of being charged with an AC level 1 charger. Conductive wireless charging has no practical application for enhancing system effectiveness. The efficiency of such systems is dependent on converters with a very high frequency.

SWC charges a vehicle when it is at a stop [71]. The vehicle is also charged while it is moving, thanks to dynamic en route charging (DWC) [72]. This was demonstrated by the proposed creation of a wireless on-the-go bus charging system in Malaga, Spain by the Endesa-led project Victoria (see also CIRCE and others) [73]. QWC, also referred to in [74] as the static en route charge method, is especially favorable for vehicles that stop at predetermined locations throughout the day, such as bus stops, traffic lights or taxi stands. Thus, wireless charging could be achieved via underground fit technology when a bus stops at a bus stop [75]. EM fields are the primary WPT system used to charge EVs. A discussion on the wireless charging modes DWC, QDC, and SWC is shown in Figure 2. Systems for wirelessly charging electric vehicles are a prospective replacement option for charging electric vehicles (EVs), potentially avoiding any plug-in issues (WEVCS). In this work, existing wireless power transfer technology that is currently available for EVs was outlined. Additionally, studied wireless transformer designs with different ferrite forms were included. Due to safety and health issues that have been brought up, WEVCS has been connected to the present expansion of international standards. The two primary application types, static WEVCS and dynamic WEVCS, were described, and the most current advancements, with components from academic and commercial research labs, were documented. Along with qualitative comparisons to other existing technologies, future concept-based WEVCS were also discussed and investigated, incorporating in-wheel wireless power systems as well as vehicle-to-grid (V2G) technologies (WCS).

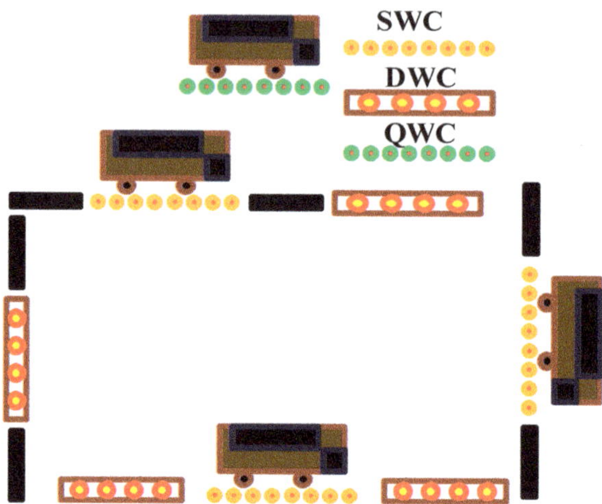

Figure 2. SWC, DWC, and QWC installation for statistical analysis.

The following various electric car (vehicle) wireless charging methods may be classified depending on their functioning principles:

1. Capacitive Wireless Charging System (CWCS);
2. Permanent Magnetic Gear Wireless Charging System (PMWC);
3. Inductive Wireless Charging System (IWC); and
4. Resonant Inductive Wireless Charging System (RIWC).

Four techniques have been utilized to develop WEVCS since wireless charging systems for EVs were first introduced: the capacitive method of wireless power transfer (CWPT), the traditional inductive method of power transfer (IPT), the resonant inductive method of wireless power transfer (RIPT), and the magnetic gear method of wireless power transfer (MGWPT) [76,77]. Currently available wireless power transfer technologies for rechargeable batteries in electric vehicles (BEVs) are listed in Table 1.

Table 1. Various methods of WPT and their parameters.

WPT Methods	The Distance between Transmitter and Receiver Circuits	Transmission of Power	Parameter Efficiency	Rate Cost	Safety and Protection
Inductive Coupling	Around a millimeter	A few watts or less	Minimum	It is economical to utilize secondhand equipment, since it is affordable and easily accessible.	It is secure from a biological perspective.
Capacitive Coupling	Multiple Kilowatts	A few Kilowatts or more	Minimum	It is less costly since the power transmission is done via aluminum plates.	Compared to the resonant approach, operation is safe.
Magnetic Resonance	A few meters	Kilowatts	Maximum	It is cost-effective, since old equipment is affordable and easily accessible.	It is secure from a biological perspective.
Microwaves	It can be produced up to 100 km.	Up to hundreds of Megawatts	Maximum	It is expensive compared to other treatments.	1 GHz to 1000 GHz high-frequency radiation is unhealthy for human health.
Optical	Using a high-intensity beam, it may be utilized across greater distances than a few meters.	Up to hundreds of Megawatts	Minimum	It has identical financial circumstances to inductive coupling.	It would be detrimental to human health.

3.1. Capacitive Wireless Power Transmission Method

For medium- and low-power implementations, like spinning machines [78], portable electronics [79], and phone chargers [80], the CWPT framework technology's cost-effectiveness and ease of use, utilizing improved mechanical configurations, as well as the geometric patterns of something like the coupling capacitors [76,77], are very advantageous. Inside the CWPT, coupling capacitors are employed to transmit power from the receiver to the source, rather than coils or magnets. Half-bridge converters receive their primary AC voltage via power quality control circuitry. The schematic of the capacitive wireless power transfer technique is presented in Figure 3.

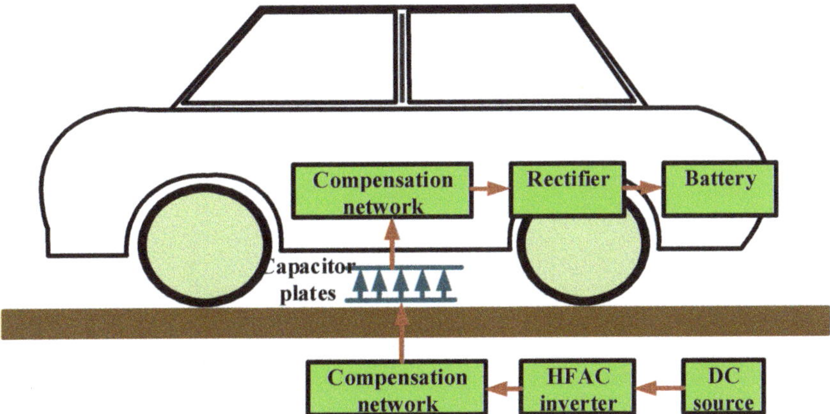

Figure 3. Schematic Representation of Capacitive Wireless Power Transmission method.

The coupling capacitors upon the receiver's side transmit the AC created by the H-bridge high-frequency. In contrast to the IPT, the CWPT runs on minimum and maximum current. Furthermore, in order to lower the range of impedance values of the transmitting and receiving sides, at the configuration of resonance, extra inductors should be coupled only with the combination present in the coupling capacitors. Given this arrangement, the circuitry could integrate soft switching. Rectifier and filter circuitry is used to change the incoming AC power to DC for either the load or the battery bank [81]. The two variables affecting power transfer levels are (1) the coupling capacitor's size and (2) the separation between its two plates. CWPT offers excellent performance and better field restrictions for small air gaps formed between the two capacitors' plates. Since its introduction, CWPT has only been partially applied to electric vehicles (EVs), owing to considerable high power level needs and air gaps. The authors of [82] offered suggestions for the rotary mechanism's high capacitance coupling designs and air-gap reduction. According to the authors of [83], a receiver might be attached to the vehicle's bumper bar to help close the air gap between the two connecting plates. A static research prototype with a power output of much more than 1 kW and approximately 83% efficiency (from the DC power supply to the battery bank) was demonstrated at the 540 kHz operating frequency.

It is possible to wirelessly transfer energy between both the transmitter and receiver sides by utilizing the displacement current that the varying electric magnetic field creates. In this case, coupling capacitors are used as the transmitter and receiver for wireless power transmission, in place of magnets or coils [84–90].

The AC voltage is sent into the power factor component adjustment circuit to improve the efficiency range, maintain voltage levels, and reduce transmission losses. The high-frequency supply of AC is then provided to the transmitting plate, producing an oscillating electric field that, via electrostatic induction, produces displacement current at the receiving plate. After that, it is provided to a half-bridge for the generation and improvement of maximum AC voltage. The receiver side AC voltage is transformed into DC and utilized to power or charge the battery throughout the BMS using filter circuits and a rectifier. The voltage, frequency, coupling capacitor dimension and size, and air gap produced between both the transmitting and receiving sides affect how much power is transmitted. It runs between 100 and 600 kHz in frequency.

3.2. Magnetic Gear Wireless Power Transmission Method

The magnetic gear WPT (MGWPT) is very different from the CWPT and IPT, as shown in Figure 4. This method uses two side-by-side synchronized permanent magnets (PM), as opposed to earlier WEVCS that relied on coaxial cable. The transmitter side winding gets the main power supply, as the current source causes the primary PM to suffer mechanical torque power. The primary PM spins and mechanically communicates with the secondary PM to apply torque, utilizing mechanical torque. The generator mode functioning is performed by the primary PM of such a combination of synchronous PMs. In contrast, the secondary PM gathers power and sends it to the battery through the power converter and BMS [91]. A 1.6-kW laboratory prototype called the MGWPT was created; it was able to provide and deliver over an air gap distance of 150 mm.

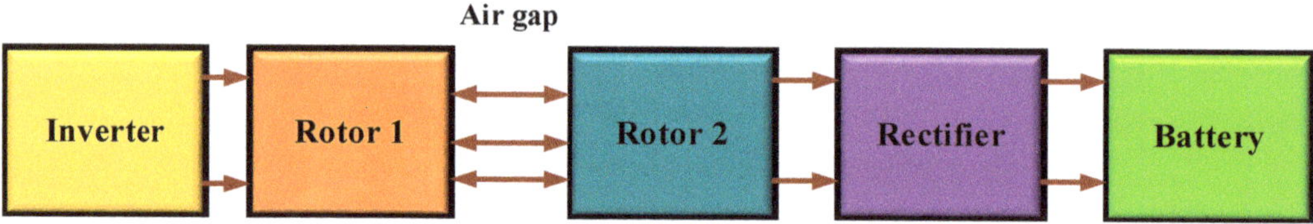

Figure 4. Magnetic gear wireless power transmission method.

Nevertheless, this approach in dynamic and static systems is fraught with several challenges. According to [92], rotators stopped synchronizing at 150 Hz, which significantly affected the transmitted power. A sophisticated feedback mechanism must be used to continuously shift the speed from the primary side to the battery side to avoid going over the upper limitation of power. As the coupling between the two synchronized windings dramatically weakens, the capacity to transmit power is inversely associated with separating the primary and secondary PMs from axis to axis. This makes it potentially beneficial for fixed WEVCS, but rather tricky for dynamic applications [93].

The armature winding and the synchronized permanent magnets make up the transmitter and receiver, respectively. Functioning at the transmitter side is comparable to motor operation [94–100]. The spinning of the transmitter magnet is caused by the mechanical tension induced on the transmitter winding when AC is applied to it. The receiver's PM is torqued by the transmitter's magnetic interaction shift, which makes the receiver magnet rotate synchronously with the transmitter magnet. The receiver now acts as a generator because the permanent magnetic field of the receiver has changed, converting mechanical power input into electrical output at the receiver winding. Permanent magnets connected by rotating gear are referred to as magnetic gear. Power converters rectify and filter the produced AC power at the receiver side before feeding it to the side of the battery.

3.3. Inductive Power Transmission Method

In 1914, Nikola Tesla developed the traditional IPT for wireless power transmission. Figure 5 shows the fundamental block diagram of the conventional IPT. Several EV charging systems have had an impact on it. IPT has been evaluated and carried out in some packages, ranging from mW to kW for the transmission of contactless strength from the supply to the receiver. In 1996, a well-known automobile manufacturer (GM) unveiled the Chevrolet S10 EV. The magnet-charge IPT (J1773) system was used to feed it, which provided stage 2 (6.6 kW) slow and degree three (50 kW) rapid charges [101]. The magnetic-number-one rate coil, a charging paddle (inductive coupler), was inserted into the auto's charging port, where the secondary coil received energy and could charge the vehicle. The University of Georgia displayed a 6.6 kW stage 2 EV charger with a seventy-seven kHz running frequency and a two hundred–four hundred V charging variety. This IPT made use of a 10-KVA coaxial winding transformer. The core premise of IWC is induction by Faraday's law. Electricity is transmitted wirelessly using magnetic field mutual induction between the transmitter and receiving coils [102–108]. When the principle AC delivery is supplied to the transmitter coil, an AC magnetic area that passes through it and transports the electrons creates AC power. The electric car's storage device is charged with this rectified and filtered AC output. The frequency, mutual inductance, and separation between the transmitter and receiver coils all affect how much electricity is sent and received. IWC makes use of a frequency variety of 19 to 50 kHz.

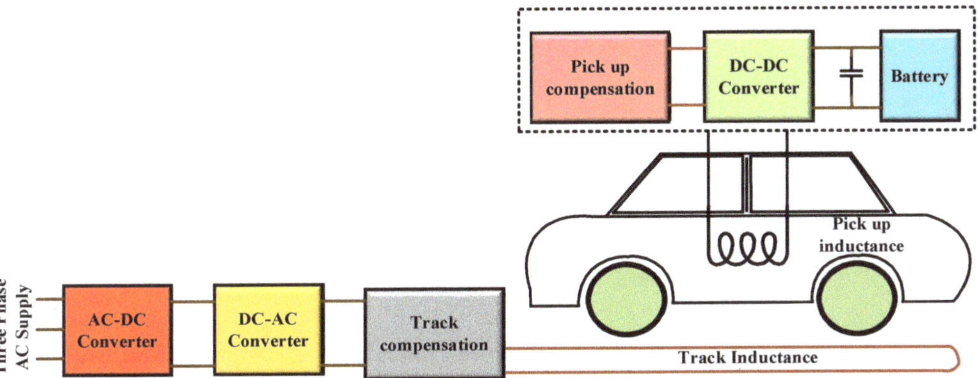

Figure 5. Inductive power transmission method.

3.4. Resonant Inductive Wireless Charging System (RIWC)

Regardless of weaker magnetic fields, resonance operation makes it possible to switch an equal amount of electricity as in IWC because resonators with excessive pleasant elements transmit electricity at a much higher charge. Power can be transmitted across long distances without the use of cables—the resonant inductive wireless charging system is shown in Figure 6.

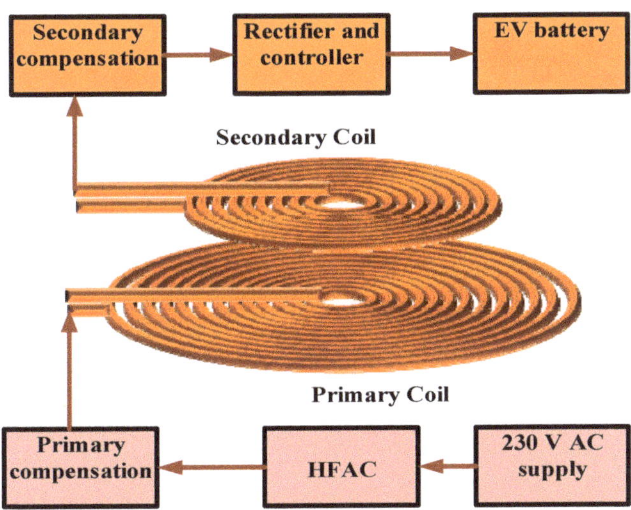

Figure 6. Resonant Inductive Wireless Charging System (RIWC).

The most significant power that may be sent over the air occurs when the resonant frequencies (bandwidth) of the sides of the propagation (transmitter) and reception (receiver) coils are matched, or when the transmitter and receiver coils are adjusted [109–112]. Additional reimbursement networks are consequently delivered in series and parallel to the transmitter and receiver coils to achieve appropriate resonance frequencies. Together with an increase in resonance frequency, these extra compensation networks also help to cut down on additional losses. The RIWC's operating frequency ranges from 10 to 150 kHz.

4. Static and Dynamic Wireless Charging

Wireless EV charging systems fall into two categories, depending on the application:

1. Static Wireless Charging;
2. Dynamic Wireless Charging.

4.1. Static Wireless Charging Method

As the name suggests, the vehicle charges every time it is in static mode. The block representation of the static wireless charging technique is shown in Figure 7. Thus, it would be easy to park the EV within a particular spot or in storage that permitted interface with the WCS. The transmitter would be located underground, while the receiver would be set up in the automobile's underside. Before getting out of the car to complete charging, the driver would align the transmitter and receiver [113–120]. The space between the edges of the transmitter and receiver, the scale of their pads, and the AC supply strength would all affect charging speed.

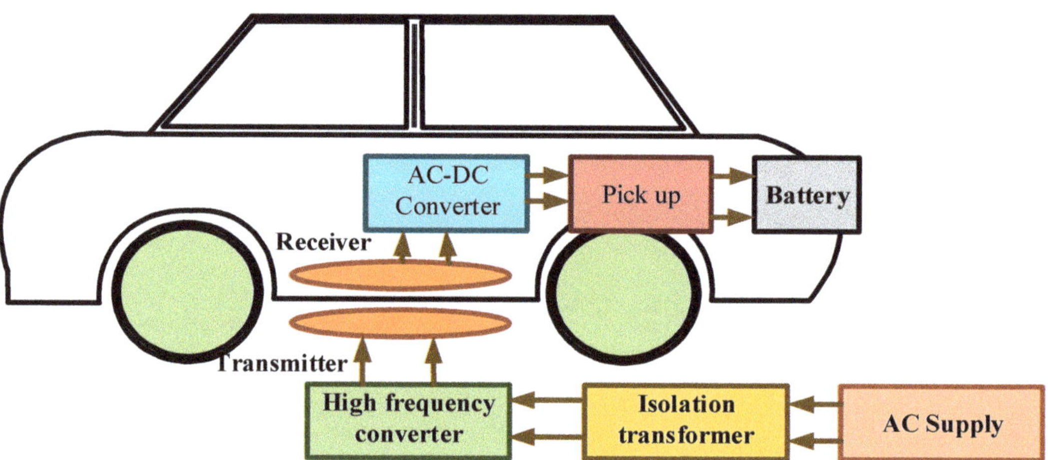

Figure 7. Static Wireless Charging Method.

The best places to build SWCs are those where EVs are routinely parked for long periods.

Wireless Charging Types and Charging Methods

These days, the world is moving toward electrified mobility, both to offer an alternative to expensive fuel for transportation and to minimize the pollution emissions created by nonrenewable fossil fuel cars. However, for electric vehicles, the two main problems preventing their adoption over conventional vehicles are their driving ranges and charging procedures.

Wired charging technology made it possible to charge your electric vehicle while parked. There is no longer a need for drawn-out charging station lines. Why can't power be delivered over the air? We are already accustomed to wireless data, audio, and video transmission [121].

The principles of transformer operation and wireless charging are identical. Wireless charging uses a transmitter and a receiver. The transmitter coil receives high-frequency alternating current from a 220 V 50 Hz AC source. The high-frequency AC creates an alternating magnetic field, which interrupts the receiver coil and enables the receiver coil to generate AC power. However, the transmitter and receiver's resonance frequency should remain constant for wireless charging to function. Compensation networks are implemented on both aspects to preserve this resonance frequency. Furthermore, the battery management system delivers this rectified DC power, generated from the receiver side which is connected to the battery side (BMS). Figures 8 and 9 show the wireless charging methods.

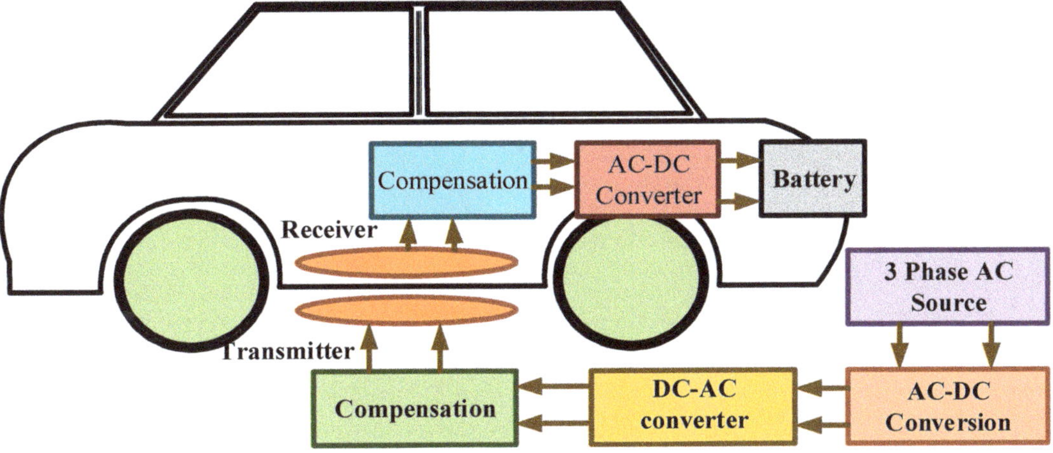

Figure 8. Wireless charging system.

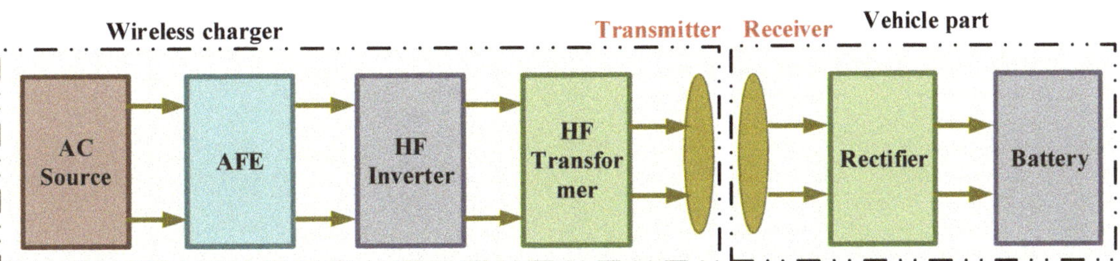

Figure 9. Block Representation of a Wireless Charging System.

The wireless charging block diagram displays the conversion processes and their overall effectiveness. Each conversion step is designed such that you the end user is able acquire the most effective efficiency [122]. The efficiency depicted in the diagram is general. More than 95% efficiency has been attained, according to Qualcomm, Oak Ridge National Laboratory (ORNL), and others [123–127]. There are several phases in a wireless charging system, and each stage varies in complexity and efficiency. When converting PS from AC to DC, Active Front End (AFE) with Power Factor Correction is used. Power transmission from IPT calls for excessive-frequency AC to be successful. As a result, the DC–AC converter transforms low-frequency DC into high-frequency AC. A remote high-frequency transformer is placed between the converter and the primary coil in order to prevent primary winding isolation breakdown. The alternating magnetic discipline produced by this excessive-frequency AC follows Ampere's equation. Due to the interplay of this magnetic discipline with the secondary coil, high-frequency AC is produced, in keeping with Faraday's law. The secondary repayment community is utilized to evolve to resonant surroundings, improving performance. A compelling rectifier is then used to feed the battery to rectify the AC energy. The three crucial components of an EV wireless charging system are the compensation network, the power electronics converters, and the remote, loosely coupled primary and secondary coils, which are covered via ferrite.

4.2. Dynamic Wireless Charging Method

Dynamic wireless charging is used to recharge EVs while they are being driven, making it unnecessary to wait while the battery charges. This theory, put forth in 1978 by J. G. Bolger et al., states that energy is transferred to the vehicle as it moves [128]. A study at KAIST has been working to develop dynamic wireless charging since 2009.

This study addressed continuous power transmission, high-frequency contemporary controlled inverters, and various electromagnetic interference parameters [129]. Choi et al. provided a beneficial analysis of OLEV. Dynamic wireless charging solves most of the problems with electric vehicles, such as battery capacity range, range anxiety, cost, etc. Inductive wireless power transfer is used by current dynamic wireless charging gadgets [130–133]. This technique is based on a pickup coil mounted in the electric vehicle (EV) that obtains an electromagnetic field, generating high-frequency current, and coils hidden beneath the road pavement. The on-road coils constantly deliver power to the pickup coil throughout a track. After being adequately prepared, the EV battery can be charged by the current captured by this coil. To transfer energy to the integrated system with a transmitter coil and several resonators, low-power wireless systems have been created. However, because they follow a path, these systems are worthless for electric vehicles (EVs) [134]. The two track types developed for DWC systems have different shapes, referred to as stretched tracks and lumped tracks. A stretched track includes a transmitter coil that is notably larger than the pickup coil of a lumped track, which contains many coils with radii that could reach near the pickup coil. Even as KAIST evolved the OLEV (electric automobile) prototype as a consequence of researching stretched tracks, a study group from Auckland University researched aggregated tracks [135,136]. Only a portion of the lumped track with the linked transmitter coil can drive the related receiver coil. This supply strategy, often referred to as segmentation, aids in increasing DWC effectiveness and reducing electromagnetic field radiation from the non-coupled rail segments. Prior research on coil sizing for static wireless charging systems has centered chiefly on sizing the coils [137] and researching axes misalignment's effects [138]. In terms of DWC systems, some researchers have looked into coil-based lumped tracks placed side by side [139], while others have assessed the appropriate coil length in a stretched track [140,141]. Galvanic isolation and user ease are two benefits of EV wireless charging over touch charging. To avoid using wires and cords and to circumvent the need for careful charging and discharging, it is possible to top off a vehicle's battery frequently. At the same time, a vehicle may be parked at different charging places, including at work, at home, at a traffic light, and while shopping. By incorporating a charging lane into motorways that would allow charging while driving, DWC could do away with fast charging infrastructures. Compared to cable charging, wireless charging has lower cost, size, manufacturing complexity, efficiency, and power density.

Table 2 shows the study's review of the works based on WPT.

EV wireless charging presents difficulties that must be considered for effective power transfer. Wireless power transfer necessitates energy conversion, which reduces the efficiency of conversion and transfer; as a result, it must to be optimized, and transfer efficiency improved. These issues have made it difficult for certain businesses to switch from conductive charging to wireless charging. Each component is a crucial research subject in the application of wireless chargers. Each factor—e.g., distance, geometry, frequency, compensation topology, coil design, and alignment—has an indirect or direct impact on the practical system's performance. DWC effectiveness is also influenced by EV speed and many underlying issues, as demonstrated in Figure 10.

Table 2. Overview of earlier works on WPT.

Problems Addressed	Performance Assessment	Key Contribution	Resolution
In an EV, using a cable circuit will harm the charger. Daily maintenance will be performed. The AC outlet requires plug-in.	Yokogawa digital power meter	Simple, cheap WPC prototype	The receiver coil attached to the battery picked up the magnetic induction created by the transmitter under the road.
Inductance coupling (efficiency improved).	Network reflection coefficient and scattering parameters	Employment of repeater concept	The power was doubled and reproduced in the air by the repeater coil, which was positioned between the transmitter and receiver.
Inhibited transmission efficiency and power loss transmission.	Unnamed optimization strategy	Phase shift amplitude control	While the active portion of the blockage was left alone on the receiving end, the sensitive portion of the obstruction had its impedance modified.

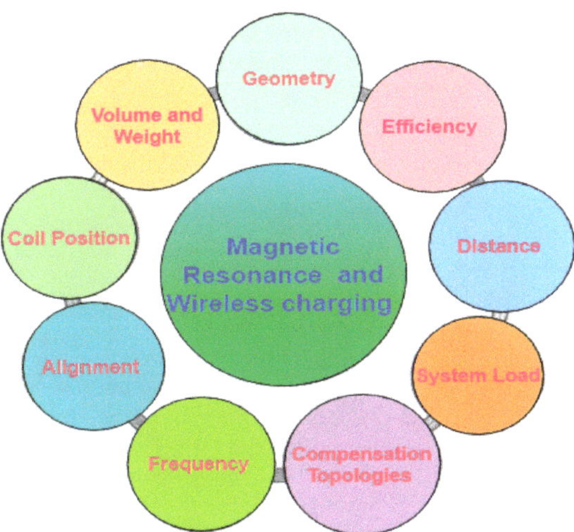

Figure 10. Demonstration of wireless charging.

5. Standards for Wireless Electric Vehicle Charging

Electric vehicles may now be wirelessly charged without the need for a plug. However, it would be unproductive or harmful if every manufacturer developed their own proprietary standard for wireless charging that could not be used in conjunction with other technologies. The user experience for wireless EV charging could be improved. The Society of Automotive Engineers (SAE), International Electro-Technical Commission (IEC), Underwriters Laboratories (UL), and Institute of Electrical and Electronics Engineers (IEEE) are a few of the international organizations working to develop standards [142–144].

SAE J2954 established the Alignment Technique for Light-Duty Plug-In EVs WPT. According to this standard, the maximum input power at level 1 is 3.7 kW, level 2 is 7.7 kW, level 3 is 11 kW, and level 4 is 22 kW. The minimal target reliability must also be higher than 85% when aligned. The side-to-side accuracy should be less than 4 inches, and the maximum permitted ground clearance should be 10 inches. The best alignment technique was determined to be magnetic triangulation, helping both human-controlled and autonomous vehicles find parking spaces.

- SAE. J1772 standard described the EV/PHEV conductive system of charge couplers.
- SAE. J2847/6 standard described the communication transmission between wireless EV Chargers and wirelessly-charged vehicles.
- SAE. J1773 standard described the EV inductive method of coupled charging.

- SAE. J2836/6 standard described the usage of a wireless charging system of communication for PEVs.
- UL subject 2750 described the investigation's general plan for WEVCS.
- IEC. 61980-1 Cor.1 Ed.1.0 described the general configuration of the EV WPT system.
- IEC. 62827-2 Ed.1.0 described the WPT-Management: Multiple Varieties of Management of Device Control.
- IEC. 63028 Ed.1.0 described the WPT-Air fuel alliance resonant baseline system requirements.

Table 3 shows the safety and technical standards for WPT.

Table 3. WPT U.S. Technical specifications and safety requirements.

Standard Developer	Standard Name	Published Year	Description
SAE	J2836/6_201305	2013	Applications for PEV wireless charging communication
SAE	J2953/1_201310	2013	Equipment for PEV compatibility with electric vehicles (EVSE)
SAE	J2953/2_201401	2014	Procedures for testing PEV compatibility with EVSE
SAE	J2953/3	2016	EVSE and PEV interoperability test scenarios
SAE	J2953/4	2020	reporting on PEV charge rates
SAE	J2847/6	2020	Wireless EV charging stations and light-duty PEVs can communicate for WPT.
SAE	J2954	2020	WPT for plug-in light-duty vehicles, as well as alignment techniques

5.1. EV Wireless Charging: Implementations and Standards

For the reliable implementation of high voltage and high power WPT, standards are necessary, since wireless charging is quickly overtaking other EV charging methods on the market. In addition to checking setup configurations for wireless charging, standardization also encompasses safety requirements, efficiency, electromagnetic restrictions, and interoperability goals as for a reliable computational design [145]. A crucial prerequisite for EVs to be practical after standardization is ubiquity. Customers should not need to worry about charging station compatibility [146–150]. The entire wireless power transfer system is contained in the IEC-61980-1 standard, from the network supply to electric vehicles (EVs) for charging the vehicle's battery, or the use of standardized equipment (or equipment parameters) at the standard power supply range of 1000 V AC or 1500 V DC. All of these were addressed using the SAE standard in SAE TIR J2954. This was the first actual wireless power transmission specification created by SAE considering EV charging. The static wireless charging industry largely followed its own trends. The owner of an EV might want to wirelessly charge their vehicle from a domestic wireless charger, workplace charger, or commercial charger, among many others, while enjoying similar charging functionality due to the frequency spectrum, interoperability, protection, coil specifications, and EMC/EMF constraints in SAE TIR J2954. The suggested frequency band per SAE 2954 is 85 kHz for any light-duty electric cars (81.39 kHz–90 kHz). Table 4 [151–158] shows the major wireless charging standards ready to be checked in the next years to be cyber-resilient [159].

Table 4. Standardization of charging power levels for light-duty EVs.

Classification	Power Level	Standard Status
WPT1	3.7 kW	SAE J2954
WPT2	7.7 kW	SAE J2954
WPT3	11 kW	SAE J2954 (WIP)
WPT4	22 kW	SAE J2954 (WIP)

A schematic diagram of a static wireless charging improvement for a stationary placement is shown in the Figure 7. A high-frequency converter is connected to the grid-side supply [160–169]. The principal pad receives this high-frequency feed (transmitter). Magnetic resonance is connected with the coils of the primary and secondary regions. An AC–DC converter is also employed on the load side to deliver power straight to the battery type. A battery management system (BMS) regulates the battery's state of charge (SOC) and overall health. The BMS is linked to the car's controller area network (CAN), which manages the vehicle's sensing component. The vehicle is connected to the management pole for wireless charging through radio indicators [17–177].

Alternately, a different DC–DC converter must be used if the vehicle's battery management system does not permit powering the battery directly. The deployment of static wireless charging is completely operator-free. With an intelligent controlling system coupled with CAN and BMS, all charging types would be feasible without human intervention. Wireless charging will only overtake the market if it provides more convenience at a lower cost [178–182] together with optimization algorithm adopted in other fields [183]. Table 5 shows the international technical standards of the EV.

Table 5. Worldwide technical standardization.

Standards Inventor	Name of the Standard	Invention Year	Description
IEC	61980-1-Ed.2.0	2020	General Requirement, Part-I, EV-WPT system
IEC	61980-3	2019	WPT System-part-3 for electric vehicles: particular specifications for entire magnetic field WPT systems
IEC	61980-2	2019	Specific criteria for wireless communication systems among electric road vehicles (EVs) and infrastructure, outlined in part two of the electric vehicle WPT systems.
IEC	61980-1:2015/COR1:2017	2017	General Requirement, Part 1 of the EV-WPT system
IEC	61980-1:2015/COR1	2017	General Requirement, Part 1 of the EV-WPT system
IEC	61980-Ed.1.0.New Addition	2015	General Requirement, Part 1 of the EV-WPT system
IEC	61980/1 AMD 1 Ed.1.0	2015	General Requirement, Part 1 of the EV-WPT system
IEC/TS	61980-2 Ed.1.0	2017	Widespread requirements for communication between electric-powered road cars and infrastructure concerning WPT devices and element 2 of EV-WPT systems
IEC/TS	61980-3 Ed.1.0	2015	Part 2 of the general necessities for the magnetic field power transmission gadget for EV-WPT systems
ISO	19363:2020	2020	Magnetic field WPT for electrically-driven road vehicles: protection and interoperability necessities
ISO	9363:2017	2017	Safety and interoperability criteria for electrically-driven road vehicles' magnetic fields

5.2. Companies Working to Develop and Improving WCS

- The Evatran Group developed plug-less charging for first-generation wireless electric vehicles such the Nissan Leaf, Chevrolet Volt, Tesla Model S, and Audi i3.
- Recently, WiTricity Corporation worked with Honda Motor Co. Ltd., Nissan, GM, Hyundai, and Furukawa Electric to create WCS for sedans and SUVs.
- Qualcomm Halo produced WCS for passenger, sport, and race cars, and Witricity Corporation obtained Qualcomm Halo.
- Hevo Power has been manufacturing WCS for a passenger automobile.
- Bombardier Primove manufactured WCS for vehicles ranging from rider automobiles to SUVs.
- Siemens and BMW have been manufacturing WCS for rider automobiles.

- Momentum Dynamic manufactured WCS for corporate and commercial fleet buses.
- Conductix-Wampfler manufactured WCS for buses and industrial fleets.

5.3. Challenges Faced by WEVCS

- The current infrastructure is insufficient for the necessary installations. Hence, developing dynamic and static wireless charging stations on highways will necessitate building new infrastructure [184–189].
- Maintaining EMC, EMI, and frequencies according to standards is necessary for human health and safety. Table 6 shows the various challenges faced by WEVCS.

Table 6. Challenges and issues faced by WEVCS.

WPT Types	Resonant Coupling	Induction Coupling	Microwave
Difficulties	Moderate	Low	High
Distance capability	Maximum of 1 km	5 mm	-
Efficiency performance	High	Low	High
Transmission of Loss	Moderate	High	Low
Number of receivers entering	Multiple receivers are applicable	Utilizing a single receiver is appropriate	Single receiver
Power wave	Fluctuating power signal	Continuous	Continuous
Radiation energy	Non-radiant power	Non-radiant power	Radiant power
Frequency range	Power transmitted at 6.7 MHz Control signals at 2.4 GHz	110–205 KHz	300 MHZ–300 GHz
Receiver	Coil of copper with few turns	Coil of copper with few turns	Rectenna with SCR
Safety considerations	Risk of sparks produced at several million volts	Considered harmless	Detrimental to living matter, like telecommunications
Transmission of energy	Electromagnetic resonance	EMI	Radiowave, microwave, and laser
Transmitter	Primary coil with a short gap and few turns. The secondary coil contained 10 times as many turns as the primary coil without a gap.	Several-turn copper coil	Antenna for transmission that uses a wave guide

6. EV-Based Vehicle-to-Grid (V2G)

Wireless charging—the need for EV charging, specifically—is one of the most significant difficulties facing the current electrical infrastructure. Vehicle-to-grid (V2G) topologies could be used to solve this problem. Vehicle-to-grid (V2G) is a well-known application for EVs, representing power delivery to the electrical grid. H. Nguyen et al. performed an in-depth analysis of V2G technology's integration and coordination with conductive charging methods [190–196]. Flexibility, automated charging, and bidirectional discharging are necessary components for a V2G integration. Most of the aforementioned criteria can be met via wireless charging. Through the electrification of transportation, fossil fuels may eventually be replaced by renewable energy. Local microgrids could gain power from V2G, and these could be combined with renewable energy systems [197–204]. Although they cannot be used to generate electricity directly, batteries are used for storage. Numerous initiatives have been made in the last ten years to increase energy conversion and lessen reliance on fossil fuels for electricity production and transportation electrification. As a result, we have seen expanding usage of electric vehicles (EVs) for mobility. In the future, it is anticipated that renewable energy sources will be used to generate power [205–214]. Because of the unpredictability of climatic conditions, renewable electricity sources (RES), especially wind and solar systems, present issues relating to the main grid's sustainability and power supply quality.

Adopting EVs on a broad scale, whether hybrid or entirely battery-based, also poses significant issues for the electrical grid [215]. The best option would be to use RES to offset the necessary demand for EVs [216]. Additionally, the power grid's stability and power quality could be improved by integrating RES with EVs, which has already been acknowledged by V2G systems. For this level of RES, EV, and grid integration, EV charging and discharging systems must be flexible, autonomous, easy, safe, and reliable. Research has placed great emphasis on design elements such as maximum efficiency, very low rate (cost), adaptability, and autonomous charging and discharging techniques. Due to automation, wireless charging of EVs could offer and produce bidirectional power flow between EVs and the grid. The comparison of interactivity between wired and wireless connections was briefed in [217,218]. According to the studies mentioned above, wireless connectivity or charging could achieve up to 65% connectivity, whereas conventional connectivity only achieved rates of 10% [219]. Several wireless communication devices have proven able to detect automatically and comply with V2G standards, helping to automate wireless charging. With the advent of wireless charging, a vehicle's interface with the grid could improve, enabling extra vehicle power whenever required.

6.1. Applications for EV Wireless Charging: LOD and FOD

Foreign Object Detection (FOD) detects foreign objects that are close to the charging pads and might or might not interfere with the transmission fields. Snow, dirt, twigs, water, oil, grease, leaves, and other items that might or might not interact with the magnetic field are considered benign objects. FOD may include magnetic materials, metal objects interacting with high-frequency fields, and animals sleeping on charging pads. Any object with metal available between the transmission and the reception sides prevents the charging process from continuing because of the power flow of eddy current through it [220]. The EV and the charging system could thus be impacted by the heating of the object that has the metal particle or any other conducting object.

Due to a strong electromagnetic field, living matter and objects are also affected. Any living thing could suffer harm during charging. Under FOD Detection Methods, numerous real-world examples of a charging apparatus coming into contact with living things for both short- and long-term exposure, such as children near the car to pick up a ball/toy, a driver reaching out to hold something felt, a pet lying still for a while, etc, were detailed.

- System variables
- Efficiency of power loss
- Actual temperature
- Image from wave-based detection
- Thermal ultrasonic radar
- Field-based laser light detection
- Resistance inductance capacitance

Most things exposed to intense magnetic fields experience both long- and short-term effects. Therefore, the International Committee on Electromagnetic Safety (ICES), Institute of Electrical and Electronics Engineers (IEEE), and International Commission on Non-Ionizing Radiation Protection (ICNIRP) recommendations [221] have determined rules governing magnetic field limits. The FOD detection methods are shown in Figure 11.

Different scientists have introduced various LOD and FOD techniques. WiTricity created an FOD technique using an overlapping coil structure that measured current, voltage, and resonators that had the phase and frequency. Another type of FOD technique is power detection, which measures a power loss brought on via the availability of a foreign object [222]. This technique is typically helpful for the minimum power transmission of wireless charging. An FOD method mainly depends on the fluctuation of quality factor (Q), as introduced by S. Fukuda et al. in [223]. Due to the position of the coils, this approach's Q could not be used for EV charging applications. Other FOD techniques, e.g., RFID, video cameras, and radar-based systems, were also considered in SAE standards [224,225]. The category-by-category classification of FOD techniques is shown in Figure 12. The benefits and drawbacks of each of the strategies mentioned above are listed in Table 7.

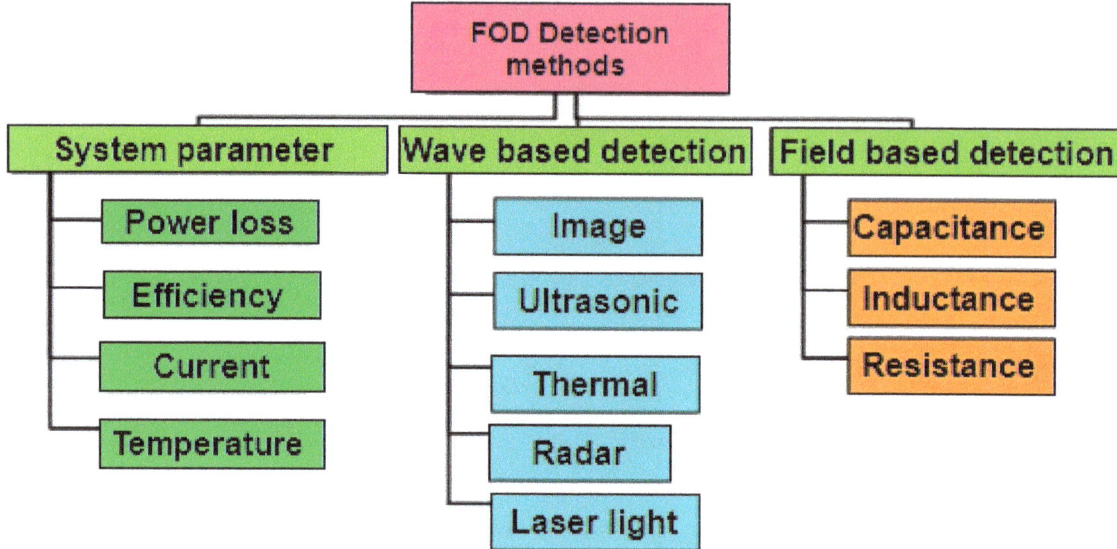

Figure 11. Categories of foreign object detection and metal object detection.

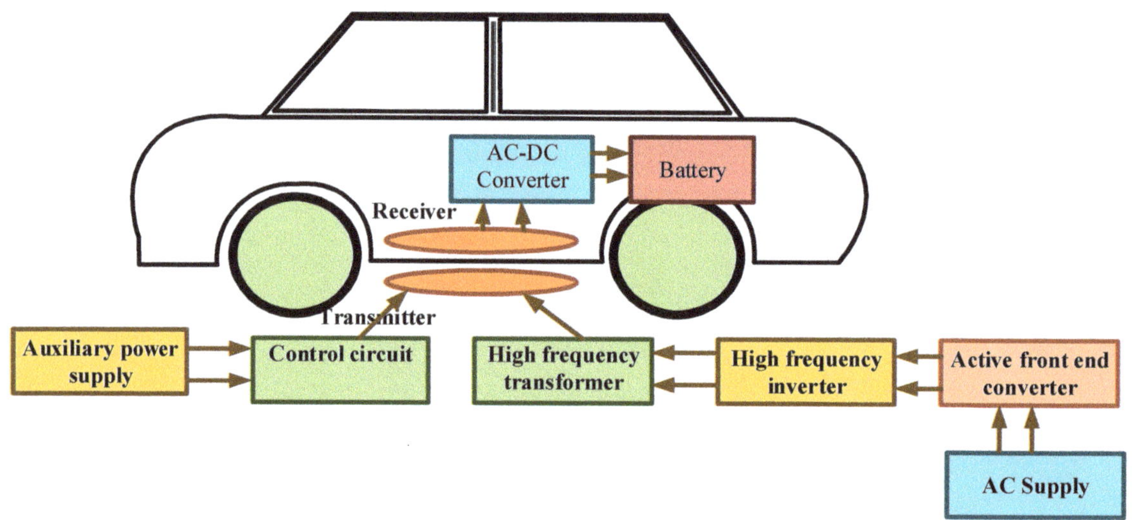

Figure 12. A wireless charging system and an FOD system.

For EV wireless charging for light-duty plug-in and alignment techniques, SAE. J2954 was created. [226–228]. This standard had to be revised to include FOD technologies. Metal objects represent a serious problem, and must be found and removed because of intense heat, and living things are a problem because magnetic radiation can distort living cells. Numerous researchers have put forth solutions to FOD (foreign object detection) and LOD (living object detection) problems.

Table 7. Advantages and disadvantages of FOD.

Methods for Detecting Foreign Objects	Benefits	Drawbacks	Comments
Method for power; oss analysis	Detects while it is powered.	High-powered wireless charging is not recommended. Only metal detection is involved, not the transmitter or receiver.	Power consumption is low.
System parameter change detection method	No extra equipment and implementation is easy.	A small metal item is hard to find. Depending on the primary power source, only metal detection occurs.	Power consumption is low.
Image, thermal and radar sensing	Can identify living things and metal.	High price, failure-pronse, and has environmental factors.	Detects both metal and living things.
Magnetic field change detection method	High-power wireless charging is acceptable, regardless of the weather.	Low-power wireless charging is challenging; detection occurs while charging.	Relates to high-power wireless charging.
Laser sensor	Able to find any object, suitable for all wireless charging levels. Reliable and simple to implement.	Costly, but very simple and robust.	Proposed laser sensor-based system.

6.2. LOD Detection Prototype Implementation

A novel method of FOD and LOD detection was put forth in this research work. First, a beam array of light was produced over the transmitter using two different multiple laser-sensor combinations. Second, a similar type of technology was also introduced; a reflecting mesh covered the transmitter, with two high-quality reflecting mirrors and two identical combinations of laser-detectors, which were used and utilized [229–231]. The configuration was presented in the suggested method of the laser-sensor-based detection system. The suggested system could be added as an accessory to any wireless charging system already in use. The laser sensor arrangement, control circuit, and auxiliary power supply made up the auxiliary system. Enclosed ferromagnetic shielding was used to protect the auxiliary control and power systems from the strong magnetic field. To make the charger installation simple and compatible with all static types of charging, the laser-sensor arrangement was placed in a physical, non-magnetic, sturdy frame that could be made according to the transmitter form and size.

Moreover, each electric vehicle's underbody system is unique, and their chassis are elevated from the ground by at least 10 cm. As a result, the auxiliary system was created to elevate 5 to 10 cm.

7. Quadruple Power Pad Coil Analysis for Wireless EV Charging

The power pad coil configuration is among the most important design considerations for EV wireless charging applications. The most crucial stage of creating an effective and trustworthy wireless power transmission system is choosing the best power pad design. Each available coil design offers benefits that are appropriate for particular applications. In this chapter, the two-objective optimization challenges, involving optimum size and design, dimension and shape, and current directions in sub-square structures of Quadruple Power Pad (QPP), were examined. Finite Element Analysis (FEA), utilizing ANSYS Maxwell®, was used to verify the dimensional design's optimization for minimum area interaction with the current directions, as well as the maximum amount of coupling coefficient with minimal interference among the corresponding coils. The outcomes of each case study were thoroughly examined and analyzed in comparison. The comparability evaluation of the structure of the QPP structure design with the Double (DD), Rectangular (D), and Double D Quadrature (DDQ) coil architectures represented another significant contribution to this chapter. Results were observed and compared, and other coil structures were used to confirm the QPP structure's computability. Figure 13 shows the quadruple power pad coil.

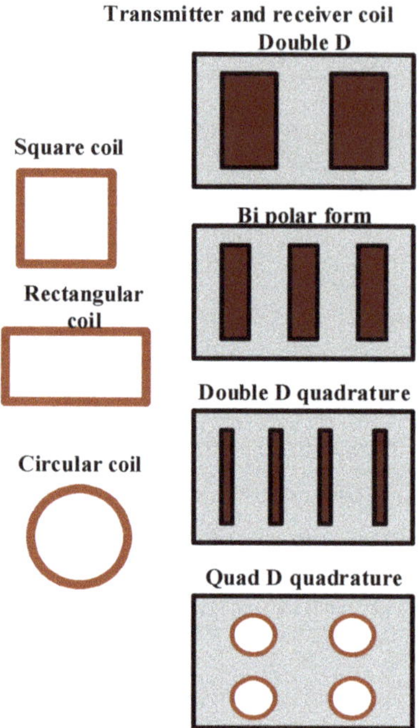

Figure 13. Quadruple Power Pad Coil.

7.1. Background

A new technique for charging electrical items was made possible by wireless power transfer. However, problems in wireless charging of EVs exist, including high-frequency power conversion converters, power pad design [232–234], electromagnetic field protection [235,236], metal object detection, and foreign object detection [237]. All of these are crucial to research and the creation of standards as in optimization [238] and structural field [239]. The main topics of this chapter were the structural study and compatibility of the QPP structure with other coil winding structures, including the D, DD, and DDQ power pads. Maximizing the value of the quality factors and the geometrical layout were the main design difficulties for the power pad.

However, the quality factor and the coupling coefficient of the transmitted and received coils directly impacts the efficiency of wireless power transmission. Unipolar, bipolar, and solenoid coil architectures are the three primary coil types used in the static method of wireless charging systems. The unipolar coil's configurations create the vertical magnetic flux and exhibit a maximum coupling coefficient due to the coil excitation, which only creates one set of polarities. Due to the primary coil excitation producing two sets of magnetic polarities, bipolar coil configurations produce the vertical magnetic flux and exhibit a low coupling coefficient [240]. Due to the double-sided magnetic flux that the solenoid coil produces, and the fact that only half of it is interconnected with the receiver coil, the solenoid construction is ineffective for EV wireless charging. DD, DDQ, BP, and solenoid pads are examples of non-polarization of power pads. Polarization coil structures, such as rectangular, circular, and square-shaped power pads, are another way to classify power pads. Non-polarized pads have one single pole and magnetic flux dispersed in all directions, with one pole located in the coil's core and the other outside of it. Two poles, north and south, are produced in the polarized power pad.

The magnetic field is localized in the central region and is parallel to the corresponding nonpolarized power pads. While at the center of the polarized power pad, the magnetic field is parallel. The interoperability between polarized power pads will, therefore, be higher when they are not in alignment, even if no evident power (VA) links appear when they are correctly aligned. Multi-coil topologies like DD and bipolar pads have many sets of mutually-disconnected coils (BP) [241,242]. The circular coils' lack of sharp edges results in a limited amount of eddy current and a dramatic peak in magnetic flux in the center of the primary coil, which is advantageous for high power transfer. The coil design structure is round and reduced, but nevertheless prone to misalignment, because of the restricted dispersion of the flux diameter [243]. D-shaped coils [244] are the ideal coil configuration, i.e., an array-type arrangement, such as in dynamic wireless charging. Unfortunately, corners with sharp edges in the coil structure are inappropriate, owing to the development of hotspots and eddy currents. Although hexagon configurations have very high maximum power transmission at the coil's core, they are inappropriate because they lose power at the periphery [245]. Better misalignment tolerance is demonstrated by the oval-shaped coil structure, although it performs less well in high-power applications. Because non-polar power pads perform poorly under horizontal misalignment, multi-coil rectangle configurations have been used to create polar pad constructions. Both single-phase and three-phase applications can benefit from multi-coil designs. Bipolar, DD, solenoid, QPP, DDQ, and Quad D Quadrature architectures are a few examples.

7.2. Analysis of the QPP Configuration

An in-depth analysis and discussion of QPP structures was provided in Figure 14. A structural investigation of the QPP structure was carried out with mathematical study, modeling, and simulation to determine its coupling coefficient. Later, interoperability testing was done to see how well the QPP structure worked with other rectangular coil topologies like D, DD, and DDQ. Geometrical diagrams of the quadruple power pad structure are illustrated in Figure 14.

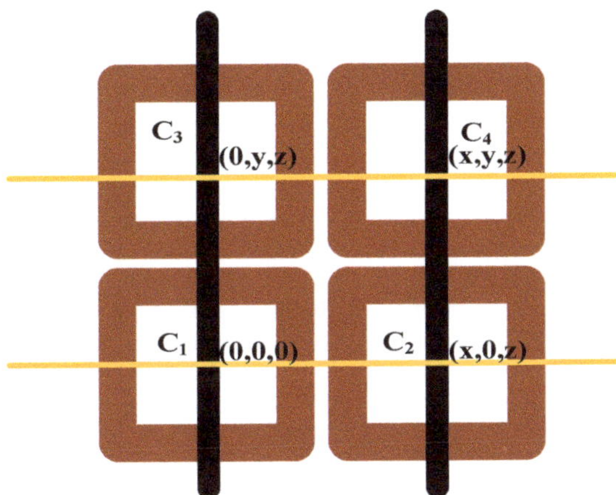

Figure 14. Geometrical diagram of quadruple power pad structure.

7.3. Misalignment Prevention for Wireless Charging Technology of Electric Vehicles: Design, Development, and Implementation

One of the primary barriers to wireless charging for electric vehicles (EVs) is the mismatch between power pads [246]. In this chapter, the intelligent alignment of the receiving coil to minimize electromagnetic leakage was discussed. The recommended remedy comprises the employment of sensors to gauge the flux intensity, both in the center and at the corners of the receiver coil. A controller circuit and a stepper motor driver are coupled to orient the receiver coil in two dimensions. To detect flux, the receiver is outfitted with a variety of Hall effect sensors, with the sensor that receives the minor flux producing the least voltage. Additionally, the controller instructs the reception coil to be moved in the direction of the sensor that registers the most excellent flux level. This chapter evaluated and confirmed the magnetic flux distribution at the optimum alignment site between the transmitter and receiver with finite element analysis (FEA) utilizing Ansys Maxwell®. The recommended modification enhanced the efficiency of wireless power transmission while reducing flux leakage. The conceptual model built the suggested system, and each system component was briefly examined and discussed. The simulation and modeling results confirmed the value of the recommended intelligent alignment.

With the improvement of EV technology's range, complex research on the wireless charging infrastructure will be required [247]. The main elements that significantly influence wireless charging include the charging coil design [248], compensation topologies [249–251], coil misalignment [252], and the frequency of wireless power transmission [253]. Coupling efficiency, however, increased at a precisely aligned location when a ferrite core was used [254,255]. Even a minimal change in the charging coils' alignment caused the efficiency to decline [256]. The charging efficiency was significantly impacted by lateral and horizontal misalignment [257]. Numerous academics have suggested compensation topologies to increase misalignment tolerance [257,258]. The effectiveness of WPT is impacted by misalignment in all directions [259]. Misalignment is inevitable if vehicles are manually parked, however, even if the suggested solutions were put into practice. Additionally, a method to self-align the transmitter or receiver is needed, which could use sensors to determine the degree of misalignment between the coils present in the region of transmission and reception. The position of the receivers or transmitters could be then modified using the proper autonomous control technique until power transmission was restored to its ideal state.

In this chapter, an ideal solution was conceived, created, and developed to solve the fundamental issue of wireless EV charging systems' power pads being out of alignment. A magnetic tracking-based automatic alignment receiver system (AARS) was suggested. AARS is a cutting-edge technique that uses two-dimensional control of the receiver, with a hall effect sensor over the transmitter's magnetic field that has been minimized to automatically align the receiving coil over the transmitter. To determine the increase in alignment efficiency, the outputs of the wireless charging system's electrical circuit analysis, modeling, and simulation were examined in Section 5.2. The recommended system's hardware implementation procedure was detailed in Section 5.3.

The projected AARS system's results, implementation issues, and solutions were covered. For short air gaps with significant magnetic field coupling, inductive energy transfer is a very effective form of wireless power transmission, but it would be difficult to use for greater power wireless charging. Moreover, even a slight misalignment would have a major detrimental effect on the effectiveness of power transfer. Below, electrical research was used to show the degree to which misalignment affected how well wireless power transfer was able to function.

7.4. Analysis of Wireless Power Transfer Efficiency Caused by EV Static Wireless Charging Misalignment

The transformer (which has no core or uses an air layer as a core) and wireless power transfer both operate on the same fundamental principles [260]. Series–series (SS) compensation topology was employed in this investigation. The operating frequency was particular for the constant current operation, where the primary side inductance and capacitance were present. This investigation was done to show the elements directly or indirectly affecting wireless power transmission effectiveness. A general block architecture of an EV wireless charging station was presented. The high-frequency converter, rectifiers, power supply, load, and coupling coils made up the static wireless charging system. To comprehend the connection between the misalignments and the effectiveness of the WPT, simple circuit analysis was carried out, as the alignment between the coils between the transmitter and receiver could alter the coefficient in the coupling.

Consequently, the coupling coefficient had a direct impact on efficiency. Inductance, resistance, capacitance, voltage, and current were the primary electrical properties on the secondary (receiver) side. The primary (transmitter) side included properties such as supply voltage, operating angular frequency, secondary side load resistance, and mutual inductance of the WPT system.

According to Equation (1), the mutual inductance and frequency of operation were connected:

$$M = \frac{V_s}{\omega I_1} \tag{1}$$

According to WPT, Equation (2) described the connection between mutual inductance and coupling coefficient:

$$M = K\sqrt{L_1 L_2} \tag{2}$$

7.5. Two Receiver Coils Were Used in a Novel Wireless Charging System Employed in Electric Vehicles

Most electric vehicle systems are designed around various components to ensure the maximum power and dependability of the automobile. The majority of these components' connections to the charging system are shown in Figure 15. Dynamic wireless power transmission could reduce the cost of onboard batteries in hybrid cars and aid with range anxiety. Pure electric vehicles have long used wireless recharging, enabling charging even while the car is moving. Analysis was difficult, nevertheless, because of the complicated working philosophy of this approach, and the existence of so many different variables and elements. Nevertheless, several characteristics, including the vehicle speed and the shape, volume, and sizes of the coil receivers, were determined by the vehicle's condition, i.e., whether it was in motion or stationary [261]. This study proposed a brand-new technique for enhancing dynamic wireless recharge system performance. The suggested technique for increasing charging power included a dynamic continuous statistical model that could characterize and analyze source-to-vehicle power transmission even when a vehicle was in motion. The suggested mathematical model presented and addressed each of the physical parameters associated with the model. The outcomes demonstrated the viability of the suggested model. Additionally, by placing two coil receivers under the car, the simulation results were validated by experimental testing [262].

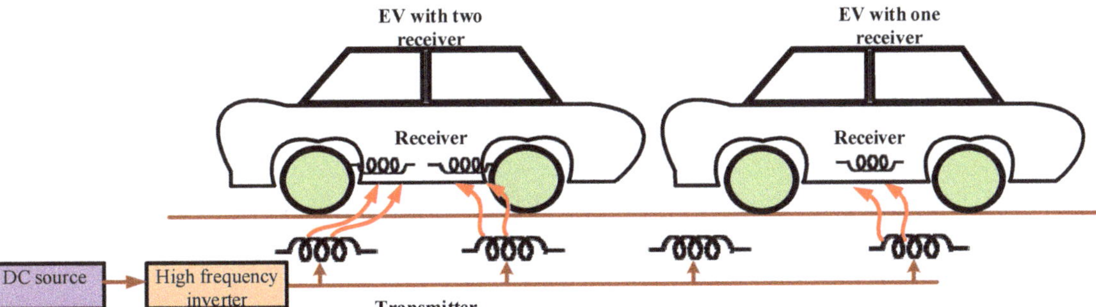

Figure 15. Composition of wireless transmitter system.

The paucity of fossil fuels and environmental concerns indicates new energy challenges. Traditional transportation accounts for a sizable share of global oil use, which generates substantial emissions [263]. The investigation and development of electric vehicle (EV) technology, as a solution to these issues, is essential and will continue to impact the automotive industry as a whole. Electric power storage is now a very popular field of study [264–266]. Electrical energy storage technology improvements have increased the power and mass of energy generated, making it possible to meet automotive demands [267–269]. The primary shortcoming of these storage options is the high cost of manufacture [270]. To minimize the overall cost of EVs, experts have been working to develop effective storage solutions and improve their reliability and charging strategies. This sector has seen the development of numerous storage systems that have been successfully incorporated into the electric power train system, improving their performance [271–275]. Therefore, losses relating to vehicle power systems could be reduced by controlling the main grid, engine, or even the rechargeable battery system flawlessly. The designers of [276] studied several control techniques to increase system efficacy. Numerous studies on the internal structures of batteries have been done to boost overall production. Numerous approaches were implemented to increase the energy efficiency of E-transportation systems [277–279]. These solutions were included in current iterations of electric cars and now have a proven track record of independence. Some researchers have concentrated on charging equipment, seeking to maximize overall performance to increase vehicle efficiency and autonomy.

In contrast, the authors presented a unique wireless charging method in [280–283]. The hybrid recharging system, which also used two types of internal power sources for the automobile, received additional consideration [284]. Additionally, PV systems have been integrated into vehicles to supply power from several sources, including blended power sources [285–289]. The main goals of the problems, addressed in the models above regarding shading impact, charging remedies, and vehicles' extra structures, included PV recharging and hybrid techniques and their remedies [290]. Research has been done on wireless charging techniques to discover more reliable and practical solutions. Accordingly, based on the literature relating to recharging strategies, more researchers have demonstrated that two components must be used, a receiver and transmitter, in parallel, for this method to be very effective.

The whole machine's performance would be constrained if the two sections were to move apart by a few inches [291]. As a result, the analysis could only be correct if the two pieces were aligned correctly and stable (i.e., not in motion). The precision of the analysis would change if one of these were still moving. Inductive power transfer, magnetic gear wireless power transfer, capacitive wireless power transfer, and inductive coupling link wireless power are just a few of the wireless energy transfer methods discussed in the literature. Of these, inductive coupling link wireless power has proven to be among the most popular. Numerous representations emerged in the literature due to the extensive discussion surrounding mathematical expressions and their representations of this recharging tool. In [292], the authors looked into static modeling to increase the effectiveness of a 50 kW, 22 kHz, 70 kHz and 85 kHz wireless charging system range for electric vehicles. This concept was only evaluated when the receiver and transmitter coils were overlaid, and it was based on mutual inductance among primary and secondary coils. The authors of [67] looked at a dynamic setting to comprehend the connection between the receiver and transmitter coil orientation deviations. Calculations were made to determine the output voltage and total efficiency factor so as to build an integrated computational framework similarly to other fields [293,294].

The analysis and mathematical model considered internal factors, including the inductance, resistance, and pitch angle among the two coils. The specifications of each of the two prior approaches were examined in [24], which also contrasted and examined the two methods. These evaluations were performed using a single receiver coil without considering the significance of the divergence speed between the receiver coils and transmitter halves. The issue, involving a two-receiver system, was not adequately examined in any current research, and its dynamic yield has not been examined.

Pitch elevation angle, resistance, coil size, inductance, spacing among the coils, and the displacement speed present in the receiver coil were assessed in the newly presented model in connection to the efficacy of the coil's recharging tool. This model helped specify the proper number of wireless coils to completely power the car while it was on a charged road. The recharging process was described in detail using a detailed mathematical model. It also offered intriguing data, showing how the physical equations worked. The tests were performed under two different circumstances: when the car was stopped and when it was in motion.

Two wireless receivers were tested in the vehicle as part of this investigation, and the outcomes demonstrated the value of the suggested model. The authors considered where the receiver was in relation to the transmitter and when the vehicle's speed changed. Additionally, the effects of the autonomous driving system were discussed. Current approaches were contrasted with these findings. The results of this study showed the benefits of two receivers below the car. The suggested concept underwent experimental testing utilizing a prototype, and the findings were confirmed.

8. Wireless Charging System—Composition

This research looked at the wireless power transfer system, which had two main parts: one on the road and one within a car. On the road, the stationary part was referred to as the transmitter. The second part, placed below the vehicle, was called the moveable receiver. Each of the two halves utilized an electronic system and was isolated from the other by a vacuum. The transmitter block generated a magnetic flux with a maximum frequency. This magnetic flux was converted into electrical energy whenever connected to the receiver coil, which was then used to recharge the EV battery. The other had two receivers, whereas the first has just one [295].

As seen in Figure 16, the transmitter component was installed on the road and coupled to various electrical components to ensure the receivers and the AC power supply were compatible. It provided the initial energy, AC power coupled to the AFE converter, which created a controlled DC voltage. This part of the transmitter block was updated by a PFC block, which kept track of the reactive power going from the source to the transmitter to preserve grid stability. After that, a strong excitation current was delivered to the transmitter coil using a high-frequency (HF) full-bridge inverter [296]. Two variables significantly influenced the entire system's profitability. The first element had to do with the compensating technique that ensured the accuracy of the current and voltage waves. The second crucial element was how the coils were made, specifically whether or not the transmitter coil surface was mainly related to the coil form's circularity. More information on these factors has been provided in the two following subsections [297].

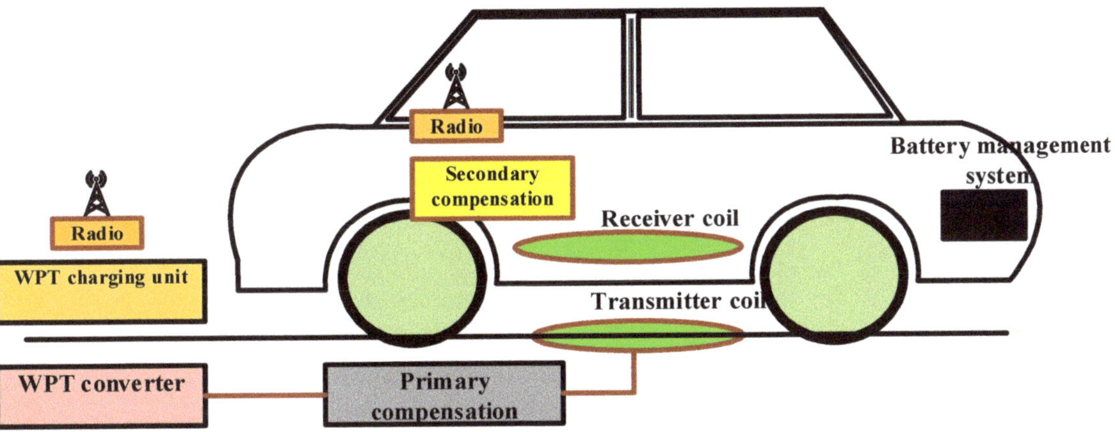

Figure 16. Two instances of electric vehicles.

8.1. Topologies for Compensation

A wireless power system transmission (WPT) system might use four resonant circuit topologies. After placing the capacitor on either side, that was indicated. The types of topologies are: series–series (SS) method, series–parallel (SP) method, parallel–series (PS) method, and parallel–parallel(PP) method, assuming that the connection and interconnection can be in series (S) type or parallel (P) type with the coil. Figure 17 provides an illustration of these plans. More information regarding these methods was provided in [298–300]. For management of the filter, the values of the first and second inductances and the capacitors (L_1, L_2) and (C_1, C_2) are fixed. To increase the transmission of power, one must reduce the apparent power supply and ensure the distribution of the active energy to the load; the coupler's primary and secondary circuits—and perhaps inductance and capacitance—are used [301]. The authors of [302,303] researched various topologies and used a few working prototypes to demonstrate their system concepts.

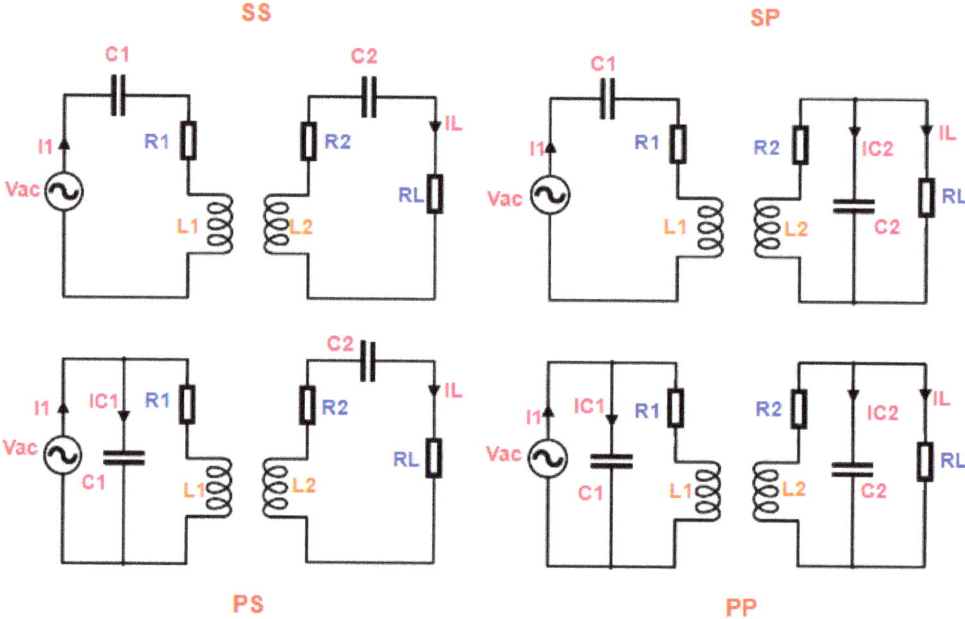

Figure 17. Compensation topologies for primary and secondary resonant circuits.

Compensation is crucial in a resonant inductive power transmission system. The system's VA ratings must consistently be decreased whenever the coupling coefficient falls under 0.3. Both parties' compensation should be reasonable and capable of performing. In the case of parasitic capacitance, the network's architecture prevents the system from resonating or receiving compensation. Additional reactive parts, like inductors or capacitors, are required to change the operating resonant frequency.

Mono resonant topology refers to the primary and basic fundamental compensation that may be achieved by interconnecting the single capacitor in series or parallel. A kind of compensation known as multi-resonant compensation affects several reactive elements. On the other hand, improper compensation results in greater reactive power and current. Reactive current enhances conduction and semiconductor losses, especially on the inverter side.

The main goals of compensation are:

- decreased reactive power;
- the feasibility of operating with a gentle duty cycle;
- avoidance of bifurcation and segmentation;
- the creation of a system able to tolerate severe misalignment; and
- to achieve optimum efficiency, bifurcation tolerance, a compact design, and cost reduction.

An automatically coupled voltage power source inverter and series compensated transmitter's coil is possible. An inductor transforms a paralleled compensatory coil winding of the inverter into an inverted current source utilizing an inductance. Secondary compensation is utilized to reduce the coil's VA rating. It is possible to offset the series network's secondary side by converting the transmitter coil's continuous output on the current side of the series network of a transmitter coil to an input source voltage.

Primarily on the secondary side, parallel networking rectification creates a current source [304]. To reduce the coil VA, Zero's Phase Angle(ϕ)(ZPA) criteria would need to be enhanced. This situation would only be feasible if the voltage, as well as the current, were in phase. This can be accomplished by fine-tuning the main capacitor using predetermined load and coupling characteristics. The primary side compensation, like the secondary side compensation, aims to achieve zero switching current (ZCS) or zero switching voltage (ZVS) if it retains a small portion of reactive or active power demand [305]. Compensation schemes can. be adjusted to the resonance frequency, also known as the ZPA frequency, due to abrupt changes in some parameters. This bifurcation occurs in the RIPT system, and the resulting parameter specification is referred to as the critical parameter. The electrical properties are modified by bifurcation. Electronic components could suffer harm as a result.

In RIPT system circumstances, basic resonant topologies exhibit bifurcation and primary level-type capacitance.

8.2. Mono-Resonant Compensation Networks

Based on the capacitor connection, there are four different compensation topologies. The two letters can then be used to address them because of the series/parallel link. As shown in Figure 17, the first sign denotes the connection of the primary side, while the subsequent (second) symbol denotes the interconnection present on the secondary side. Series–series type (SS), series–parallel type (SP), parallel–series type (PS), and parallel–parallel type (PP) are the four variations. QS must be established as a secondary quality criterion in order to get major reimbursement $Q_s = \frac{\omega_0 L_P}{R_L}$ for series-type compensation, and $Q_s = R_L / \omega_0 L_P$ for parallel-type compensation, where $in\omega\_0$ represents the frequency in resonance. The quality factor (QF) is the proportion of reactive power to active power. Table 8 displays the primary capacitances of fundamental compensatory techniques.

Table 8. Requirements for averting the basic compensations' bifurcation occurrence.

	SS	SP	PS	PP
Primary Capacitance	$\dfrac{1}{\omega^2 \cdot L_p}$	$\dfrac{1}{\omega^2 \cdot L_p \cdot \frac{M^2}{L_S}}$	$\dfrac{L_P}{\left(\frac{\omega^2 \cdot M^2}{R}\right)^2 + L_p^2 \omega^2}$	$\dfrac{L_p - \frac{M^2}{L_s}}{\left(\frac{R \cdot M^2}{L_s^2}\right)^2 + \omega^2 \left(L_p - \frac{M^2}{L_s}\right)^2}$
Bifurcation	$Q_p > \dfrac{4 Q_s^3}{4 Q_s^2 - 1}$	$Q_p > Q_s + \dfrac{1}{Q_s}$	$Q_p > Q_s$	$Q_p > Q_s + \dfrac{1}{Q_s}$

Because they increase performance, compensations like SS-type and SP-type are frequently employed in these applications. The capacitance levels are not impacted by changes in load thanks to the their compensatory networks. Additionally, SS compensation is unaffected by the main coupling coefficient of the network's capacitance. Owing to the uniqueness of the network's coupling coefficient, this leads to reduced susceptibility to misalignment. This requirement is typically included in the approach known as DWPT. Additionally, as SP compensation type is dependent only on some of the available coefficients of the coupling range, a higher primary capacitance value is required for solid magnetic coupling [306]. In an SP topology, the mutual inductance squared equals the main side transmission impedance. In this circumstance, putting DWC into practice is rather tricky. Two additional networking topologies, namely the PP type and PS type, have different properties, based only on the compensation network's resistive load and coupling coefficient range. Current source converters power these systems. The primary capacitance value needed for PP topology is higher than for PS [307]. Table 9 shows the analysis of the review study of the resonant power transmission.

Table 9. Analysis of review study on resonant power transmission.

Complication to be Resolved	Involvement in This Paper	Resolution	Performance
A single base station for power delivery and data collecting in WSN	Fully automated recharging of mobile vehicles.	The driving assistance (automotive vehicle) travels along defined routes according to the blueprint, and OPT-4 transfers electricity to the necessary nodes.	OPT-5 OPT-1
Multi-frequency	Multi-frequency unwired power transmission system.	Certain electrically-powered equipment could only receive electricity from a predetermined frequency channel.	—
The transmitter circuit's nearby and reserved loads each received the same amount of energy.	Technique for electrical circuit separation matching of impedance	Employment of several repeaters and resonators to create arithmetic derivations was suggested.	Power division method

For series–series compensation, the PF is the low coupling coefficient that causes unity and maximum accuracy. In the presence of a receiver, the associated impedance only goes to zero at the resonance frequency. However, the rated current is constrained by parasitic impedance [308], which makes operation potentially dangerous. Additionally, SP-type compensation is influenced by the more extensive primary (main) capacitance ranging value, and the coupling coefficient is required for robust electromagnetic coupling [309]. In adding more topologies, such as PP and PS types, coupling coefficient ranges and load resistances dictate capacitance values. In such systems, current source converters are employed. SS-compensating secondary sides have become a well-liked alternative for bidirectional wireless chargers due to their symmetry, which facilitates the construction of identical control topologies. The overall impedance for the four topologies is shown in Table 10. Research [310] claimed the following to be possible descriptions of the mutual inductance between two coils:

$$M = \pi \mu_0 r^4 N^2 / 2D^3 \tag{3}$$

where μ_0 represents the vacuum permeability, the spacing between the two coaxial coils is shown by the letters D, N, and r, which represent the coil's radius, turns, and number. The load is given, by transmitted power, as:

$$P = \frac{\omega_0 M^2 Q_s}{L_S} * I_p^2 \tag{4}$$

Table 10. Overall impedance of compensation topologies.

Symbol Representation	Equation
Z_{T-SS}	$\left[R_P + J\left(\omega L_P - \frac{1}{\omega C_P}\right)\right] + \dfrac{\omega^2 M^2}{[R_S + R_L + J\left(\omega L_S - \frac{1}{\omega C_S}\right)]}$
Z_{T-SP}	$\left[R_P + J\left(\omega L_P - \frac{1}{\omega C_P}\right)\right] + \dfrac{\omega^2 M^2}{[R_S + J\omega L_S + \frac{R_L}{1 + JR_L C_S \omega}]}$
Z_{T-PS}	$\dfrac{1}{(R_P + J\omega L_P) + \dfrac{\omega^2 M^2}{(R_S + R_L + J\left(\omega L_S - \frac{1}{\omega C_S}\right))}} + J\omega L_P$
Z_{T-PP}	$\dfrac{1}{(R_P + J\omega L_P) + \dfrac{\omega^2 M^2 (1 + JR_L C_S \omega)}{(R_L + ((R_S + J\omega L_S)(1 + JR_L C_S \omega)))}} + J\omega C_P$

Misalignment reduces mutual inductance, which changes the impedance range of the entire system. Thus, according to Equations (1) and (2), the power transfer enhances power production and effectiveness and is approximately proportionate to the transmission signal. The total compensation, at the most basic level, and how they relate to mutual inductance, misalignment, and whole impedance, including current mutual inductance for output power and beneficial transmission effects, were discussed.

The average total impedance falls when the ratio of current to load rises under both the series–series-type topology and the series–parallel-type topology compensatory design. Thus, the total value of the impedance rate would progressively grow along with the misalignment under the topologies of parallel–series type, as well as the parallel–parallel type compensations, resulting in a sudden reduction in the value of the current [311]. PS-type and PP-type coils' compensations would offer a maximum power factor value (PF) value and very high efficacy at low mutual inductances, and thus, the broader value spectrum of mutual inductance variations and load fluctuation [312]. Table 10 shows the detailed symbol representation of the topologies.

The value in the power factor (PF) of the PP-type layout is minimal; at the same time, the parallel main (primary) loads require a maximal current rate; primary parallel loads require a maximum reference voltage [313]. The primary side input impedance range produced by series-type compensation, presenting mostly on the secondary side of the network connection (SS-type or PS-type), has a significantly and comparatively minimal value compared to that produced by compensation for parallel networks (mostly on the secondary side in the network connection (SP-type or PP-type) [56]. A quick comparison of many fundamental network topologies is shown in Table 11. The positives and negatives of each strategy are listed.

Table 11. Comparison table of the Network Topologies.

Characteristics of Topology	SS-Type Topology	SP-Type Topology	PS-Type Topology	PP-Type Topology
The primary compensation capacitance found in the load condition, which a significant impact on topology.	-	-	Interdependent	Interdependent
The circuit equivalent impedance at resonance	Minimum	Minimum	Maximum	Maximum
The AC power supply type that will be utilized to transfer a large amount of power	Voltage power source	Voltage power source	Current generator or power source with very high voltage	Current generator or power source with very high voltage
At the stable current source (SS, SP), energy is transmitted (PS, PP)	Lower	Higher	Lower	Higher
Peak performance of efficiency	High	Low	High	Low
Power factor tolerance for changing frequency	Lower	Greater	Lower	Greater
The capability of power transmission	Maximum	Maximum	Minimum	Minimum
As a function of distance, power factor sensitivity	Minimum	Minimum	Medium	Medium
Alignment tolerances	Maximum	Maximum	Medium	Minimum
The impedance range at the resonance state	Minimum	Minimum	Maximum	Maximum
Suitability for use in electric vehicles (EV)	Maximum	Maximum	Medium	Medium

8.3. Coil Design

WPT makes it possible for electrical power to go from the source to the receiver by utilizing an air-core wireless transformer architecture [1].

In Figure 18, many planar coil designs for WPT systems are depicted, including rectangular, circular, and hybrid forms. These are applied to boost output and fix transmitter and receiver misalignment problems similarly to structural issues [314]. Additionally, each model's associated benefits and drawbacks are listed in a similar table [315]. The literature review assessed several WPT architectures' viability and magnetic coupling for automotive applications. These studies mainly concentrated on circularly-shaped structures. Inductive power transfer for a 2 kW circular planar construction was recently tested in [316,317]. It was proven that this model's null zone was the lowest. This design was chosen for this research as a result. Two coils, attached and allowing for the transmission of electricity through a magnetic field, made up the system's core component for wireless power transfer (WPT). In WPT systems, an electrical current discharge among the principal (primary) side coil creates a changing magnetic field over time. Whenever the secondary winding of the coil, which is in the reception (receiving) part, is near to that same primary (main) side coil, present in the propagation (transmitter) section, voltage is produced as the magnetic field is halted. Several variables, including the separation among the two winding of coils, the intensity of the electromagnet field, and the number of coiled turns provided throughout the time, affect the expected size of the induced emf voltage. Due to this voltage, the secondary coil of the receiver will have current flowing through it.

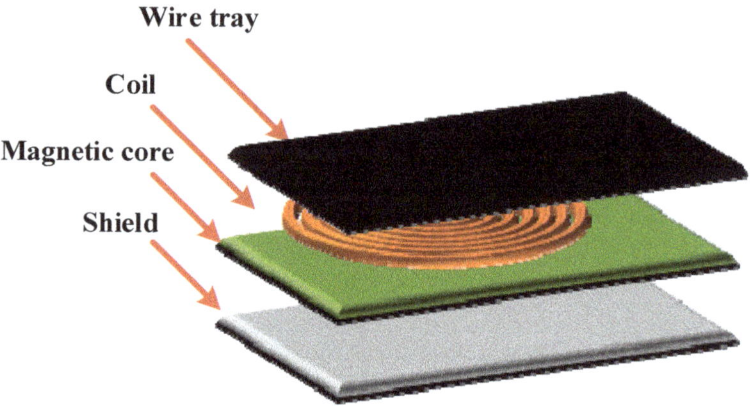

Figure 18. Exploded view of the power pad.

Power transmission resistance may not have been provided by the permeability and electrical flux channel that linked the coil that had the windings used to create the transformer's loosely coupling coil. When every coil was interconnected via the appropriate compensating communication network, the resonance movements enhances the electrical charging current, which traveled on the coil side. The Q factor of certain coil winding and the availability of coefficient in coupling (k)—which were connected toward the optimum permitted tolerance to boost the wide-angle length of aperture and lateral displacement in the longitudinal/lateral interfaces—were important criteria utilized for the overall process of designing the coil pads included within the primary (active) and secondary (passive) coils.

Another way to improve coil design in the coupling is to progressively increase coil size or decrease the air gap size of the aperture. Thus, the width present in the air gap, or the size of the coil, is determined by the implementation, and EV charging is prohibited. The coupling coefficient can be enhanced utilizing same-sized coils, e.g., a mutual inductance range for a 0.1 m aperture measurement vs. the proportion of the receiver coil (Rx) to the transmitter coil (Tx) radius.

In an EV application, with utilization of coils of the same size, both the current in the eddy coil to the vehicle chassis and the permeability of the magnetic field around the coil field are reduced [318]. When the volume and size of the receiver's coil are lessened, it is easier to install and more suitable for use in vehicles. There are more advantages in terms of weight reduction, size, and dimension when automobiles are equipped with a secondary side circuit. Additionally, each pad costs less, since less ferrite is used. These pot designs are consequently shifted to discs, rods, or plates that are evenly distributed across the coil [319]. The pad architecture created by Budhia et al. saved ferrite while maintaining the crucial connection between the circuits' primary (main) coil and secondary coil [157,320]. The authors, Budhia et al., separated the coils present in the winding to produce two different coils, interconnected in series mode. Better coil topologies have been studied and investigated due to their capacity to create a unidirectional flux. The DD and DDQ pads [321], bipolar-type pad [321], tripolar-type pad [322], and zigzag design [323] are well-known examples. The structural component of the core's top and also the side faces are wrapped in the shape of a round helical coil to form an electrical flux pipe-like pad known as a DD-type pad. This creates paths in the electromagnetic flux, which then turns away from the coil winding formed around the core material and returns to its initial position. Moreover, the backside of the coil creates zero magnetic flux with this arrangement.

The formation of the flux in the electromagnetic field connecting the main coil side accessible in the primary (active) and the secondary (passive) side available in the reception (receiver) pad is caused by the coefficient in the coupling between most of the two coils housed on a single pad.

Dimensions of the flux pipe need to be constructed in order to evaluate the coupling effect. Thus, only lateral flux is combined in the DD design, which is a downside. The quadrature coil described in [324] could be connected to the vertical parts to create the DDQ structure. The DDQ pad is another example which calls for extra wires, because it joins two windings to create a circular coil with nothing more than a double flux height. According to the authors, Bosshard et al. [325], the parameter performance of the DD charge methods and rectangle approaches were distinct. In contrast to the two independent coils utilized in a DD pad, tripolar design variants needed three separate coils associated with the DDQ coil. In an egg-like shape and size, three different kinds of coils imbricated and detached from each other. It was feasible to separate the coils by diminishing the magnetic flux in the neighboring coil winding and altering the fabricating region [326]. This kind of coil winding pad offered a wide quasi-misarrangement tolerance and, as a result, a more effective and appropriate design space for distributing the chosen, rated amount of power, by allowing significantly more winding of coils [327].

Moreover, the leakage field was somewhat decreased compared to a circular pad. The requirement for a complex control strategy design, however, was a significant drawback. A separate inverter used each coil independently, which resulted in a very high cost. An alternative layout was proposed that used three kinds of coils for each pad, with one larger group of magnetic coils twisted into a rectangle next to smaller rectangular coils [328]. The smaller coils were able to precisely and uniformly control the magnetic flux.

The load condition received power from the coil and made tiny adjustments across a significant misalignment region. Most of the time, a considerable amount of wire cable was needed. In order to increase the transmission distance, inter-junction coils could be placed concurrently among the transmitting coils' main side and the receiving winding coil [329]. Additionally, this investigation did not understand the future coils or outsourcing [330]. Whenever the layout design was separated in the power utility grid, the conclusion of the comparison of the achievability of different coil designs (Table 12) was that there was zero tolerance for the misalignment rate or the altitude of the magnetic flux path. This is crucial for any further innovations in the system [331] or cyber-resilience [332–334]. Additionally, other data-driven processes for design and optimization must be considered [335,336].

Table 12. Comparative evaluation of various coils' design methodologies.

Design Specification of Coils	Rate Range of Misalignment	Tolerance for the Flux Path's Height
Circular type	Zero at 40–50% range in diameter.	1/4 amount of a coil's diameter
Magnetic Flux pipe/flat solenoid	A great step toward tolerance.	1/2 amount of a coil's length
DD coil type	Null at 35% of the length of the pad (x-direction).	1/2 the amount of the coil's length
DDQ coil type	Around 96% of the length, null (x-direction).	2 circular times
Bipolar coil type	Approximately 96% of the length was null (x-direction).	2 circular times
Tripolar coil type	Non-symmetrical type of tolerance.	N/A
Zigzagcoil type	No null and empty values are present in this.	1/(2.5) amount of the coil's length

8.4. Batteries and Electric Vehicles

Wireless charging gives EVs, both pure and hybrid, greater independence. As such, it has become much more prevalent on motorways. Additionally, since a battery-powered electric vehicle (BEV) cannot currently be charged when moving, advancements in wireless recharging methods might be helpful. Hence, it will be crucial for manufacturers to disclose BEV's interior designs. The basic variants for the primary electric motor consist of battery systems coupled to an inverter [160]. The entire system is managed or supervised by a control unit. Because specific current systems do not allow for wireless charging, the preceding mention solely refers to the vehicle-to-grid (V2G) classification's basic form. The size of the motor must be maximized in recently- and soon-to-be-introduced EV models, battery technology must progress, and several easy charging methods must be supported in order to contribute to the reduction of pollutants [337–339]. It is commonly known that the battery, which supplies the necessary energy to run the system, is among an electric car's crucial components (EV). This component type was modeled in this work to examine power flow and demonstrate how well wireless recharging tools work [340,341]. The charging/discharging voltage of lithium-ion batteries is determined by devices comprising a DC voltage source and a series of variable resistance. It is influenced by several factors, as shown in the Equations and the recommended methods, considering related assumptions, that must be adopted in order to optimize battery function and performance. [342–344].

9. Conclusions

In this work, we compared and contrasted several types of power pads utilized for the wireless charging types of electric vehicles. Thanks to mutual induction, electric vehicle batteries may be charged wirelessly, without physical connections. Critical obstacles to EV wireless charging include power transmission frequency, power pad design, space between transmission of coils, and alignment of transfer coils. These were reviewed in this paper. This paper also discussed power pad design, involving economic analysis, in addition to optimization of the coil and core size, material, and shape for rapid prototyping.

The article evaluated existing coil shapes and designs, using a ferrite core across the coils, to create a powerful power pad for wireless EV charging. Because of the unusual way the flow was distributed, analysis was done using the 3D FEA. Only three kinds of coils—D, DD, and DDQ—were used to examine the impact of the magnetic ferrite core. The comparison was based on data imported from various findings, and magnetic flux patterns and simulation results were evaluated. The magnetic configuration of the power pad coils was simulated using Ansys 3D Maxwell simulation software. The findings showed that the DD model coil had the largest coupling coefficient, best magnetic fields, and greatest tolerance for misalignment, as explained in detail in this paper. The ferrite core, inserted across coils, somewhat improved the coupling coefficient and aligned the magnetic flux pattern, which was also discussed.

The development of EVs and HEVs is inevitable, as a result of growing worries over the energy crisis and energy usage. While an overview of new technologies could be helpful to many stakeholders, many fascinating innovations that have been developed in the previous few decades were also discussed herein. This paper sought to provide engineers, researchers, and academics eager to pursue their interests in this field with a place to start. The latest EV/HEV models, electrochemical energy sources, wireless charging infrastructures, and electric vehicles, in general, represent some of the major subjects covered in this article. As such, this work was designed to give readers a road map, so they may start their own fieldwork.

Author Contributions: Conceptualization, G.P. and U.S.; methodology, G.P.; software, U.S.; validation, G.P., U.S. and P.V.; formal analysis, G.P. and U.S.; investigation, B.N.; resources, B.N.; data curation, U.S.; writing—original draft preparation, G.P and U.S.; writing—review and editing, G.P., U.S., P.V. and B.N.; visualization, B.N.; supervision, U.S.; project administration, U.S.; funding acquisition, B.N. All authors have read and agreed to the published version of the manuscript.

Funding: This research received no external funding.

Conflicts of Interest: The authors declare no conflict of interest.

References

1. Dai, J.; Ludois, D.C. A Survey of Wireless Power Transfer and a Critical Comparison of Inductive and Capacitive Coupling fo r Small Gap Applications. *IEEE Trans. Power Electron.* **2015**, *30*, 6017–6029. [CrossRef]
2. Miller, J.M.; Onar, O.C.; Chinthavali, M. Primary-Side Power Flow Control of Wireless Power Transfer for Electric Vehicle Charging. *IEEE J. Emerg. Sel. Top. Power Electron.* **2015**, *3*, 147–162. [CrossRef]
3. Qiu, C.; Chau, K.T.; Liu, C.; Chan, C.C. Overview of wireless power transfer for electric vehicle charging. *Electr. Veh. Symp. Exhib.* **2013**, *7*, 1–9.
4. Choi, S.Y.; Gu, B.W.; Jeong, S.Y.; Rim, C.T. Advances in wireless power transfer systems for roadway-powered electric vehicles. *IEEE J. Emerg. Sel. Top. Power Electron.* **2015**, *3*, 18–36. [CrossRef]
5. Mi, C.C.; Buja, G.; Choi, S.Y.; Rim, C.T. Modern Advances in Wireless Power Transfer Systems for Roadway Powered Electric Vehicles. *IEEE Trans. Ind. Electron.* **2016**, *63*, 6533–6545. [CrossRef]
6. Li, S.; Liu, Z.; Zhao, H.; Zhu, L.; Shuai, C.; Chen, Z. Wireless Power Transfer by Electric Field Resonance and its Application in Dynamic Charging. *IEEE Trans. Ind. Electron.* **2016**, *63*, 6602–6612. [CrossRef]
7. Choi, S.; Huh, J.; Lee, W.Y.; Lee, S.W.; Rim, C.T. New cross-segmented power supply rails for roadway-powered electric vehicles. *IEEE Trans. Power Electron.* **2013**, *28*, 5832–5841. [CrossRef]
8. Bosshard, R.; Kolar, J.W. Multi-Objective Optimization of 50 kW/85 kHz IPT System for Public Transport. *IEEE J. Emerg. Sel. Top. Power Electron.* **2016**, *4*, 1370–1382. [CrossRef]
9. Budhia, M.; Covic, G.; Boys, J. A new IPT magnetic coupler for electric vehicle charging systems. In Proceedings of the IECON 2010—36th Annual Conference on IEEE Industrial Electronics Society, Glendale, AZ, USA, 7–10 November 2010; pp. 2487–2492.
10. Budhia, M.; Boys, J.T.; Covic, G.A.; Huang, C.Y. Development of a single-sided flux magnetic coupler for electric vehicle IPT charging systems. *IEEE Trans. Ind. Electron.* **2013**, *60*, 318–328. [CrossRef]
11. Nagendra, G.R.; Covic, G.A.; Boys, J.T. Determining the physical size of inductive couplers for IPT EV systems. In Proceedings of the Conference Proceedings-IEEE Applied Power Electronics Conference and Exposition-APEC, Fort Worth, TX, USA, 16–20 March 2014; pp. 3443–3450.
12. Wireless Power Consortium Products. Available online: https://www.wirelesspowerconsortium.com/products (accessed on 10 May 2019).
13. Wireless Chargers Archives-Qi Wireless Charging. Available online: http://www.qiwireless.com/category/wirelesschargers (accessed on 10 May 2019).
14. Barnard, J.M.; Ferreira, J.A.; Van Wyk, J.D. Sliding transformers for linear contactless power delivery. *IEEE Trans. Ind. Electron.* **1997**, *44*, 774–779. [CrossRef]
15. Lu, F.; Zhang, H.; Hofmann, H.; Mi, C. A high efficiency 3.3 kW loosely-coupled wireless power transfer system without magnetic material. In Proceedings of the 2015 IEEE Energy Conversion Congress and Exposition, ECCE 2015, Montreal, QC, Canada, 20–24 September 2015; pp. 2282–2286.
16. Wu, H.H.; Gilchrist, A.; Sealy, K.D.; Bronson, D. A high efficiency 5 kW inductive charger for EVs using dual side control. *IEEE Trans. Ind. Inform.* **2012**, *8*, 585–595. [CrossRef]
17. Li, W. High Efficiency Wireless Power Transmission at Low Frequency Using Permanent Magnet Coupling. Master's Thesis, University of British Columbia (Vancouver), Kelowna, BC, Canada, 2009.
18. Ahmad, A.; Alam, M.S.; Chaban, R.C. Efficiency enhancement of wireless charging for Electric vehicles through reduction of coil misalignment. In Proceedings of the 2017 IEEE Transportation Electrification Conference and Expo (ITEC), Chicago, IL, USA, 22–24 June 2017; pp. 21–26.
19. Mizuno, T.; Yachi, S.; Kamiya, A.; Yamamoto, D. Improvement in efficiency of wireless power transfer of magnetic resonant coupling using magneto plated wire. *IEEE Trans. Magn.* **2011**, *47*, 4445–4448. [CrossRef]
20. Sakamoto, H.; Harada, K.; Washimiya, S.; Takehara, K.; Matsuo, Y.; Nakao, F. Large air gap coupler for inductive charger. *IEEE Trans. Magn.* **1999**, *35 Pt 2*, 3526–3528. [CrossRef]
21. Takanashi, H.; Sato, Y.; Kaneko, Y.; Abe, S.; Yasuda, T. A large air gap 3 kW wireless power transfer system for electric vehicles. In Proceedings of the 2012 IEEE Energy Conversion Congress and Exposition (ECCE), Raleigh, NC, USA, 15–20 September 2012; pp. 269–274.
22. Duan, C.; Jiang, C.; Taylor, A.; Bai, K. Design of a zero-voltage-switching large air-gap wireless charger with low electric stress for electric vehicles. *IET Power Electron.* **2013**, *6*, 1742–1750. [CrossRef]
23. Narayanamoorthi, R. Cross Interference Free Dual Frequency Wireless Power Transfer Using Frequency Bifurcation for Dynamic Biomedical Implants. *IEEE Trans. Electromagn. Compat.* **2021**, *63*, 286–293. [CrossRef]
24. Kurs, A.; Karalis, A.; Moffatt, R.; Joannopoulos, J.D.; Fisher, P.; Soljacic, M. Wireless power transfer via strongly coupled magnetic resonances. *Science* **2007**, *317*, 83–86. [CrossRef]
25. Zhang, Y.; Zhao, Z.; Chen, K. Frequency decrease analysis of resonant wireless power transfer. *IEEE Trans. Power Electron.* **2014**, *29*, 1058–1063. [CrossRef]
26. Villa, J.L.; Sallán, J.; Osorio, J.F.S.; Llombart, A. High-misalignment tolerant compensation topology for ICPT systems. *IEEE Trans. Ind. Electron.* **2012**, *59*, 945–951. [CrossRef]
27. Li, S.; Li, W.; Deng, J.; Nguyen, T.D.; Mi, C.C. A Double-Sided LCC Compensation Network and Its Tuning Method for Wireless Power Transfer. *IEEE Trans. Veh. Technol.* **2015**, *64*, 2261–2273. [CrossRef]

28. Yvkoff, L. Will DC Fast Charging Harm Electric Car Batteries? 2010. Available online: https://www.cnet.com/roadshow/news/will-dc-fast-charging-harm-electric-car-batteries/ (accessed on 15 May 2019).
29. A Simple Guide to DC Fast Charging. Available online: http://www.fleetcarma.com/dcfast-charging-guide/ (accessed on 15 May 2019).
30. Lukic, S.; Pantic, Z. Cutting the Cord: Static and Dynamic Inductive Wireless Charging of Electric Vehicles. *IEEE Electrif. Mag.* **2013**, *1*, 57–64. [CrossRef]
31. Yilmaz, M.; Krein, P.T. Review of Charging Power Levels and Infrastructure for Plug-In Electric and Hybrid Vehicles and Commentary on Unidirectional Charging. *IEEE Int. Electr. Veh. Conf. IEVC* **2012**, *28*, 2151–2169.
32. Khaligh, A.; Dusmez, S. Comprehensive topological analysis of conductive and inductive charging solutions for plug-in electric vehicles. *IEEE Trans. Veh. Technol.* **2012**, *61*, 3475–3489. [CrossRef]
33. J1772: SAE Electric Vehicle and Plug in Hybrid Electric Vehicle Conductive Charge Chapter 2 53 Coupler-SAE International. Available online: http://standards.sae.org/j1772_201602/ (accessed on 16 May 2019).
34. Ahmad, F.; Alam, M.S.; Asaad, M. Developments in xEVs charging infrastructure and energy management system for smart microgrids including xEVs. *Sustain. Cities Soc.* **2017**, *35*, 552–564. [CrossRef]
35. Porsche Panamera S E-Hybrid PluginCars.com. Available online: http://www.plugincars.com/porsche-panamera-s-e-hybrid (accessed on 15 May 2019).
36. 2017 Audi A3 Sportback e-tron® I Audi USA. Available online: https://www.audiusa.com/models/audi-a3-sportback-e-tron/2017 (accessed on 20 May 2019).
37. Specs I Cadillac ELR Forum. Available online: http://www.myelr.com/cadillac-elrspecs (accessed on 20 May 2019).
38. Chevrolet Pressroom-United States-Spark EV. Available online: http://media.chevrolet.com/media/us/en/chevrolet/vehicles/spark-ev/2016.tab1.html (accessed on 20 May 2019).
39. 2017 Ford®C-MAX Energi SE Plug-In Hybrid I Model Highlights I Ford.com. Available online: https://www.ford.com/cars/c-max/2017/models/c-max-energi-se/ (accessed on 21 May 2019).
40. Mercedes S550 Plug-in Hybrid I PluginCars.com. Available online: http://www.plugincars.com/mercedes-s550-plug-hybrid (accessed on 21 May 2019).
41. Mercedes-Benz C350 Plug-In Hybrid-EVBox. Available online: http://www.evbox.com/go-electric/electric-cars/mercedes-benz/mercedes-benz-c350-plug-in-hybrid/ (accessed on 21 May 2019).
42. Smart Electric Drive I PluginCars.com. Available online: http://www.plugincars.com/smart-ed (accessed on 22 May 2019).
43. The Toyota Prius Plug-in Hybrid I PluginCars.com. Available online: http://plugincars.com/toyota-prius-plugin-hybrid (accessed on 30 May 2019).
44. Specifications I i-MiEV I MITSUBISHI MOTORS. Available online: http://www.mitsubishi-motors.com/en/showroom/i-miev/specifications/ (accessed on 30 May 2019).
45. Nissan LEAF Will Include Fast Charge Capability and Emergency Charging Cable at Launch-Gas 2. Available online: http://gas2.org/2010/05/27/nissan-leaf-will-includefast-charge-capability-and-emergency-charging-cable-at-launch/ (accessed on 30 May 2019).
46. Review and Pictures of Porsche Cayenne S E-Hybrid I PluginCars.com. Available online: http://www.plugincars.com/porsche-cayenne-s-e-hybrid (accessed on 30 May 2019).
47. 2017 Volkswagen e-Golf Specifications. Available online: http://www.neftinvw.com/blog/2017-volkswagen-e-golf-specifications/ (accessed on 30 May 2019).
48. Specs I Ford Focus Electric Forum, My Focus Electric. Available online: http://www.myfocuselectric.com/specs/ (accessed on 30 May 2019).
49. Fiat 500e I PluginCars.com. Available online: http://www.plugincars.com/fiat-500e (accessed on 5 June 2019).
50. 2017 Kia Soul EV Specifications. Available online: http://www.kiamedia.com/us/en/models/soul-ev/2017/specifications (accessed on 5 June 2019).
51. Honda Accord Plug-in Hybrid I PluginCars.com. Chapter 2 54. Available online: http://www.plugincars.com/honda-accord-plug-hybrid (accessed on 5 June 2019).
52. "Chevrolet Spark EV I PluginCars.com". Available online: https://www.ford.com/cars/cmax/2017/models/c-max-energi-se/ (accessed on 5 June 2019).
53. 2017 BMW i3 (94 Ah) Release Date, Price and Specs-Roadshow. Available online: https://www.cnet.com/roadshow/auto/2017-bmw-i3/preview/ (accessed on 5 June 2019).
54. Hui, S.Y. Technology for Portable Electronic Products and Qi. *Proc. IEEE* **2013**, *101*, 1290–1301. [CrossRef]
55. Chen, W.; Liu, C.; Lee, C.H.T.; Shan, Z. Cost-effectiveness comparison of coupler designs of wireless power transfer for electric vehicle dynamic charging. *Energies* **2016**, *9*, 906. [CrossRef]
56. Transfer, P. *SAgE Singapore Scholarships Dynamic Charging for Electric Vehicles (EV) by Wireless*; SAE International: Warrendale, PA, USA, 2022.
57. Bhattacharya, S.; Tan, Y.K. Design of static wireless charging coils for integration into electric vehicle. In Proceedings of the 2012 IEEE Third International Conference on Sustainable Energy Technologies (ICSET), Kathmandu, Nepal, 24–27 September 2012; pp. 146–151.
58. Miller, J.M.; Jones, P.T.; Li, J.M.; Onar, O.C. ORNL experience and challenges facing dynamic wireless power charging of EV's. *IEEE Circuits Syst. Mag.* **2015**, *15*, 40–53. [CrossRef]

59. Moschoyiannis, S.; Maglaras, L.; Jiang, J.; Topalis, F.; Maglaras, A. Dynamic wireless charging of electric vehicles on the move with Mobile Energy Disseminators. *Int. J. Adv. Comput. Sci. Appl.* **2015**, *6*, 239–251.
60. Laccone, F.; Malomo, L.; Froli, M.; Cignoni, P.; Pietroni, N. Automatic Design of Cable-Tensioned Glass Shells. *Comput. Graph. Forum* **2020**, *39*, 260–273. [CrossRef]
61. Jang, Y.J.; Jeong, S.; Lee, M.S. Initial energy logistics cost analysis for stationary, quasi-dynamic, and dynamic wireless charging public transportation systems. *Energies* **2016**, *9*, 483. [CrossRef]
62. Mohamed, A.A.S.; Lashway, C.R.; Mohammed, O. Modeling and Feasibility Analysis of Quasi-dynamic WPT System for EV Applications. *IEEE Trans. Transp. Electrif.* **2017**, *3*, 343–353. [CrossRef]
63. Ojika, S.; Miura, Y.; Ise, T. Evaluation of Inductive Contactless Power Transfer Outlet with Coaxial Coreless Transformer. *Electr. Eng. Jpn.* **2016**, *195*, 57–67, (English Transl. Denki Gakkai Ronbunshi). [CrossRef]
64. Esteban, B.; Sid-Ahmed, M.; Kar, N.C. A Comparative Study of Power Supply Architectures in Wireless EV Charging Systems. *IEEE Trans. Power Electron.* **2015**, *30*, 6408–6422. [CrossRef]
65. SLi; Mi, C.C. Wireless power transfer for electric vehicle applications. *IEEE J. Emerg. Sel. Top. Power Electron.* **2015**, *3*, 4–17.
66. Chao, Y.-H.; Shieh, J.-J. Series-parallel loosely coupling power supply with primary-side control. In Proceedings of the 2012 IEEE Third International Conference on Sustainable Energy Technologies (ICSET), Kathmandu, Nepal, 24–27 September 2012; pp. 352–356.
67. Park, C.; Lim, S.; Shin, J.; Lee, C.-Y. How much hydrogen should be supplied in the transportation market? Focusing on hydrogen fuel cell vehicle demand in South Korea: Hydrogen demand and fuel cell vehicles in South Korea. *Technol. Forecast. Soc. Change* **2022**, *181*, 121750. [CrossRef]
68. Huh, J.; Lee, S.W.; Lee, W.Y.; Cho, G.H.; Rim, C.T. Narrow-width inductive power transfer system for online electrical vehicles. *IEEE Trans. Power Electron.* **2011**, *26*, 3666–3679. [CrossRef]
69. Kashani, S.A.; Soleimani, A.; Khosravi, A.; Mirsalim, M. State-of-the-Art Research on Wireless Charging of Electric Vehicles Using Solar Energy. *Energies* **2023**, *16*, 282. [CrossRef]
70. Kim, J.H.; Lee, B.S.; Lee, J.H.; Lee, S.H.; Park, C.B.; Jung, S.M.; Lee, S.G.; Yi, K.P.; Baek, J. Development of 1-MW Inductive Power Transfer System for a High-Speed Train. *IEEE Trans. Ind. Electron.* **2015**, *62*, 6242–6250. [CrossRef]
71. Julio, A. Ruiz ITS systems developing in Malaga. 2nd Congress EU Core Net Cities, 2014. Available online: https://intellias.com/intellias-opens-its-first-spanish-officein-malaga/ (accessed on 19 February 2023).
72. INTIS-Integrated Infrastructure Solutions. Available online: http://www.intis.de/intis/downloads_e.html (accessed on 7 June 2019).
73. Umenei, A.E. Understanding Low Frequency Non-Radiative Power Transfer. WHITEPAPER-Fult. Innov. LLC.; Wirel. Power Consortium . . . , no. June 2011. Available online: https://www.semanticscholar.org/paper/UNDERSTANDING-LOW-FREqUENCY-NON-RADIATIVE-POWER-Umenei/777695524c40f6d352130a71b047ea611d0ee86a (accessed on 30 December 2022).
74. Sazonov, E.; Neuman, M.R. *Wearable Sensors: Fundamentals, Implementation and Applications*; Academic Press: Cambridge, MA, USA, 2014.
75. Sun, T.; Xie, X.; Wang, Z. *Wireless Power Transfer for Medical Microsystems*; Springer: New York, NY, USA, 2013.
76. Chabalko, M.J.; Besnoff, J.; Ricketts, D.S. Magnetic Field Enhancement in Wireless Power with Metamaterials and Magnetic Resonant Couplers. *IEEE Antennas Wirel. Propag. Lett.* **2016**, *15*, 452–455. [CrossRef]
77. Agbinya, J.I. *Wireless Power Transfer*; River Publishers: Aalborg, Denmark, 2012; Chapter 4; p. 119.
78. Dashora, H.K.; Bertoluzzo, M.; Buja, G. Reflexive properties for different pick-up circuit topologies in a distributed IPT track. In Proceedings of the 2015 IEEE International Conference on Industrial Informatics, INDIN 2015, Cambridge, UK, 22–24 July 2015; pp. 69–75.
79. Kazmierkowski, M.P.; Moradewicz, A.J. Unplugged but connected: Review of contactless energy transfer systems. *IEEE Ind. Electron. Mag.* **2012**, *6*, 47–55. [CrossRef]
80. Maxwell, J.C. Summary for Policymakers. *Treatise Electr. Magn.* **1954**, *53*, 1–30.
81. Ampere's Law-Reference Notes. Available online: http://notes.tyrocity.com/ampereslaw/ (accessed on 10 June 2019).
82. Lopez-Ramos, A.; Menendez, J.R.; Pique, C. Conditions for the Validity of Faraday's Law of Induction and Their Experimental Confirmation. *Eur. J. Phys.* **2008**, *29*, 1069–1076. [CrossRef]
83. Justin, A. Biot-Savart Law. *Int. J. Res.* **2015**, *2*, 2348–6848.
84. Maxwell, J. A Dynamical Theory of the Electromagnetic Field. *Proc. R. Soc.* **1863**, 459–512.
85. Lu, X.; Wang, P.; Niyato, D.; Kim, D.I.; Han, Z. Wireless Charging Technologies: Fundamentals, Standards, and Network Applications. *IEEE Commun. Surv. Tutor.* **2016**, *18*, 1413–1452. [CrossRef]
86. Leblanc, M.; Hutin, M. Transformer System for Electric Railways. U.S. Patent 527,857, 1894.
87. Wireless Transmission of Energy. Available online: https://teslaresearch.jimdo.com/wireless-transmission-of-energy-1/ (accessed on 10 June 2019).
88. JBolger, G.; Kirsten, F.A.; Ng, L.S. Inductive power coupling for an electric highway system. In Proceedings of the 28th IEEE Vehicular Technology Conference, Denver, CO, USA, 22–24 March 1978; pp. 137–144.
89. Zell, C.E.; Bolger, J.G. Development of an engineering prototype of a roadway powered electric transit vehicle system: A public/private sector program. In Proceedings of the 32nd IEEE Vehicular Technology Conference, San Diego, CA, USA, 23–26 March 1982; Volume 32, pp. 35–38.

90. California PATH Program Roadway Powered Electric Vehicle Project Track Construction And Testing Program Phase 3D. Traffic, 1994. Available online: https://escholarship.org/uc/item/1jr98590Accessed (accessed on 20 December 2022).

91. Shinohara, N. Wireless power transmission progress for electric vehicle in Japan. *IEEE Radio Wirel. Symp. RWS* **2013**, 109–111. Available online: https://onlinelibrary.wiley.com/doi/10.1002/tee.22340 (accessed on 20 December 2022).

92. Nikitin, P.V.; Rao, K.V.S.; Lazar, S. An Overview of Near Field UHF RFID. In Proceedings of the 2007 IEEE International Conference on RFID, Grapevine, TX, USA, 26–28 March 2007; pp. 167–174.

93. Power Supply-Is There Any Difference between Induction and Resonant Wireless Energy Transfer-Electrical Engineering Stack Exchange. Available online: https://electronics.stackexchange.com/questions/25176/is-there-any-difference-betweeninduction-and-resonant-wireless-energy-trans (accessed on 15 June 2019).

94. Jung, G.; Jeon, S.; Cho, D.; Member, S. Design and Implementation of Shaped MagneticResonance-Based Wireless Power Transfer System for Roadway-Powered Moving Electric Vehicles. *IEEE Trans. Ind. Electron.* **2014**, *61*, 1179–1192.

95. Seungyoung, A.; Joungho, K. Magnetic field design for high efficient and low EMF wireless power transfer in on-line electric vehicle. In Proceedings of the 5th European Conference on Antennas and Propagation (EUCAP), Rome, Italy, 11–15 April 2011; pp. 3979–3982.

96. Nam, P.S.; Dong, H.C. *The On-Line Electric Vehicle: Wireless Electric Ground Transportation Systems*; Springer: Berlin, Germany, 2017.

97. Hori, Y. Novel EV society based on motor/capacitor/wireless; Application of electric motor, supercapacitors, and wireless power transfer to enhance operation of future vehicles. In Proceedings of the 2012 IEEE MTT-S International Microwave Workshop Series on Innovative Wireless Power Transmission: Technologies, Systems, and Applications, Kyoto, Japan, 10–11 May 2012; pp. 3–8.

98. Imura, T.; Okabe, H.; Uchida, T.; Hori, Y. Study on open and short end helical antennas with capacitor in series of wireless power transfer using magnetic resonant couplings. In Proceedings of the IECON Proceedings (Industrial Electronics Conference), Porto, Portugal, 3–5 November 2009; pp. 3848–3853.

99. Covic, G.A.; Boys, J.T.; Budhia, M.; Huang, C. Electric vehicles—Personal transportation for the future. *World Electr. Veh. J.* **2010**, *4*, 693–704. [CrossRef]

100. Garnica, J.; Chinga, R.A.; Lin, J. Wireless Power Transmission: From Far Field to Near Field. *Proc. IEEE* **2013**, *101*, 1321–1331. [CrossRef]

101. Matsumoto, H. Research on solar power satellites and microwave power transmission in Japan. *IEEE Microw. Mag.* **2002**, *3*, 36–45. [CrossRef]

102. Shinohara, N.; Kubo, Y. Wireless Charging for Electric Vehicle with Microwaves. In Proceedings of the 2013 3rd International Electric Drives Production Conference (EDPC), Nuremberg, Germany, 29–30 October 2013.

103. Brown, W.C. The History of Power Transmission by Radio Waves. *IEEE Trans. Microw. Theory Tech.* **1984**, *32*, 1230–1242. [CrossRef]

104. Range, S. Beam Efficiency of Wireless Power Transmission via Radio Waves from Short Range to Long Range. *J. Electromagn. Eng. Sci.* **2010**, *10*, 4–10. [CrossRef]

105. Kapranov, V.V.; Matsak, I.S.; Tugaenko, V.Y.; Blank, A.V.; Suhareva, N.A. Atmospheric turbulence effects on the performance of the laser wireless power transfer system. In *Free-Space Laser Communication and Atmospheric Propagation XXIX*; SPIE: San Francisco, CA, USA, 2017; p. 100961E.

106. Dickinson, R.M.M. Performance of a High-Power, 2.388-GHz Receiving Array in Wireless Power Transmission Over 1.54 km. In *MTT-S International Microwave Symposium Digest*; IEEE: Piscataway, NJ, USA, 1976; Volume 76, pp. 139–141.

107. Shinohara, N. Wireless charging system of electric vehicle with GaNSchottky diodes. In Proceedings of the IMS2011 Workshop WFA "Wireless Power Transmission", Baltimore, MD, USA, 10 June 2011; p. CD-ROM.

108. Tritschler, J.; Reichert, S.; Goeldi, B. A practical investigation of a high power, bidirectional charging system for electric vehicles. In Proceedings of the 16th European Conference on Power Electronics and Applications, Lappeenranta, Finland, 26–28 August 2014; pp. 1–7.

109. Triviño, A.; González-González, J.M.; Aguado, J.A. Wireless Power Transfer Technologies Applied to Electric Vehicles: A Review. *Energies* **2021**, *14*, 1547. [CrossRef]

110. Wireless Electric Vehicle Charging Technology | Halo & Power Transfer | Qualcomm. Available online: https://www.qualcomm.com/solutions/automotive/wevc (accessed on 15 June 2019).

111. R. Schuylenbergh, Koeranadvan; Puers, Inductive Powering: Basic Theory and Application to Biomedical System, no. 1. 2014. Available online: https://link.springer.com/book/10.1007/978-90-481-2412-1 (accessed on 19 February 2023).

112. Kamineni, A.; Covic, G.A.; Boys, J.T. Analysis of Coplanar Intermediate Coil Structures in Inductive Power Transfer Systems. *IEEE Trans. Power Electron.* **2015**, *30*, 6141–6154. [CrossRef]

113. Nguyen, T.-D.D.; Li, S.; Li, W.; Mi, C.C. Feasibility study on bipolar pads for efficient wireless power chargers. In Proceedings of the 2014 IEEE Applied Power Electronics Conference and Exposition-APEC 2014, Fort Worth, TX, USA, 16–20 March 2014; pp. 1676–1682.

114. Zhang, W.; Wong, S.-C.; Tse, C.K.; Chen, Q. An Optimized Track Length in Roadway Inductive Power Transfer Systems. *IEEE J. Emerg. Sel. Top. Power Electron.* **2014**, *2*, 598–608. [CrossRef]

115. Zhang, Z.; Chau, K.T. Homogeneous Wireless Power Transfer for Move-and-Charge. *IEEE Trans. Power Electron.* **2015**, *30*, 6213–6220. [CrossRef]

116. Li, W.; Zhao, H.; Li, S.; Deng, J.; Kan, T.; Mi, C.C. Integrated LCC Compensation Topology for Wireless Charger in Electric and Plug-in Electric Vehicles. *IEEE Trans. Ind. Electron.* **2015**, *62*, 4215–4225. [CrossRef]

117. Chen, L.; Liu, S.; Zhou, Y.C.; Cui, T.J. An optimizable circuit structure for highefficiency wireless power transfer. *IEEE Trans. Ind. Electron.* **2013**, *60*, 339–349. [CrossRef]
118. Bertoluzzo, M.; Buja, G.; Dashora, H.K. Lumped Track Layout Design for Dynamic Wireless Charging of Electric Vehicles. *IEEE Trans. Ind. Electron.* **2016**, *63*, 6631–6640.
119. Venkatesan, M.; Rajamanickam, N.; Vishnuram, P.; Bajaj, M.; Blazek, V.; Prokop, L.; Misak, S. A Review of Compensation Topologies and Control Techniques of Bidirectional Wireless Power Transfer Systems for Electric Vehicle Applications. *Energies* **2022**, *15*, 7816. [CrossRef]
120. Zaheer, A.; Hao, H.; Covic, G.A.; Kacprzak, D. Investigation of multiple decoupled coil primary pad topologies in lumped IPT systems for interoperable electric vehicle charging. *IEEE Trans. Power Electron.* **2015**, *30*, 1937–1955. [CrossRef]
121. Bertoluzzo, M.; Di Barba, P.; Forzan, M.; Mognaschi, M.E.; Sieni, E. Optimization of Compensation Network for a Wireless Power Transfer System in Dynamic Conditions: A Circuit Analysis Approach. *Algorithms* **2022**, *15*, 261. [CrossRef]
122. Ongayo, D.; Hanif, M. Comparison of Circular and Rectangular Coil Transformer Parameters for Wireless Power Transfer Based on Finite. In Proceedings of the 2015 IEEE 13th Brazilian Power Electronics Conference and 1st Southern Power Electronics Conference (COBEP/SPEC), Fortaleza, Brazil, 29 February 2016.
123. Mahmud, M.H.; Elmahmoud, W.; Barzegaran, M.R.; Brake, N. Efficient Wireless Power Charging of Electric Vehicle by Modifying the Magnetic Characteristics of the Transmitting Medium. *IEEE Trans. Magn.* **2017**, *63*, 6631–6640. [CrossRef]
124. Sekiya, N.; Monjugawa, Y. A Novel REBCO Wire Structure That Improves Coil Quality Factor in MHz Range and its Effect on Wireless Power Transfer Systems. *IEEE Trans. Appl. Supercond.* **2017**, *27*, 1–5. [CrossRef]
125. Yilmaz, T.; Hasan, N.; Zane, R.; Pantic, Z. FbifMulti-Objective Optimization of Circular Magnetic Couplers for Wireless Power Transfer Applications. *IEEE Trans. Magn.* **2017**, *53*, 1–12. [CrossRef]
126. Boys, J.T.; Covic, G.A. Inductive Power Transfer Systems (IPT) Fact Sheet: No. 1-Basic Concepts. Available online: http://www.qualcomm.com/media/documents/ (accessed on 16 February 2019).
127. Hwang, K.; Cho, J.; Kim, D.; Park, J.; Kwon, J.H.; Kwak, S.I.; Park, H.H.; Ahn, S. An autonomous coil alignment system for the dynamic wireless charging of electric vehicles to minimize lateral misalignment. *Energies* **2017**, *10*, 315. [CrossRef]
128. Liu, N.; Habetler, T.G. Design of a Universal Inductive Charger for Multiple Electric Vehicle Models. *IEEE Trans. Power Electron.* **2015**, *30*, 6378–6390. [CrossRef]
129. Vaka, R.; Keshri, R.K. Review on Contactless Power Transfer for Electric Vehicle Charging. *Energies* **2017**, *10*, 636. [CrossRef]
130. Ni, W.; Collings, I.B.; Wang, X.; Liu, R.P.; Kajan, A.; Hedley, M.; Abolhasan, M. Radio alignment for inductive charging of electric vehicles. *IEEE Trans. Ind. Informatics* **2015**, *11*, 427–440. [CrossRef]
131. Ko, Y.D.; Jang, Y.J. The Optimal System Design of the Online Electric Vehicle Utilizing Wireless Power Transmission Technology. *Intell. Transp. Syst. IEEE Trans.* **2013**, *14*, 1255–1265. [CrossRef]
132. Choi, S.Y.; Huh, J.; Lee, W.Y.; Rim, C.T. Asymmetric coil sets for wireless stationary EV chargers with large lateral tolerance by dominant field analysis. *IEEE Trans. Power Electron.* **2014**, *29*, 6406–6420. [CrossRef]
133. Choi, S.Y.; Jeong, S.Y.; Gu, B.W.; Lim, G.C.; Rim, C.T. Ultraslim S-Type Power Supply Rails for Roadway-Powered Electric Vehicles. *IEEE Trans. Power Electron.* **2015**, *30*, 6456–6468. [CrossRef]
134. ORNL Surges Forward with 20-Kilowatt Wireless Charging for Vehicles | ORNL. 2016. Available online: https://www.ornl.gov/news/ornl-surges-forward-20-kilowatt-wirelesscharging-vehicles (accessed on 15 June 2019).
135. Wu, H.H.; Masquelier, M.P. An overview of a 50kW inductive charging system for electric buses. In Proceedings of the 2015 IEEE Transportation Electrification Conference and Expo, ITEC 2015, Dearborn, MI, USA, 14–17 June 2015.
136. Fisher, T.M.; Farley, K.B.; Gao, Y.; Bai, H.; Tse, Z.T.H. Electric vehicle wireless charging technology: A state-of-the-art review of magnetic coupling systems. *Wirel. Power Transf.* **2014**, *1*, 87–96. [CrossRef]
137. Bojarski, M.; Asa, E.; Colak, K.; Czarkowski, D. A 25 kW industrial prototype wireless electric vehicle charger. In Proceedings of the Conference Proceedings-IEEE Applied Power Electronics Conference and Exposition-APEC, Long Beach, CA, USA, 20–24 March 2016; Volume 2016, pp. 1756–1761.
138. Bojarski, M.; Asa, E.; Colak, K.; Czarkowski, D. Analysis and Control of Multiphase Chapter 2 59 Inductively Coupled Resonant Converter for Wireless Electric Vehicle Charger Applications. *IEEE Trans. Transp. Electrif.* **2017**, *3*, 312–320. [CrossRef]
139. Bosshard, R. *Multi-Objective Optimization of Inductive Power Transfer Systems for EV Charging*; ETH Zurich: Zürich, Switzerland, 2015.
140. "Charging Electric Buses Quickly and Efficiently: Bus Stops Fitted with Modular Components Make "Charge & Go" Simple to Implement. 2013. Available online: http://www.conductix.us/en/news/2013-05-29/charging-electric-buses-quickly-andefficiently-bus-stops-fitted-modular-components-make-charge-go (accessed on 15 June 2019).
141. Sato, F.; Morita, J.; Takura, T.; Sato, T.; Matsuki, H. Research on Highly Efficient Contactless Power Station System using Meander Coil for Moving Electric Vehicle Model. *J. Magn. Soc. Jpn.* **2012**, *36*, 249–252. [CrossRef]
142. Chigira, M.; Nagatsuka, Y.; Kaneko, Y.; Abe, S.; Yasuda, T.; Suzuki, A. Small-size lightweight transformer with new core structure for contactless electric vehicle power transfer system. In Proceedings of the IEEE Energy Conversion Congress and Exposition: Energy Conversion Innovation for a Clean Energy Future, ECCE 2011, Proceedings, Phoenix, AZ, USA, 17–22 September 2011; pp. 260–266.
143. Yang, Y.; Cui, J.; Cui, X. Design and Analysis of Magnetic Coils for Optimizing the Coupling Coefficient in an Electric Vehicle Wireless Power Transfer System. *Energies* **2020**, *13*, 4143. [CrossRef]

144. Horiuchi, T.; Kawashima, K. Study on Planar Antennas for Wireless Power Transmission of Electric Vehicles. *IEEJ. Trans. Ind. Appl.* **2010**, *130*, 1371–1377. [CrossRef]

145. Laccone, F.; Malomo, L.; Pérez, J.; Pietroni, N.; Ponchio, F.; Bickel, B.; Cignoni, P. A bending-active twisted-arch plywood structure: Computational design and fabrication of the FlexMaps Pavilion. *SN Appl. Sci.* **2020**, *2*, 1505. [CrossRef]

146. Developments in Wireless Power Transfer Standards and Regulations | IEEE Standards University. Available online: http://www.standardsuniversity.org/e-magazine/june2016/selected-developments-wireless-power-transfer-standards-regulations/ (accessed on 13 June 2019).

147. Alam, M.M.; Mekhilef, S.; Bassi, H.; Rawa, M.J.H. Analysis of LC-LC2 Compensated Inductive Power Transfer for High Efficiency and Load Independent Voltage Gain. *Energies* **2018**, *11*, 2883. [CrossRef]

148. Cho, J.-H.; Lee, B.-H.; Kim, Y.-J. Maximizing Transfer Efficiency with an Adaptive Wireless Power Transfer System for Variable Load Applications. *Energies* **2021**, *14*, 1417. [CrossRef]

149. Elliott, G.A.J.; Covic, G.A.; Kacprzak, D.; Boys, J.T. A new concept: Asymmetrical pick-ups for inductively coupled power transfer monorail systems. *IEEE Trans. Magn.* **2006**, *42*, 3389–3391. [CrossRef]

150. Keeling, N.A.; Covic, G.A.; Boys, J.T. A unity-power-factor IPT pickup for highpower appl5ications. *IEEE Trans. Ind. Electron.* **2010**, *57*, 744–751. [CrossRef]

151. Green, A.W. 10 kHz inductively coupled power transfer-concept and control. In Proceedings of the 5th International Conference on Power Electronics and Variable-Speed Drives, London, UK, 26–28 October 1994; pp. 694–699.

152. Okasili, I.; Elkhateb, A.; Littler, T. A Review of Wireless Power Transfer Systems for Electric Vehicle Battery Charging with a Focus on Inductive Coupling. *Electronics* **2022**, *11*, 1355. [CrossRef]

153. Zhang, W.; Member, S.; Mi, C.C. Compensation Topologies of High-Power Wireless Chapter 2 60 Power Transfer Systems. *IEEE Trans. Veh. Technol.* **2016**, *65*, 4768–4778. [CrossRef]

154. Tho, H.N.; Zhang, C.; Zhang, J.; Lee, S.B.; Jang, I.G. Layout Optimization of the Receiver Coils for Transfer Systems. *IEEE J. Emerg. Sel. Top. POWER Electron.* **2017**, *5*, 1311–1321.

155. Throngnumchai, K.; Kai, T.; Minagawa, Y. A study on receiver circuit topology of a cordless battery charger for electric vehicles. In Proceedings of the IEEE Energy Conversion Congress and Exposition: Energy Conversion Innovation for a Clean Energy Future, ECCE 2011, Proceedings, Phoenix, AZ, USA, 17–22 September 2011; pp. 843–850.

156. IEC 61980-1:2015 | IEC Webstore. Available online: https://webstore.iec.ch/publication/22951 (accessed on 15 June 2019).

157. IEC 61980-1:2015/COR1:2017 | IEC Webstore. Available online: https://webstore.iec.ch/publication/59640 (accessed on 10 June 2019).

158. IEC 61980-1-Electric Vehicle Wireless Power Transfer (WPT) Systems-Part 1: General Requirements | Engineering360. Available online: http://standards.globalspec.com/std/10072168/iec-61980-1 (accessed on 15 February 2019).

159. Annarelli, A.; Nonino, F.; Palombi, P. Understanding the management of cyber resilient systems. *Comput. Ind. Eng.* **2020**, *149*, 106829. [CrossRef]

160. Wireless Power Transfer for Light-Duty Plug-in/Electric Vehicles and Alignment Methodology. Available online: https://www.sae.org/standards/content/j2954_202010/ (accessed on 17 June 2019).

161. J1773A: SAE Electric Vehicle Inductively Coupled Charging-SAE International. Available online: http://standards.sae.org/j1773_201406/ (accessed on 17 June 2019).

162. J2847/6: Communication between Wireless Charged Vehicles and Wireless EV Chargers-SAE International. Available online: http://standards.sae.org/j2847/6_201508/ (accessed on 11 June 2019).

163. J2931/6: Signaling Communication for Wirelessly Charged Electric Vehicles-SAE International. Available online: http://standards.sae.org/j2931/6_201508/ (accessed on 15 February 2019).

164. Charging Stations | Industries | UL. Available online: http://industries.ul.com/energy/e-mobility/charging-stations (accessed on 17 February 2019).

165. Jeong, S.; Jang, Y.J.; Kum, D. Economic Analysis of the Dynamic Charging Electric Vehicle. *IEEE Trans. Power Electron.* **2015**, *30*, 6368–6377. [CrossRef]

166. Giler, E. WiTricity. Available online: https://www.ted.com/talks/eric_giler_demos_wireless_electricity (accessed on 10 February 2019).

167. ICNIRP. Available online: http://www.icnirp.org/ (accessed on 17 June 2019).

168. Mazzeo, D.; Matera, N.; Oliveti, G. Interaction Between a Wind-PV-Battery-Heat Pump Trigeneration System and Office Building Electric Energy Demand Including Vehicle Charging. In Proceedings of the 2018 IEEE International Conference on Environment and Electrical Engineering and 2018 IEEE Industrial and Commercial Power Systems Europe (EEEIC/I&CPS Europe), Palermo, Italy, 12–15 June 2018; pp. 1–5. [CrossRef]

169. Nguyen, H.N.T.; Zhang, C.; Mahmud, A. Optimal Coordination of G2V and V2G to Support Power Grids With High Penetration of Renewable Energy. *IEEE Trans. Transp. Electrif.* **2015**, *1*, 188–195. [CrossRef]

170. Tho, H.N.; Zhang, C.; Zhang, J. Dynamic Demand Control of Electric Vehicles to Support Power Grid with High Penetration Level of Renewable Energy. *IEEE Trans. Transp. Electrif.* **2016**, *2*, 66–75.

171. Wei, W.; Liu, F.; Mei, S.; Hou, Y. Robust energy and reserve dispatch under variable renewable generation. *IEEE Trans. Smart Grid* **2015**, *6*, 369–380. [CrossRef]

172. Jain, P.; Jain, T. Impacts of G2V and V2G power on electricity demand profile. In Proceedings of the 2014 IEEE International Electric Vehicle Conference, IEVC 2014, Florence, Italy, 17–19 December 2014.

173. Huang, X.; Qiang, H.; Huang, Z.; Sun, Y.; Li, J. The interaction research of smart grid and EV based wireless charging. In Proceedings of the 2013 IEEE Vehicle Power and Propulsion Conference (VPPC), Beijing, China, 15–18 October 2013; pp. 354–358.
174. Brooks, A.N. Final Report Grid Regulation Ancillary Service. *Regulation* **2002**, *1*, 61.
175. Arif, S.M.; Lie, T.T.; Seet, B.C.; Ayyadi, S.; Jensen, K. Review of Electric Vehicle Technologies, Charging Methods, Standards and Optimization Techniques. *Electronics* **2021**, *10*, 1910. [CrossRef]
176. Song, K.; Lan, Y.; Zhang, X.; Jiang, J.; Sun, C.; Yang, G.; Yang, F.; Lan, H. A Review on Interoperability of Wireless Charging Systems for Electric Vehicles. *Energies* **2023**, *16*, 1653. [CrossRef]
177. Jeong, S.Y.; Kwak, H.G.; Jang, G.C.; Choi, S.Y.; Rim, C.T. Dual-Purpose Nonoverlapping Coil Sets as Metal Object and Vehicle Position Detections for Wireless Stationary EV Chargers. *IEEE Trans. Power Electron.* **2018**, *33*, 7387–7397. [CrossRef]
178. Kuyvenhoven, N.; Dean, C.; Melton, J.; Schwannecke, J.; Umenei, A.E. Development of a foreign object detection and analysis method for wireless power systems. In Proceedings of the ISPCE 2011–2011 IEEE Symposium on Product Compliance Engineering, Proceedings, San Diego, CA, USA, 10–12 October 2011.
179. Fukuda, S.; Nakano, H.; Murayama, Y.; Murakami, T.; Kozakai, O.; Fujimaki, K. A novel metal detector using the quality factor of the secondary coil for wireless power transfer systems. In Proceedings of the 2012 IEEE MTT-S International Microwave Workshop Series on Innovative Wireless Power Transmission: Technologies, Systems, and Applications, IMWSIWPT 2012-Proceedings, Kyoto, Japan, 10–11 May 2012.
180. Kato, T.; Ninomiya, Y.; Masaki, I. An Obstacle Detection Method by Fusion of Radar and Motion Stereo. *IEEE Trans. Intell. Transp. Syst.* **2002**, *3*, 182–188. [CrossRef]
181. Xu, Q.; Ning, H.; Chen, W. Video-based foreign object debris detection. In Proceedings of the 2009 IEEE International Workshop on Imaging Systems and Techniques, IST 2009-Proceedings, Shenzhen, China, 11–12 May 2009.
182. Futatsumori, S.; Morioka, K.; Kohmura, A.; Yonemoto, N. Design and measurement of W-band offset stepped parabolic reflector antennas for airport surface foreign object debris detection radar systems. In Proceedings of the 2014 International Workshop on Antenna Technology: Small Antennas, Novel EM Structures and Materials, and Applications, iWAT 2014, Sydney, Australia, 4–6 March 2014.
183. Laccone, F.; Malomo, L.; Pérez, J.; Pietroni, N.; Ponchio, F.; Bickel, B.; Cignoni, P. FlexMaps Pavilion: A twisted arc made of mesostructured flat flexible panels. In Proceedings of the IASS Symposium 2019—60th Anniversary Symposium of the International Association for Shell and Spatial Structures, 2019; Structural Membranes 2019—9th International Conference on Textile Composites and Inflatable Structures, FORM and FORCE, Barcelona, Spain, 7–10 October 2019; pp. 509–515.
184. Ahmad, A.; Alam, M.S.; Varshney, Y.; Khan, R.H. A state of the Art review on Wireless Power Transfer a step towards sustainable mobility. In Proceedings of the 2017 14th IEEE India Council International Conference (INDICON), Roorkee, India, 15–17 December 2017; pp. 1–6.
185. M. S. P Wireless charging 2020–interoperable and standardized Technological and standardization challenges Status and goals of standardization International projects and timeline Activities and status of project STILLE Project goals, structure and activities. 2017; pp. 1–13.
186. Xue, Z.; Candemir, S.; Antani, S.; Long, L.R.; Jaeger, S.; Demner-Fushman, D.; Thoma, G.R. Foreign object detection in chest X-rays. In Proceedings of the 2015 IEEE International Conference on Bioinformatics and Biomedicine, BIBM 2015, Washington, DC, USA, 9–12 November 2015.
187. Jang, G.C.; Jeong, S.Y.; Kwak, H.G.; Rim, C.T. Metal object detection circuit with non-overlapped coils for wireless EV chargers. In Proceedings of the 2016 IEEE 2nd Annual Southern Power Electronics Conference, SPEC 2016, Auckland, New Zealand, 5–8 December 2016.
188. ICNIRP. Guidelines for limiting exposure to time-varying electric and magnetic fields (1 Hz to 100 kHz). *Int. Comm. NON-IONIZING Radiat. Prot.-Health Phys.* **2010**.
189. *IEEE Std C95.1-2005 (Revision IEEE Std C95.1-1991)*; IEEE Standard for Safety Levels With Respect to Human Exposure to Radio Frequency Electromagnetic Fields, 3 kHz to 300 GHz. IEEE: Piscataway, NJ, USA, 2005.
190. Ziegelberger, G. ICNIRP statement on the, guidelines for limiting exposure to timevarying electric, magnetic, and electromagnetic fields (UP to 300 GHz). *Health Phys.* **2009**, *97*, 257–258.
191. Poguntke, T.; Schumann, P.; Ochs, K. Radar-based living object protection for inductive charging of electric vehicles using two-dimensional signal processing. *Wirel. Power Transf.* **2017**, *4*, 88–97. [CrossRef]
192. Wang, Y.; Chiang, C. Foreign Metal Detection by Coil Impedance for EV Wireless Charging System. In Proceedings of the 28th International Electric Vehicle Symposium and Exhibition 2015, Goyang, Republic of Korea, 3–6 May 2015; pp. 1–4.
193. Shahjalal, M.; Shams, T.; Tasnim, M.N.; Ahmed, M.R.; Ahsan, M.; Haider, J. A Critical Review on Charging Technologies of Electric Vehicles. *Energies* **2022**, *15*, 8239. [CrossRef]
194. Amjad, M.; Farooq-i-Azam, M.; Ni, Q.; Dong, M.; Ansari, E.A. Wireless charging systems for electric vehicles. *Renew. Sustain. Energy Rev.* **2022**, *167*, 112730. [CrossRef]
195. Savari, G.F.; Sathik, M.J.; Raman, L.A.; El-Shahat, A.; Hasanien, H.M.; Almakhles, D.; Aleem, S.H.A.; Omar, A.I. Assessment of charging technologies, infrastructure and charging station recommendation schemes of electric vehicles: A review. *Ain Shams Eng. J.* **2023**, *14*, 101938. [CrossRef]
196. Alam, M.S.; Ahmad, A.; Khan, Z.A.; Rafat, Y.; Chabaan, R.C.; Khan, I.; Al-Shariff, S.M. A Bibliographical Review of Electrical Vehicles (xEVs) Standards. *SAE Int. J. Altern. Powertrains* **2018**, *7*, 63–98. [CrossRef]

197. Ahmad, A.; Alam, M.S. Magnetic Analysis of Copper Coil Power Pad with Ferrite Core for Wireless Charging Application. *Trans. Electr. Electron. Matererials* **2019**, *20*, 165–173. [CrossRef]
198. ElGhanam, E.; Hassan, M.; Osman, A.; Kabalan, H. Design and Performance Analysis of Misalignment Tolerant Charging Coils for Wireless Electric Vehicle Charging Systems. *World Electr. Veh. J.* **2021**, *12*, 89. [CrossRef]
199. Mwasilu, F.; Justo, J.J.; Kim, E.K.; Do, T.D.; Jung, J.W. Electric vehicles and smart grid interaction: A review on vehicle to grid and renewable energy sources integration. *Renew. Sustain. Energy Rev.* **2014**, *34*, 501–516. [CrossRef]
200. Khan, W.; Ahmad, A.; Ahmad, F.; SaadAlam, M. A Comprehensive Review of Fast Charging Infrastructure for Electric Vehicles. *Smart Sci.* **2018**, *6*, 256–270. [CrossRef]
201. Debbou, M.; Colet, F. Inductive wireless power transfer for electric vehicle dynamic charging. In Proceedings of the IEEE PELS Workshop on Emerging Technologies: Wireless Power, WoW 2016, Knoxville, TN, USA, 4–6 October 2016; pp. 118–122.
202. Kim, K.R.; Kim, D.H.; Kim, H.J. Magnetic resonance wireless power transmission using a LLC resonant circuit for a locomotion robot's battery charging. In Proceedings of the Intelligent Robotics and Applications: 6th International Conference, ICIRA 2013, Busan, South Korea, 25–28 September 2013.
203. Uddin, M.K.; Ramasamy, G.; Mekhilef, S.; Ramar, K.; Lau, Y.C. A review on highfrequency resonant inverter technologies for wireless power transfer using magnetic resonance coupling. In Proceedings of the 2014 IEEE Conference Energy Conversion, CENCON 2014, Johor Bahru, Malaysia, 13–14 October 2014; pp. 412–417.
204. Mou, X.; Gladwin, D.T.; Zhao, R.; Sun, H. Survey on magnetic resonant coupling wireless power transfer technology for electric vehicle charging. *IET Power Electron.* **2019**, *12*, 3005–3020. [CrossRef]
205. Mousa, A.G.E.; Abdel Aleem, S.H.E.; Ibrahim, A.M. Mathematical Analysis of Maximum Power Points and Currents Based Maximum Power Point Tracking in Solar Photovoltaic System: A Solar Powered Water Pump Application. *Int. Rev. Electr. Eng.* **2016**, *11*, 97. [CrossRef]
206. Asna, M.; Shareef, H.; Achikkulath, P.; Mokhlis, H.; Errouissi, R.; Wahyudie, A. Analysis of an Optimal Planning Model for Electric Vehicle Fast-Charging Stations in Al Ain City, United Arab Emirates. *IEEE Access* **2021**, *9*, 73678–73694. [CrossRef]
207. Mohamed, N.; Aymen, F.; Issam, Z.; Bajaj, M.; Ghoneim, S.S.M. The Impact of Coil Position and Number on Wireless System Performance for Electric Vehicle Recharging. *Sensors* **2021**, *21*, 4343. [CrossRef]
208. Ma, G.; Kamaruddin, M.H.; Kang, H.S.; Goh, P.S.; Kim, M.H.; Lee, K.Q.; Ng, C.Y. Watertight integrity of underwater robotic vehicles by self-healing mechanism. *Ain. Shams Eng. J.* **2021**, *12*, 1995–2007. [CrossRef]
209. Younes, Z.; Alhamrouni, I.; Mekhilef, S.; Reyasudin, M. A memory-based gravitational search algorithm for solving economic dispatch problem in micro-grid. *Ain. Shams Eng. J.* **2021**, *12*, 1985–1994. [CrossRef]
210. Hussien, A.M.; Hasanien, H.M.; Mekhamer, S.F. Sunflower optimization algorithmbased optimal PI control for enhancing the performance of an autonomous operation of a microgrid. *Ain. Shams Eng. J.* **2021**, *12*, 1883–1893. [CrossRef]
211. Sobhy, M.A.; Abdelaziz, A.Y.; Hasanien, H.M.; Ezzat, M. Marine predators algorithm for load frequency control of modern interconnected power systems including renewable energy sources and energy storage units. *Ain. Shams Eng. J.* **2021**, *12*, 3843–3857. [CrossRef]
212. Savari, G.F.; Krishnasamy, V.; Sathik, J.; Ali, Z.M.; Abdel Aleem, S.H.E. Internet of Things based real-time electric vehicle load forecasting and charging station recommendation. *ISA Trans.* **2020**, *97*, 431–447. [CrossRef]
213. Mostafa, M.H.; Aleem, S.H.E.A.; Ali, S.G.; Abdelaziz, A.Y.; Ribeiro, P.F.; Ali, Z.M. Robust energy management and economic analysis of microgrids considering different battery characteristics. *IEEE Access* **2020**, *8*, 54751–54775. [CrossRef]
214. Kawasan, M.; Zobaa, A.F.; Hasanien, H.M.; Aleem, S.H.A.; Ali, Z.M. Towards accurate calculation of supercapacitor electrical variables in constant power applications using new analytical closed-form expressions. *J. Energy Storage* **2021**, *42*, 102998.
215. Rawa, M.; Abusorrah, A.; Bassi, H.; Mekhilef, S.; Ali, Z.M.; Aleem, S.H.A.; Hasanien, H.M.; Omar, A.I. Economical-technical-environmental operation of power networks with windsolar-hydropower generation using analytic hierarchy process and improved grey wolf algorithm. *Ain. Shams Eng. J.* **2021**, *12*, 2717–2734. [CrossRef]
216. Erdogan, A.; Kizilkan, O.; Colpan, C.O. Thermodynamic performance assessment of solar based closed brayton cycle for different supercritical fluids. In Proceedings of the 2019 4th International Conference on Smart and Sustainable Technologies (SpliTech), Split, Croatia, 18–21 June 2019; pp. 1–4. [CrossRef]
217. Jahangir, H.; Tayarani, H.; Ahmadian, A.; Golkar, M.A.; Miret, J.; Tayarani, M.; Gao, H.O. Charging demand of Plug-in Electric Vehicles: Forecasting travel behavior based on a novel Rough Artificial Neural Network approach. *J. Clean Prod.* **2019**, *229*, 1029–1044. [CrossRef]
218. Majhi, R.C.; Ranjitkar, P.; Sheng, M. Assessment of dynamic wireless charging based electric road system: A case study of Auckland motorway. *Sustain. Cities Soc.* **2022**, *84*, 104039. [CrossRef]
219. Sudimac, B.; Ugrinović, A.; Jurčević, M. The application of photovoltaic systems in sacred buildings for the purpose of electric power production: The case study of the Cathedral of St. Michael Archangel Belgrade. *Sustainability* **2020**, *12*, 1408. [CrossRef]
220. Ji, B.; Song, X.; Cao, W.; Pickert, V.; Hu, Y.; Mackersie, J.W.; Pierce, G. In situ diagnostics and prognostics of solder fatigue in IGBT modules for electric vehicle drives. *IEEE Trans. Power Electron.* **2015**, *30*, 1535–1543. [CrossRef]
221. Triviño-Cabrera, A.; Ochoa, M.; Fernández, D.; Aguado, J.A. Independent primaryside controller applied to wireless chargers for electric vehicles. In Proceedings of the 2014 IEEE International Electric Vehicle Conference (IEVC), Florence, Italy, 17–19 December 2014. [CrossRef]

222. Castilla, M.; Miret, J.; Matas, J.; de Vicuña, L.G.; Guerrero, J.M. Control design guidelines for single-phase grid-connected photovoltaic inverters with damped resonant harmonic compensators. *IEEE Trans. Ind. Electron.* **2009**, *56*, 4492–4501. [CrossRef]
223. Gandoman, F.H.; Van Mierlo, J.; Ahmadi, A.; Abdel Aleem, S.H.E.; Chauhan, K. Safety and reliability evaluation for electric vehicles in modern power system networks. Distrib. Energy Resour. *Microgrids* **2019**, 389–404. [CrossRef]
224. Sharaf, A.M.; Omar, N.; Gandoman, F.H.; Zobaa, A.F.; Abdel Aleem, S.H.E. Electric and Hybrid Vehicle Drives and Smart Grid Interfacing. *Adv. Renew. Energies Power Technol.* **2018**, *2*, 413–439. [CrossRef]
225. Sarkar, J.; Bhattacharyya, S. Application of graphene and graphene-based materials in clean energy-related devices Minghui. *Arch. Thermodyn* **2012**, *33*, 23–40. [CrossRef]
226. Naoui, M.; Flah, A.; Ben hamed, M. Inductive charger efficiency under internal and external parameters variation for an electric vehicle in motion. *Int. J. Powertrains* **2019**, *8*, 343–358. [CrossRef]
227. Guerrero, J.M.; de Vicuna, L.G.; Matas, J.; Castilla, M.; Miret, J. A wireless controller to enhance dynamic performance of parallel inverters in distributed generation systems. *IEEE Trans. Power Electron.* **2004**, *19*, 1205–1213. [CrossRef]
228. Rosu, S.G. A Dynamic Wireless Charging System for Electric Vehicles Based on DC/AC Converters with SiC MOSFET-IGBT Switches and Resonant Gate-Drive. In Proceedings of the IECON 2016-42nd Annual Conference of the IEEE Industrial Electronics Society, Florence, Italy, 23–26 October 2016; pp. 4465–4470.
229. Jang, Y.J.; Ko, Y.D.; Jeong, S. Optimal design of the wireless charging electric vehicle. In Proceedings of the 2012 IEEE International Electric Vehicle Conference, Greenville, SC, USA, 4–8 March 2012; pp. 1–5. [CrossRef]
230. Bellocchi, S.; Colbertaldo, P.; Manno, M.; Nastasi, B. Assessing the effectiveness of hydrogen pathways: A techno-economic optimisation within an integrated energy system. *Energy* **2023**, *263 Pt E*, 126017. [CrossRef]
231. Colak, K.; Asa, E.; Bojarski, M.; Czarkowski, D.; Onar, O.C. A Novel Phase-Shift Control of Semibridgeless Active Rectifier for Wireless Power Transfer. *IEEE Trans. Power Electron.* **2015**, *30*, 6288–6297. [CrossRef]
232. Shin, Y.; Park, J.; Kim, H.; Woo, S.; Park, B.; Huh, S.; Lee, C.; Ahn, S. Design Considerations for Adding Series Inductors to Reduce Electromagnetic Field Interference in an Over-Coupled WPT System. *Energies* **2021**, *14*, 2791. [CrossRef]
233. Narayanamoorthi, R. Modeling of Capacitive Resonant Wireless Power and Data Transfer to Deep Biomedical Implants. *IEEE Trans. Compon. Packag. Manuf. Technol.* **2019**, *9*, 1253–1263. [CrossRef]
234. Musavi, F.; Eberle, W. Overview of wireless power transfer technologies for electric vehicle battery charging. *IET Power Electron.* **2014**, *7*, 60–66. [CrossRef]
235. Haque, M.S.; Mohammad, M.; Pries, J.L.; Choi, S. Comparison of 22 kHz and 85 kHz 50 kW Wireless Charging System Using Si and SiC Switches for Electric Vehicle. In Proceedings of the 2018 IEEE 6th Workshop on Wide Bandgap Power Devices and Applications (WiPDA) 2018, Atlanta, GA, USA, 31 October 2018–2 November 2018; Volume 2018, pp. 192–198. [CrossRef]
236. Mohamed, N.; Aymen, F.; Ben Hamed, M.; Lassaad, S. Analysis of battery-EV state of charge for a dynamic wireless charging system. *Energy Storage* **2019**, *5*, e117. [CrossRef]
237. Joseph, P.K.; Devaraj, E.; Gopal, A. Overview of wireless charging and vehicletogrid integration of electric vehicles using renewable energy for sustainable transportation. *IET Power Electron.* **2019**, *12*, 627–638. [CrossRef]
238. Cristofari, A. Active-set identification with complexity guarantees of an almost cyclic 2-coordinate descent method with armijo line search. *SIAM J. Optim.* **2022**, *32*, 739–764. [CrossRef]
239. Corio, E.; Laccone, F.; Pietroni, N.; Cignoni, P. Conception and parametric design workflow for a timber large-spanned reversible grid shell to shelter the archaeological site of the Roman shipwrecks in Pisa. *Int. J. Comp. Meth. Exp. Meas.* **2017**, *5*, 551–561. [CrossRef]
240. Moosavi, S.A.; Mortazavi, S.S.; Namadmalan, A.; Iqbal, A.; Al-Hitmi, M. Design and Sensitivity Analysis of Dynamic Wireless Chargers for Efficient Energy Transfer. *IEEE Access* **2021**, *9*, 16286–16295. [CrossRef]
241. Yilmaz, M.; Krein, P.T. Review of battery charger topologies, charging power levels, and infrastructure for plug-in electric and hybrid vehicles. *IEEE Trans. Power. Electron.* **2013**, *28*, 2151–2169. [CrossRef]
242. Ghate, K.; Dole, L. A review on magnetic resonance based wireless power transfer system for electric vehicles. In Proceedings of the 2015 International Conference on Pervasive Computing (ICPC), Pune, India, 8–10 January 2015; pp. 1–3. [CrossRef]
243. Kalwar, K.A.; Aamir, M.; Mekhilef, S. Inductively coupled power transfer (ICPT) for electric vehicle charging-A review. *Renew. Sustain. Energy Rev.* **2015**, *47*, 462–475. [CrossRef]
244. Chopra, S.; Bauer, P. Analysis and design considerations for a contactless power transfer system. *INTELEC. Int. Telecommun. Energy Conf.* **2011**, 1–6. [CrossRef]
245. García, X.D.T.; Vázquez, J.; Roncero-Sánchez, P. Design, implementation issues and performance of an inductive power transfer system for electric vehicle chargers with series-series compensation. *IET Power Electron.* **2015**, *8*, 1920–1930. [CrossRef]
246. Zhao, J.; Cai, T.; Duan, S.; Feng, H.; Chen, C.; Zhang, X. A General Design Method of Primary Compensation Network for Dynamic WPT System Maintaining Stable Transmission Power. *IEEE Trans. Power Electron.* **2016**, *31*, 8343–8358. [CrossRef]
247. Mollaei, M.S.M.; Jayathurathnage, P.; Tretyakov, S.A.; Simovski, C.R. High-Impedance Wireless Power Transfer Transmitter Coils for Freely Positioning Receivers. *IEEE Access* **2021**, *9*, 42994–43000. [CrossRef]
248. Shi, X.; Qi, C.; Qu, M.; Ye, S.; Wang, G.; Sun, L.; Yu, Z. Effects of coil shapes on wireless power transfer via magnetic resonance coupling. *J. Electromagn. Waves Appl.* **2014**, *28*, 1316–1324. [CrossRef]
249. Ahmad, A.; Alam, M.S.; Mohamed, A.A.S. Design and Interoperability Analysis of Quadruple Pad Structure for Electric Vehicle Wireless Charging Application. *IEEE Trans. Transp. Electrif.* **2019**, *5*, 934–945. [CrossRef]

250. Budhia, M.; Covic, G.A.; Boys, J.T. Design and optimization of circular magnetic structures for lumped inductive power transfer systems. *IEEE Trans. Power Electron.* **2011**, *26*, 3096–3108. [CrossRef]
251. Member, S.; Covic, G.A.; Boys, J.T. Design and Optimisation of Magnetic Structures for Lumped Inductive Power Transfer Systems. In Proceedings of the 2009 IEEE Energy Conversion Congress and Exposition 2009, San Jose, CA, USA, 20–24 September 2009; pp. 2081–2088.
252. Flah, A.; Khan, I.A.; Agarwal, A.; Sbita, L.; Simoes, M.G. Field-oriented control strategy for double-stator single-rotor and double-rotor single-stator permanent magnet machine: Design and operation. *Comput. Electr. Eng.* **2021**, *90*, 1–15. [CrossRef]
253. Naoui, M.; Flah, A.; Ben Hamed, M.; Sbita, L. Review on autonomous charger for EV and HEV. In Proceedings of the 2017 International Conference on Green Energy Conversion Systems (GECS), Hammamet, Tunisia, 23–25 March 2017; pp. 1–6. [CrossRef]
254. Aymen, F.; Mahmoudi, C. A Novel Energy Optimization Approach for Electrical Vehicles in a Smart City. *Energies* **2019**, *12*, 929. [CrossRef]
255. Rawat, T.; Niazi, K.R.; Gupta, N.; Sharma, S. Impact assessment of electric vehicle charging/discharging strategies on the operation management of grid N. Mohamed, F. Aymen, M. Alqarni et al. Ain Shams Engineering Journal 13 (2022) 101569 14 accessible and remote microgrids. *Int. J. Energy Res.* **2019**, *43*, 9034–9048. [CrossRef]
256. Hwang, J.J.; Kuo, J.K.; Wu, W.; Chang, W.R.; Lin, C.H.; Wang, S.E. Lifecycle performance assessment of fuel cell/battery electric vehicles. *Int. J. Hydrog. Energy* **2013**, *38*, 3433–3446. [CrossRef]
257. Lee, J.-Y.; Han, B.-M. A Bidirectional Wireless Power Transfer EV Charger Using Self-Resonant PWM. *IEEE Trans. Power Electron.* **2015**, *30*, 1784–1787. [CrossRef]
258. Shanmugam, Y.; Sathik, J.; Almakhles, D.J. A Comprehensive Review of the On-Road Wireless Charging System for E-Mobility Applications. *Front. Energy Res.* **2022**, *10*, 926270. [CrossRef]
259. Xie, K.; Xu, J.; Pan, Z. Research and application of anti-offset wireless charging plant protection UAV. *Electr. Eng.* **2020**, *102*, 2529–2537. [CrossRef]
260. Raju, S.; Wu, R.; Chan, M.; Yue, C.P. Modeling of mutual coupling between planar inductors in wireless power applications. *IEEE Trans. Power Electron.* **2014**, *29*, 481–490. [CrossRef]
261. Mohamed, N.; Aymen, F.; Lassaad, S.; Mouna, B.H. Practical validation of the vehicle speed influence on the wireless recharge system efficiency. *Int. Energy Conf.* **2020**, 372–376. [CrossRef]
262. Wang, H.; Cheng, K.W.E. An improved and integrated design of segmented dynamic wireless power transfer for electric vehicles. *Energies* **2021**, *14*, 1975. [CrossRef]
263. Lu, F.; Member, S.; Zhang, H.; Member, S. A Dynamic Charging System With Reduced Output Power Pulsation for Electric Vehicles. *IEEE Trans. Ind. Electron.* **2016**, *63*, 6580–6590. [CrossRef]
264. Zhou, S.; Chris Mi, C. Multi-Paralleled LCC Reactive Power Compensation Networks and Their Tuning Method for Electric Vehicle Dynamic Wireless Charging. *IEEE Trans. Ind. Electron.* **2016**, *63*, 6546–6556. [CrossRef]
265. Alphones, A.; Jayathurathnage, P. Review on wireless power transfer technology (invited paper). In Proceedings of the IEEE Asia Pacific Microwave Conference (APMC), Kuala Lumpur, Malaysia, 13–16 November 2017; pp. 326–329. [CrossRef]
266. Salau, A.O.; Marriwala, N.; Athaee, M. Data Security in Wireless Sensor Networks: Attacks and Countermeasures. In *Mobile Radio Communications and 5G Networks: Proceedings of MRCN 2020*; Springer: Singapore, 2020. [CrossRef]
267. Bi, Z.; Kan, T.; Mi, C.C.; Zhang, Y.; Zhao, Z.; Keoleian, G.A. A review of wireless power transfer for electric vehicles: Prospects to enhance sustainable mobility. *Appl. Energy* **2016**, *179*, 413–425. [CrossRef]
268. Ahmad, A.; Motors, R.C.H. *Comparative Analysis of Power Pad for Wireless Charging of Electric Vehicles*; SAE: Warrendale, PA, USA, 2019; pp. 1–7. [CrossRef]
269. Ahmad, A.; Alam, M.S.; Rafat, Y.; Shariff, S.M.; Al-Saidan, I.S.; Chabaan, R.C. Foreign Object Debris Detection and Automatic Elimination for Autonomous Electric Vehicles Wireless Charging Application. *SAE Int. J. Electrified Veh.* **2020**, *9*, 93–110. [CrossRef]
270. Shanmugam, Y.; Narayanamoorthi, R.; Vishnuram, P.; Bajaj, M.; Aboras, K.M.; Thakur, P. A Systematic Review of Dynamic Wireless Charging System for Electric Transportation. *IEEE Access* **2022**, *10*, 133617–133642. [CrossRef]
271. Musavi, F.; Edington, M.; Eberle, W. Wireless power transfer: A survey of EV battery charging technologies. In Proceedings of the 2012 IEEE Energy Conversion Congress and Exposition (ECCE), Raleigh, NC, USA, 15–20 September 2012; Volume 2012, pp 1804–1810. [CrossRef]
272. Vishnuram, P.; Alagarsamy, S.; Krishnasamy, V.; Bajaj, M.; Khurshaid, T.; Nauman, D.; Kamel, S. A Comprehensive Review on EV Power Converter Topologies Charger Types Infrastructure and Communication Techniques. *Front. Energy Res.* **2023**, *11*, 101. [CrossRef]
273. Ashok, J.; Thirumoorthy, P. Design considerations for implementing an optimal battery management system of a wireless sensor node. *Indian J. Sci. Technol.* **2014**, *7*, 1255–1259. [CrossRef]
274. Wang, G.; Sun, J. Improved Magnetic Coupling Resonance Wireless Power Transfer System. *Chin. Control Conf. CCC* **2020**, *2020*, 5317–5321. [CrossRef]
275. Adaramola, B.A.; Salau, A.O.; Adetunji, F.O.; Fadodun, O.G.; Ogundipe, A.T. Adetunji Development and Performance Analysis of a GPS-GSM Guided System for Vehicle Tracking. In Proceedings of the International Conference on Computation, Automation and Knowledge Management (ICCAKM), Dubai, United Arab Emirates, 9–10 January 2020; pp. 286–290. [CrossRef]

276. Khan, S.; Ahmad, A.; Ahmad, F.; Shafaati Shemami, M.; Saad Alam, M.; Khateeb, S. A Comprehensive Review on Solar Powered Electric Vehicle Charging SystemA Comprehensive Review on Solar Powered Electric Vehicle Charging System. *Smart Sci.* **2018**, *6*, 54–79. [CrossRef]

277. Ahmad, A.; Khan, Z.A.; Saad Alam, M.; Khateeb, S. A Review of the Electric Vehicle Charging Techniques, Standards, Progression and Evolution of EV Technologies in Germany. *Smart Sci.* **2018**, *6*, 36–53. [CrossRef]

278. Khan, S.; Shariff, S.; Ahmad, A.; Saad Alam, M. A comprehensive review on level 2 charging system for electric vehicles. *Smart Sci.* **2018**, *6*, 271–293. [CrossRef]

279. Apparatus for transmitting electrical energy. *IEEE Trans. Circuits Syst. I Regul. Pap.* **2013**, *1*, 1–4.

280. Huang, R.; Zhang, B.; Qiu, D.; Zhang, Y. Frequency splitting phenomena of magnetic resonant coupling wireless power transfer. *IEEE Trans. Magn.* **2014**, *50*, 1–4. [CrossRef]

281. Xu, H.; Wang, C.; Xia, D.; Liu, Y. Design of magnetic coupler for wireless power transfer. *Energies* **2019**, *15*, 3000. [CrossRef]

282. Pinto, R.; Lopresto, V.; Genovese, A. A numerical study for the design of a new DD coil prototype for dynamic wireless charging of electric vehicles. *IET Conf. Publ.* **2018**, *2018*, 2–6. [CrossRef]

283. Wang, Q.; Li, H. Research on the wireless power transmission system based on coupled magnetic resonances. In Proceedings of the 2011 International Conference on Electronics, Communications and Control (ICECC), Ningbo, China, 9–11 September 2011; pp. 2255–2258. [CrossRef]

284. Dashora, H.K.; Buja, G.; Bertoluzzo, M.; Pinto, R.; Lopresto, V. Analysis and design of DD coupler for dynamic wireless charging of electric vehicles. *J. Electromagn. Waves Appl.* **2018**, *32*, 170–189. [CrossRef]

285. Ichikawa, K.; Bondar, H. Power Transfer System. 2012, pp. 255–259. Available online: https://www.google.ch/patents/US20120 299392 (accessed on 27 December 2022).

286. Joy, E.R.; Kushwaha, B.K.; Rituraj, G.; Kumar, P. Analysis and comparison of four compensation topologies of contactless power transfer system. In Proceedings of the 2015 4th International Conference on Electric Power and Energy Conversion Systems (EPECS), Sharjah, United Arab Emirates, 24–26 November 2015. [CrossRef]

287. Campbell, S. *Oak Ridge National Laboratory Wireless Charging of Electric Vehicles–CRADA Report*; Oak Ridge National Lab.(ORNL): Oak Ridge, TN, USA, 2016.

288. Chen, Q.; Wong, S.C.; Tse, C.K.; Ruan, X. Analysis, design, and control of a transcutaneous power regulator for artificial hearts. *IEEE Trans. Biomed. Circuits Syst.* **2009**, *3*, 23–31. [CrossRef]

289. Mazzeo, D.; Matera, N.; De Luca, R.; Musmanno, R. A smart algorithm to optimally manage the charging strategy of the Home to Vehicle (H2V) and Vehicle to Home (V2H) technologies in an off-grid home powered by renewable sources. *Energy Syst.* **2022**, 1–38. [CrossRef]

290. Di Tommaso, A.O.; Genduso, F.; Miceli, R. A small power transmission prototype for electric vehicle wireless battery charge applications. In Proceedings of the 2012 International Conference on Renewable Energy Research and Applications (ICRERA), Nagasaki, Japan, 11–14 November 2012. [CrossRef]

291. Phaebua, K.; Lertwiriyaprapa, T.; Phongcharoenpanich, C. Study of a repeater Tx antenna concept of a portable device wireless battery charging system. In Proceedings of the The 20th Asia-Pacific Conference on Communication (APCC2014), Pattaya, Thailand, 1–3 October 2014; Volume 2015, pp. 442–445. [CrossRef]

292. Berger, A.; Agostinelli, M.; Vesti, S.; Oliver, J.A.; Cobos, J.A.; Huemer, M. A Wireless Charging System Applying Phase-Shift and Amplitude Control to Maximize Efficiency and Extractable Power. *IEEE Trans. Power Electron.* **2015**, *30*, 6338–6348. [CrossRef]

293. Cristofari, A.; De Santis, M.; Lucidi, S.; Rinaldi, F. Minimization over the ℓ_1 -ball using an active-set non-monotone projected gradient. *Comput. Optim. Appl.* **2022**, *83*, 693–721. [CrossRef]

294. Laccone, F.; Malomo, L.; Pietroni, N.; Cignoni, P.; Schork, T. Integrated computational framework for the design and fabrication of bending-active structures made from flat sheet material. *Structures* **2021**, *34*, 979–994, ISSN 2352-0124. [CrossRef]

295. Redder, D.A.G.; Brown, A.D.; Skinner, J.A. A contactless electrical energy transmission system. *IEEE Trans. Ind. Electron.* **1999**, *46*, 23–30. [CrossRef]

296. Lee, S.; Huh, J.; Park, C.; Choi, N.S.; Cho, G.H.; Rim, C.T. OnLine Electric Vehicle using inductive power transfer system. In Proceedings of the 2010 IEEE Energy Conversion Congress and Exposition, Atlanta, GA, USA, 12–16 September 2010; pp. 1598–1601. [CrossRef]

297. Rahulkumar, J.; Narayanamoorthi, R.; Vishnuram, P.; Bajaj, M.; Blazek, V.; Prokop, L.; Misak, S. An Empirical Survey on Wireless Inductive Power Pad and Resonant Magnetic Field Coupling for In-Motion EV Charging System. *IEEE Access* **2023**, *11*, 4660–4693. [CrossRef]

298. Nagatsuka, Y.; Ehara, N.; Kaneko, Y.; Abe, S.; Yasuda, T. Compact contactless power transfer system for electric vehicles. In Proceedings of the The 2010 International Power Electronics Conference-ECCE ASIA-IPEC, Sapporo, Japan, 21–24 June 2010; Volume 2010, pp. 807–813. [CrossRef]

299. Ullah, R.; Khan, S.; Khan, N.A.; Tahir, M.; Ahmad, N. Effect of replacement of soybean meal by silkworm meal on growth performance, apparent metabolizable energy and nutrient digestibility in broilers at day 28 post hatch. *J. Anim. Plant Sci.* **2018**, *28*, 1239–1246.

300. Wang, C.S.; Stielau, O.H.; Covic, G.A. Design considerations for a contactless electric vehicle battery charger. *IEEE Trans. Ind. Electron.* **2005**, *52*, 1308–1314. [CrossRef]

301. Covic, G.A.; Member, S.; Boys, J.T. Modern Trends in Inductive Power Transfer for Transportation Applications. *IEEE J. Emerg. Sel. Top. Power Electron.* **2013**, *1*, 28–41. [CrossRef]
302. Pinuela, M.; Yates, D.C.; Lucyszyn, S.; Mitcheson, P.D. Maximizing DC-to-load efficiency for inductive power transfer. *IEEE Trans. Power Electron.* **2013**, *28*, 2437–2447. [CrossRef]
303. Alam, B.; Nusrat, M.; Sarwer, Z.; Zaid, M.; Sarwar, A. A General Review of the Recently Proposed Asymmetrical Multilevel Inverter Topologies. In *Innovations in Cyber Physical Systems: Select Proceedings of ICICPS 2020*; Springer: Berlin, Germany, 2021; Zaid, M.
304. Zhang, W.; Wong, S.C.; Chi, K.T.; Chen, Q. Analysis and Comparison of Secondary Series- and Parallel-Compensated Inductive Power Transfer Systems Operating for Optimal Efficiency and LoadIndependent Voltage-Transfer Ratio. *IEEE Trans. Power Electron.* **2013**, *29*, 2979–2990. [CrossRef]
305. Pantic, Z.; Bai, S.; Lukic, S.M. ZCS LCC-compensated resonant inverter for inductive-power-transfer application. *IEEE Trans. Ind. Electron.* **2011**, *58*, 3500–3510. [CrossRef]
306. Fu, M.; Tang, Z.; Ma, C. Analysis and Optimized Design of Compensation Capacitors for a Megahertz WPT System Using Full-Bridge Rectifier. *IEEE Trans. Ind. Inform.* **2019**, *15*, 95–104. [CrossRef]
307. Vishnuram, P.; Nastasi, B. Wireless Chargers for Electric Vehicle: A Systematic Review on Converter Topologies, Environmental Assessment, and Review Policy. *Energies* **2023**, *16*, 1731. [CrossRef]
308. Jamal, N.; Saat, S.; Yusmarnita, Y.; Zaid, T.; Isa, M.S.M.; Isa, A.A.M. Investigations on capacitor compensation topologies effects of different inductive coupling links configurations. *Int. J. Power Electron. Drive Syst.* **2015**, *6*, 274–281. [CrossRef]
309. Shevchenko, V.; Husev, O.; Strzelecki, R.; Pakhaliuk, B.; Poliakov, N.; Strzelecka, N. Compensation topologies in IPT systems: Standards, requirements, classification, analysis, comparison and application. *IEEE Access* **2020**, *7*, 120559–120580. [CrossRef]
310. Li, J.; Kang, J.; Tian, C.; Tian, D.; Xie, T. Study on Wireless Power Transfer Technology with Series-Series Type of Magnetic Coupling Resonance Model. *DEStech Trans. Comput. Sci. Eng.* **2017**, 225–232. [CrossRef]
311. Zhao, Q.; Wang, A.; Wang, H. Structure analysis of magnetic coupling resonant for wireless power transmission system. *Futur. Energy Environ. Mater. II* **2015**, *1*, 63–70. [CrossRef]
312. Wang, C.S.; Covic, G.A.; Stielau, O.H. General stability criterions for zero phase angle controlled loosely coupled inductive power transfer systems. In Proceedings of the IECON'01. 27th Annual Conference of the IEEE Industrial Electronics Society (Cat. No.37243), Denver, CO, USA, 29 November 2001–2 December 2001; Volume 2, pp. 1049–1054. [CrossRef]
313. Bosshard, R.; Kolar, J.W.; Mühlethaler, J.; Stevanović, I.; Wunsch, B.; Canales, F. Modeling and η-α-pareto optimization of inductive power transfer coils for electric vehicles. *IEEE J. Emerg. Sel. Top. Power Electron.* **2015**, *3*, 50–64. [CrossRef]
314. Froli, M.; Laccone, F. Experimental static and dynamic tests on a large-scale free-form Voronoi grid shell mock-up in comparison with finite-element method results. *Int. J. Adv. Struct. Eng.* **2017**, *9*, 293–308. [CrossRef]
315. Mohamed, A.A.S.; Shaier, A.A.; Metwally, H.; Selem, S.I. Interoperability of the Universal WPT3 Transmitter with Different Receivers for Electric Vehicle Inductive Charger Interoperability of the universal WPT3 transmitter with different receivers for electric vehicle inductive charger. *eTransportation* **2021**, *6*, 100084. [CrossRef]
316. Zhang, Z. Energy Systems for Electric and Hybrid Vehicles. In *Energy Cryptography for Wireless Charging of Electric Vehicles*; The Institution of Engineering and Technology: London, UK, 2016; pp. 319–417.
317. Ahmad, A.; Alam, M.S.; Chabaan, R. A Comprehensive Review of Wireless Charging Technologies for Electric Vehicles. *IEEE Trans. Transp. Electrif.* **2017**, *4*, 38–63. [CrossRef]
318. Li, Y.; Lin, T.; Mai, R.; Huang, L.; He, Z. Compact DoubleSided Decoupled Coils-Based WPT Systems for High-Power Applications: Analysis, Design, and Experimental Verification. *IEEE Trans. Transp. Electrif.* **2017**, *4*, 64–75. [CrossRef]
319. Electric Vehicle Wireless Power Transfer (WPT) Systems–Part 3: Specific Requirements for the Magnetic Field Wireless Power Transfer Systems. 2019. Available online: https://webstore.iec.ch/publication/27435 (accessed on 18 March 2019).
320. Electric Vehicle Wireless Power Transfer (WPT) Systems–Part 2: Specific Requirements for Communication between Electric Road Vehicle (EV) and Infrastructure. 2019. Available online: https://webstore.iec.ch/publication/31050 (accessed on 18 March 2021).
321. *Standard IEC 61980-1*; Electric Vehicle Wireless Power Transfer (WPT) Systems—Part I: Gen_eral Requirements | Engineering360. International Standard IEC: Geneva, Switzerland, 2015.
322. IEC—TC 69 Dashboard > Documents: Working Documents, Other Documents, Support Documents. 2017. Available online: http://www.iec.ch/dyn/www/f?p=103:30:0::::FSP%0A_ORG_ID,FSP_LANG_ID:1255,2 (accessed on 13 March 2021).
323. Electrically Propelled Road Vehicles? Magnetic Field Wireless Power Transfer? Safety and Interoperability Requirements. 2020. Available online: https://www.iso.org/standard/73547.html (accessed on 18 March 2021).
324. Electrically Propelled Road Vehicles? Magnetic Field Wireless Power Transfer? Safety and Interoperability Requirements. 2017. Available online: https://www.iso.org/standard/64700.html (accessed on 18 March 2021).
325. Use Cases for Wireless Charging Communication for Plug-in Electric Vehicles. 2013. Available online: https://www.sae.org/standards/content/j2836/6_201305/ (accessed on 18 March 2021).
326. Plug-In Electric Vehicle (PEV) Interoperability with Electric Vehicle Supply Equipment (EVSE). Available online: https://www.sae.org/standards/content/j2953/1_201310/ (accessed on 18 March 2021).
327. Test Cases for the Plug-In Electric Vehicle (PEV) Interoperability with Electric Vehicle Supply Equipment (EVSE). 2016. Available online: https://www.sae.org/standards/content/j2953/3/ (accessed on 18 March 2021).

328. Plug-in Electric Vehicle (PEV) Charge Rate Reporting. 2020. Available online: https://www.sae.org/standards/content/j2953/4/ (accessed on 18 March 2021).
329. Communication for Wireless Power Transfer between Light-Duty Plug-in Electric Vehicles and Wireless EV Charging Stations. Available online: https://www.sae.org/standards/content/j2847/6_202009/ (accessed on 18 March 2021).
330. Annarelli, A.; Fonticoli, L.F.; Nonino, F.; Palombi, G. An Evaluation Model Supporting IT Outsourcing Decision for Organizations. In *Intelligent Computing. SAI 2022*; Lecture Notes in Networks and Systems; Arai, K., Ed.; Springer: Cham, Switzerland, 2022; Volume 508. [CrossRef]
331. Nastasi, B.; Di Matteo, U. Innovative Use of Hydrogen in Energy Retrofitting of Listed Buildings. *Energy Procedia* **2017**, *111*, 435–441. [CrossRef]
332. Annarelli, A.; Palombi, G. Digitalization Capabilities for Sustainable Cyber Resilience: A Conceptual Framework. *Sustainability* **2021**, *13*, 13065. [CrossRef]
333. Annarelli, A.; Clemente, S.; Nonino, F.; Palombi, G. Effectiveness and Adoption of NIST Managerial Practices for Cyber Resilience in Italy. In *Intelligent Computing*; Lecture Notes in Networks and Systems; Arai, K., Ed.; Springer: Cham, Switzerland, 2021; Volume 285.
334. Annarelli, A.; Colabianchi, S.; Nonino, F.; Palombi, G. The Effectiveness of Outsourcing Cybersecurity Practices: A Study of the Italian Context. In Proceedings of the Future Technologies Conference (FTC) 2021; Lecture Notes in Networks and Systems. Arai, K., Ed.; Springer: Cham, Switzerland, 2022; Volume 360.
335. Cristofari, A.; Dehghan Niri, T.; Lucidi, S. On global minimizers of quadratic functions with cubic regularization. *Optim. Lett.* **2019**, *13*, 1269–1283. [CrossRef]
336. Credo, A.; Cristofari, A.; Lucidi, S.; Rinaldi, F.; Romito, F.; Santececca, M.; Villani, M. Design Optimization of Synchronous Reluctance Motor for Low Torque Ripple. In *A View of Operations Research Applications in Italy 2018*; Dell'Amico, M., Gaudioso, M., Stecca, G., Eds.; AIRO Springer Series; Springer: Cham, Switzerland, 2019; Volume 2. [CrossRef]
337. Pelliccioni, A.; Cristofari, A.; Lamberti, M.; Gariazzo, C. PAHs urban concentrations maps using support vector machines. *Int. J. Environ. Pollut.* **2017**, *61*, 1–12. [CrossRef]
338. Pelliccioni, A.; Cristofari, A.; Silibello, C.; Gherardi, M.; Cecinato, A.; Lamberti, M. Estimation of PAHs concentration fields in an urban area by means of support vector machines. In Proceedings of the 7th International Congress on Environmental Modelling and Software: Bold Visions for Environmental Modeling, iEMSs 2014, San Diego, CA, USA, 15–19 June 2014; Volume 2, pp. 987–994.
339. Armando, P.; Cristofari, A.; Mafalda, L.; Claudio, G. Pahs urban concentrations maps using support vector machine. In Proceedings of the HARMO 2014-16th International Conference on Harmonisation within Atmospheric Dispersion Modelling for Regulatory Purposes, Varna, Bulgaria, 8–11 September 2014; pp. 510–514.
340. Sanguesa, J.A.; Torres-Sanz, V.; Garrido, P.; Martinez, F.J.; Marquez-Barja, J.M. A Review on Electric Vehicles: Technologies and Challenges. *Smart Cities* **2021**, *4*, 372–404. [CrossRef]
341. Goel, S.; Sharma, R.; Rathore, A.K. A review on barrier and challenges of electric vehicle in India and vehicle to grid optimization. *Transp. Eng.* **2021**, *4*, 100057. [CrossRef]
342. Cristofari, A. An almost cyclic 2-coordinate descent method for singly linearly constrained problems. *Comput. Optim. Appl.* **2019**, *73*, 411–452. [CrossRef]
343. Cristofari, A.; Rinaldi, F. A Derivative-Free Method for Structured Optimization Problems. *SIAM J. Optim.* **2021**, *31*, 1. [CrossRef]
344. Cristofari, A.; Rinaldi, F.; Tudisco, F. Total Variation Based Community Detection Using a Nonlinear Optimization Approach. *SIAM J. Appl. Math.* **2020**, *80*, 15. [CrossRef]

MDPI

St. Alban-Anlage 66

4052 Basel

Switzerland

Tel. +41 61 683 77 34

Fax +41 61 302 89 18

www.mdpi.com

Energies Editorial Office

E-mail: energies@mdpi.com

www.mdpi.com/journal/energies